VOLUME ONE HUNDRED AND SIXTY SIX

INTERNATIONAL REVIEW OF NEUROBIOLOGY

Current Challenges in Cell Therapy for Neurodegenerative Diseases

INTERNATIONAL REVIEW OF NEUROBIOLOGY

VOLUME 166

VOLUME ONE HUNDRED AND SIXTY SIX

INTERNATIONAL REVIEW OF NEUROBIOLOGY

Current Challenges in Cell Therapy for Neurodegenerative Diseases

Edited by

EMMA L. LANE

School of Pharmacy and Pharmaceutical Sciences, Brain Repair and Intracranial Neurotherapeutics Unit (BRAIN), Cardiff University, Cardiff, United Kingdom

CHENEY J.G. DREW

Centre For Trials Research;
Brain Repair and Intracranial Neurotherapeutics Unit (BRAIN), College of Biomedical and Life Sciences, Cardiff University, Cardiff, United Kingdom

MARIAH J. LELOS

Brain Repair Group, School of Biosciences, Cardiff University, Cardiff, United Kingdom

ELSEVIER

ACADEMIC PRESS

An imprint of Elsevier

Academic Press is an imprint of Elsevier
50 Hampshire Street, 5th Floor, Cambridge, MA 02139, United States
525 B Street, Suite 1650, San Diego, CA 92101, United States
The Boulevard, Langford Lane, Kidlington, Oxford OX5 1GB, United Kingdom
125 London Wall, London, EC2Y 5AS, United Kingdom

First edition 2022

ISBN: 978-0-323-98937-4
ISSN: 0074-7742

For information on all Academic Press publications
visit our website at https://www.elsevier.com/books-and-journals

Publisher: Zoe Kruze
Acquisitions Editor: Mariana Kuhl
Developmental Editor: Federico Paulo Mendoza
Production Project Manager: Abdulla Sait
Cover Designer: Mark Rogers

Typeset by STRAIVE, India

Contents

Contributors

Oliver J.M. Bartley
Brain Repair Group, School of Biosciences, Cardiff University, Cardiff, United Kingdom

Gerhard Bauer
University of CA Davis, School of Medicine, Stem Cell Program, Institute for Regenerative Cures, GMP Facility, Sacramento, CA, United States

Karim Ben M'Barek
INSERM U861, I-Stem, AFM, Institute for Stem cell Therapy and Exploration of Monogenic diseases; Paris Saclay Evry U861; CECS, Corbeil-Essonnes, France

Monica Busse
Centre For Trials Research; Brain Repair and Intracranial Neurotherapeutics Unit (BRAIN), College of Biomedical and Life Sciences, Cardiff University, Cardiff, United Kingdom

Monville Christelle
INSERM U861, I-Stem, AFM, Institute for Stem cell Therapy and Exploration of Monogenic diseases; Paris Saclay Evry U861, Corbeil-Essonnes, France

Cheney J.G. Drew
Centre For Trials Research; Brain Repair and Intracranial Neurotherapeutics Unit (BRAIN), College of Biomedical and Life Sciences, Cardiff University, Cardiff, United Kingdom

Tara Fortino
Drexel University; Marion Murray Spinal Cord Research Center, Drexel University, Philadelphia, PA, United States

Brian Fury
University of CA Davis, School of Medicine, Stem Cell Program, Institute for Regenerative Cures, GMP Facility, Sacramento, CA, United States

Patricia Garcia Jareño
Brain Repair Group, School of Biosciences, Cardiff University, Cardiff, United Kingdom

Adam Hall
Drexel University; Marion Murray Spinal Cord Research Center, Drexel University, Philadelphia, PA, United States

James S. Harrop
Department of Neurological and Orthopedic Surgery, Thomas Jefferson University, Philadelphia, PA, United States

Lyndsey Isaacs
Cure Parkinson's, London, United Kingdom

Emma L. Lane
School of Pharmacy and Pharmaceutical Sciences, Cardiff University, Cardiff, United Kingdom

Michael A. Lane
Drexel University; Marion Murray Spinal Cord Research Center, Drexel University, Philadelphia, PA, United States

Mariah J. Lelos
Brain Repair Group, School of Biosciences, Cardiff University, Cardiff, United Kingdom

Morizur Lise
INSERM U861, I-Stem, AFM, Institute for Stem cell Therapy and Exploration of Monogenic diseases; Paris Saclay Evry U861; CECS, Corbeil-Essonnes, France

Soania Mathur
Unshakeable MD, Toronto, ON, Canada

Alessia Niceforo
Drexel University; Marion Murray Spinal Cord Research Center, Drexel University, Philadelphia, PA, United States

Patricia E. Phelps
Department of Integrative Biology & Physiology, UCLA, Los Angeles, CA, United States

James B. Phillips
UCL School of Pharmacy, University College London; UCL Centre for Nerve Engineering, London, United Kingdom

Sophie V. Precious
Brain Repair Group, School of Biosciences, Cardiff University, Cardiff, United Kingdom

Catherine A. Priest
Neurona Therapeutics, South San Francisco, CA, United States

Victoria H. Roberton
UCL School of Pharmacy, University College London; UCL Centre for Nerve Engineering, London, United Kingdom

Anne E. Rosser
Brain Repair Group, School of Biosciences, Cardiff University; Cardiff University Neuroscience and Mental Health Research Institute; Brain Repair and Intracranial Neurotherapeutics (B.R.A.I.N.) Biomedical Research Unit, College of Biomedical and Life Sciences, Cardiff University, Cardiff, United Kingdom

Victoria Spruance
Drexel University; Marion Murray Spinal Cord Research Center, Drexel University, Philadelphia, PA; Division of Kidney, Urologic, & Hematologic Diseases, National Institutes of Health, Bethesda, MD, United States

Lyandysha V. Zholudeva
Drexel University; Marion Murray Spinal Cord Research Center, Drexel University, Philadelphia, PA; Gladstone Institutes, San Francisco, CA, United States

Preface

The concept of implanting new cells to replace those that are old and dying in the CNS has been part of medical science since the 1890s. The possibility of restoring function to those with traumatic brain injury, stroke, or diseases such as retinal degeneration and Parkinson's has kept many busy at their laboratory bench. These early activities had limited successes, but through time, fundamental changes in our understanding have allowed major breakthroughs and improved the likelihood of cells surviving in their new environments to produce functional benefits. The first major evolution in this journey was in the 1940s with the use of fetal cells, which, although ethically and practically very challenging, provided meaningful proof of principle that the approach could be successful. Improvements snowballed from the 1970s onward, not just in our use of animal models of the diseases but also in analytical techniques such as histochemistry, permitting a more in-depth insight into the success criteria for implanted cells. The most recent decades have seen further evolution of stem cell technologies with various stem cell sources enabling the possibility of large-scale, high-quality cell products. Advances in transcriptomics and proteomics have also provided insights into both cell products and specific diseases like never before. However, the evolution has not just been apparent in science, but, in concert with these technological advances, the opinions of the public have also changed, alongside a set shift by the medical and scientific establishments in the appreciation of who the stakeholders are in these studies. Although the scientific and clinical teams have always been at the forefront, the design of trials is now weighing the balance of the demands between scientific rigor and patient centricity. A major contributor to this is the oft-overlooked voices of those living with diseases and the ultimate recipients of the newly developed interventions.

This volume focuses on the most recent challenges that scientists, clinicians, triallists, and those living with these diseases face as we move toward an era where cell therapy for neurodegenerative diseases may become a clinical reality. Pioneering work focused on developing cell therapies for Parkinson's disease has been at the forefront of progress in the cell therapy field. With early success using fetal cells for transplantation in clinical trials it, along with degenerative conditions of the eye, has been one of the first neurodegenerative conditions to reach clinical trials of a stem cell-derived product. Here, we not only discuss some of the challenges specific to individual disease states in defining and refining their clinical products but also address more general considerations and common problems pertinent to all the different disease areas, including selection of animal models, the best biomaterial support for newly transplanted cells, and how we can achieve the necessary production standards required of an experimental cell product in human clinical trials. This volume goes on to address the bigger picture: how can we tackle the ethical and practical challenges of clinical trials in progressive degenerative disease states, which often include cognitive impairment, and when and how should we factor in the perspectives of those who have lived with the disease and are willing to be the brave test pilots embarking on an unknown journey with these new cells.

Editors
Emma L. Lane
Cheney J.G. Drew
Mariah J. Lelos

CHAPTER ONE

Challenges in progressing cell therapies to the clinic for Huntington's disease: A review of the progress made with pluripotent stem cell derived medium spiny neurons

Patricia Garcia Jareño[a,*], Oliver J.M. Bartley[a], Sophie V. Precious[a], Anne E. Rosser[a,b,c], and Mariah J. Lelos[a]
[a]Brain Repair Group, School of Biosciences, Cardiff University, Cardiff, United Kingdom
[b]Cardiff University Neuroscience and Mental Health Research Institute, Cardiff, United Kingdom
[c]Brain Repair and Intracranial Neurotherapeutics (B.R.A.I.N.) Biomedical Research Unit, College of Biomedical and Life Sciences, Cardiff University, Cardiff, United Kingdom
*Corresponding author: e-mail address: garciap@cardiff.ac.uk

Contents

International Review of Neurobiology, Volume 166
ISSN 0074-7742
https://doi.org/10.1016/bs.irn.2022.09.003

Abstract

Huntington's disease (HD) is a hereditary, neurodegenerative disorder characterized by a triad of symptoms: motor, cognitive and psychiatric. HD is caused by a genetic mutation, expansion of the CAG repeat in the huntingtin gene, which results in loss of medium spiny neurons (MSNs) of the striatum. Cell replacement therapy (CRT) has emerged as a possible therapy for HD, aiming to replace those cells lost to the disease process and alleviate its symptoms. Initial pre-clinical studies used primary fetal striatal cells to provide proof-of-principal that CRT can bring about functional recovery on some behavioral tasks following transplantation into HD models. Alternative donor cell sources are required if CRT is to become a viable therapeutic option and human pluripotent stem cell (hPSC) sources, which have undergone differentiation toward the MSNs lost to the disease process, have proved to be strong candidates. The focus of this chapter is to review work conducted on the functional assessment of animals following transplantation of hPSC-derived MSNs. We discuss different ways that graft function has been assessed, and the results that have been achieved to date. In addition, this chapter presents and discusses challenges that remain in this field.

1. Introduction

As life expectancy increases, neurodegenerative diseases, such as Alzheimer's disease, Amyotrophic Lateral Sclerosis, Parkinson's disease (PD) and Huntington's disease (HD), are becoming more prevalent, reducing late age quality-of-life and imposing a greater demand on healthcare resources (Hou et al., 2019; Prince et al., 2013). While the pathophysiology of each of these diseases is different, they share the loss of neuronal tissue, which cannot be naturally regenerated (Vasic, Barth, & Schmidt, 2019). Clinical diagnosis typically occurs after symptom onset, which is often after substantial neuronal loss has already occurred. One possible therapeutic approach is to directly replace the cells lost to the disease in order to restore damaged tissues and alleviate symptoms. Cell replacement therapy (CRT) aims to replenish cell populations lost to a disease process by transplantation of live cells directly into the parenchyma, to induce functional recovery through reconstruction of damaged tissues (Lindvall & Bjorklund, 2004). For many neurodegenerative conditions, the classes of neural cells affected, and the pattern of loss is known, even if the detailed pathogenesis is not fully understood. In this context, replacing degenerated cells with new donor cells is both pertinent and possible.

In this chapter, we will focus on HD, an autosomal dominant genetic disorder caused by an unstable expansion mutation of the CAG trinucleotide repeat in the huntingtin (HTT) gene. This results in degeneration of the caudate and putamen (neostriatum), which leads to a triad of motor, cognitive and psychiatric symptoms that gradually worsen as the disease progresses (Vonsattel et al., 1985). Individuals suffer great loss to their quality of life as they eventually become bed-bound and require full nursing care. It is the most common single gene neurodegenerative disorder and a particularly suitable condition in which to explore CRT, as (i) single gene disorders with full penetrance are particularly powerful paradigms for clinical translation, and patients can be identified before onset of degeneration through genetic testing; (ii) early degeneration is predominantly in the striatum, providing a localized target for CRT; and (iii) the obstacles which must be overcome to develop a CRT product for clinical use are likely to be similar for other neurodegenerative diseases.

Over the last few decades, there has been substantial progress in developing CRT for HD. However, the mechanism(s) underpinning the efficacy of this approach has not been definitively established. One of the key concepts being explored is the idea that reconstruction of critical elements of host neural circuitry is key to achieve neural repair. Circuit reconstruction is based on the idea that the transplanted cell product will receive afferents and project efferents to relevant areas of the brain and establish functional synapses between these regions. For HD, the main goal is to reconstruct the dysfunctional cortico-striatal-pallidal loops, by replacing the degenerated striatal medium spiny neurons (MSNs) that are lost through the disease process (Tabrizi et al., 2012). Homotopic transplantation of MSNs into the striatal nucleus offers the opportunity for newly transplanted cells to form synaptic connections with biologically relevant brain regions and reform these lost circuits.

The potential of CRT for HD has been established using rodent and human fetal striatal tissues in preclinical studies (Mazzocchi-Jones, Dobrossy, & Dunnett, 2009; Pauly, Piroth, Dobrossy, & Nikkhah, 2012). Additionally, early clinical trials have demonstrated safety and feasibility, with some evidence of clinical benefit (Bachoud-Levi et al., 2006; Gaura et al., 2004; Rosser & Bachoud-Levi, 2012). These have provided a useful proof-of-concept and evidence that transplants can achieve circuit reconstruction with subsequent recovery of motor and cognitive dysfunctions (Brasted, Watts, Robbins, & Dunnett, 1999; Brasted, Watts, Torres, Robbins, & Dunnett, 1999; Brasted, Watts, Torres, Robbins, & Dunnett, 2000; Dobrossy & Dunnett, 2005, 2007; Mazzocchi-Jones et al., 2009).

However, there are significant constraints associated with the use of fetal-derived donor cells (Boer, 1994; Hurelbrink et al., 2003), and for CRT to transition to the clinic, thus a donor cell source is required that can circumvent these limitations. Human pluripotent stem cells (hPSC) are a renewable cell source which can, theoretically, be directed to specific cell fates and undergo long-term cryo-storage. This provides the opportunity to undertake quality control testing, thus making them a more readily applicable and attractive donor cell source.

In this chapter, we consider the current progress in the field, with a specific focus on the extent to which hPSCs can differentiate into MSN-like progenitors and whether hPSC-derived transplants in HD models have been shown to ameliorate functional deficits and reconstruct neural circuits. Specifically, we will focus on the available data reporting functional integration of hPSC-derived striatal grafts and critically consider the evidence associated with neural circuit reconstruction. In addition, we will discuss the remaining challenges facing the field for developing a cell therapy product for neural repair in HD.

2. Derivation of MSN-like cells from hPSCs: What has been shown *in vitro* and *in vivo*?

The cell type most affected in HD is the striatal medium spiny neuron (MSN), thereby making them the most suitable target cell type for CRT. MSNs are a heterogenous population of GABAergic neurons which represent ~90% of the striatal neuron population (Anderson & Reiner, 1991; Ouimet, Langley-Gullion, & Greengard, 1998); they commonly express CTIP2, DARPP32 and a range of other subtype-specific markers. Achieving the derivation of MSNs from hPSCs has been a central issue for CRT and the wider HD field and much time has been invested developing *in vitro* differentiation protocols. Developmental knowledge has provided a foundation for the creation of these MSN differentiation protocols, leading to the first reported success around 15 years ago by Aubry et al. (Aubry et al., 2008).

Aubry and colleagues generated neural rosettes and treated them with developmentally relevant factors (Sonic hedgehog (SHH), Dickkopf-1 (DKK1, a WNT inhibitor) and brain-derived neurotrophic factor (BDNF)) to promote a lateral ganglionic eminence (LGE) fate (the area of the developing brain where striatal MSNs originate). They validated their differentiations by exploring the expression of genes associated with the LGE and striatal MSNs, successfully demonstrating that exposure to SHH and DKK1 increased the expression of subpallial markers DLX2 and GSX2. Across the course of the differentiation, they observed a variety of other key striatal markers including FOXG1, SOX1, PAX6, SIX3, GAD67, CALB1, and DARPP32. Toward the end of the maturation process (62–72 days *in vitro* (DIV)) they reported 22% of the population achieved a neuronal phenotype, as indicated by expression of the neuronal marker MAP2. Of these MAP2 positive cells, approximately 53% expressed the hallmark MSN marker DARPP32, but only 36% of MAP2 positive cells were found to be GABAergic. The group then trialed the transplantation of these cells across five stages of maturation into the quinolinic acid (QA) lesioned rat brain, to determine the most suitable stage for transplantation. Subsequent quantification of the grafts revealed that approximately 21% of the transplanted population which had committed to a neuronal phenotype also expressed DARPP32. There was evidence of substantial overgrowth, and while no pluripotent markers were observed, the graft was still highly proliferative with both Ki67 and nestin positive cells identifiable. These large grafts overfilled much of the native striatum, had defined borders indicating limited integration with host tissues and were detrimental to the health of the rats. No functional assessments were conducted, but much was learned from this study, including the importance of considering the optimum differentiation stage for transplantation and assessing key markers in order to understand the cell therapy product more comprehensively.

Since that first publication, the field has made several advancements to the methods used to generate hPSC derived MSNs. Here, we will summarize the outputs of key papers which aimed to (i) directly differentiate hPSCs toward MSN progenitor phenotypes, (ii) transplant those cell therapy products into HD models, and (iii) assess features of graft functionality. This section aims to provide relevant information about the differentiation protocols used to develop and characterize the resulting cell products from these works. *In vivo* functional assessments will be discussed in Section 3.

2.1 Ma et al. (2012) optimized SHH exposure to yield high proportions of DARPP32 expressing cells from human embryonic stem cells (hESCs)

In 2012, Ma and colleagues (Ma et al., 2012) promoted a ventral telencephalic fate by activating the SHH pathway in forebrain progenitors (either using SHH or the SHH agonist purmorphamine) to generate progenitors that were comparable to those found in the LGE. They modulated the dose of SHH to optimize the expression of two genes associated with the subpallium and LGE fates; GSX2 and MEIS2. Following maturation, they reported a highly neuronal population (93% positive for βIII-tubulin), of which the majority were GABAergic (90%), and of these most were also positive for DARPP32 (89%, approximately 75% of the total population). High-performance liquid chromatography (HPLC) at 90 DIV identified release of the neurotransmitter GABA, and patch-clamp recordings following 70 DIV confirmed the presence of neurons able to evoke action potentials. Despite reporting a relatively pure population *in vitro*, grafts of the same cells yielded only 59% DARPP32-positive neurons at 4 months post-transplantation into the QA lesioned mouse neostriatum, and the intensity of DARPP32 staining was noted to be lower than that observed in endogenous DARPP32 positive neurons. Nevertheless, this remains one of the higher proportions of DARPP32 positive cells reported in grafts to date. The remaining graft-derived cells were generally still relevant to the striatum, with low numbers of a variety of striatal interneuron markers reported (e.g., PVALB and STT). In addition to identifying cell types, the group also observed graft derived projections to the host anterior substantia nigra, and found that these cells were positive for DARPP32 and TAC1, but not PENK, suggesting they were a specific subtype of MSN (e.g., in this case D1 subtype). Finally, the graft images show large grafts and indeed, despite transplanting 100,000 cells per animal, final numbers of donor-derived cells within the grafts were reported in the range of ~3 million cells. However, although proliferating cells were present at 1-week post-transplantation as shown by the expression of Ki67 and nestin, both these markers were absent at 4 months post-transplantation.

2.2 Jeon et al. (2012) generated LGE like progenitors from mutant and healthy HD human induced pluripotent stem cells (hiPSC) lines

Using PA6 stromal cell co-culture and BDNF, Jeon and colleagues (Jeon et al., 2012) generated forebrain and GABAergic neural precursors from a hESC line and two hiPSC lines, one of which was derived from a patient with juvenile HD carrying 72 CAG repeats (HD-mutant hiPSCs). They reported variability in the formation of neural rosette structures, indicative of the capacity of these lines to commit to a neuroepithelial fate. Following manual selection of rosettes, no further differences in cell lines were observed, and they achieved a high proportion of neurons with 89% of cells expressing MAP2, 75% expressing GABA and 27% of cells expressing DARPP32. Additionally, they reported some DARPP32 positive cells with LGE markers GSX2 and DLX2, indicating successful derivation of an MSN-like phenotype. Patch-clamp electrophysiological recordings confirmed the cells were GABA sensitive as GABA exposure evoked whole cell currents which were blocked when the GABA antagonist picrotoxin was applied. To assess their *in vivo* potential, 200,000 precursor cells were transplanted into the QA-lesioned rat striatum and histologically characterized after 12 weeks. Grafted cells expressed DARPP32, GAD65/67, GABA, and synaptophysin, indicating that some of the engrafted cells expressed MSN relevant genes, were GABAergic, and were capable of forming synapses. However, no quantification of these markers was reported, and the authors note that it is unclear if functional recovery is due to connectivity or other factors.

2.3 Delli Carri et al. (2013) used dual SMAD inhibition before directing to an LGE like fate

In 2013, Delli Carri et al. (Delli Carri et al., 2013) used dual SMAD inhibition (*via* inhibition of bone morphogenic protein (BMP) and transforming growth factor beta (TGFβ) pathways, (Chambers et al., 2009)) to generate a highly pure neural culture, which was then patterned to an MSN fate also using SHH and DKK1. Cells were differentiated to 80 DIV, at which time the authors reported that 51% of cells expressed the mature neuronal marker MAP2, and of this neuronal population, ~20% expressed DARPP32, ~60% expressed CTIP2 and ~78% expressed GABA, with electrophysiological assessment revealing functionally active neurons. In this study, cells were transplanted at 38 DIV into the striata of QA-lesioned rats. At 3 weeks post-transplantation, proliferating cells, identified by immuno-positive staining of Ki67 and nestin, were seen with no expression of striatal markers. Although graft images showed large grafts that appeared to increase in size over time and filled the host striatum, the expression of proliferative markers decreased over time and by 6 weeks post-transplantation neuronal markers βIII-tubulin and MAP2 were detected in the grafts. Striatal markers,

FOXP1, FOXP2 and DARPP32, were evident in graft-derived cells (evidenced by co-localization with human nuclei immuno-staining) at 9 weeks post-transplantation, although the proportions of cells expressing striatal genes are not reported.

2.4 Arber et al. (2015) introduced the manipulation of TGFβ pathways with Activin A

In 2015, Arber and colleagues (Arber et al., 2015) published a protocol also based on dual SMAD inhibition for initial neural induction, but which differed from previous works by its unique use of the TGFβ family protein Activin A to promote an LGE-like phenotype, instead of SHH and DKK1 mediated WNT inhibition. With this approach, authors observed a rapid increase over 9 days in LGE markers CTIP2, NOLZ1 (ZNF503), and FOXP2, but no change in medial ganglionic eminence (MGE) markers NKX2.1 and LHX8, suggesting this approach may specifically upregulate genes known to be involved in MSN specific fates without also activating MGE-like fates. Following maturation to DIV 43, ~40% of these cells expressed DARPP32 and 80% expressed CTIP2. The group also identified other striatal genes in these cultures including ARPP21, CALB1, PENK, TAC1, and GAD1. Furthermore, GABAergic neurons generated from this protocol were reported to be electrophysiologically active. These cells were further characterized following transplantation into a QA lesion rat model of HD. At 8 weeks post-transplantation, low levels of DARPP32 were observed, but by 16 weeks this had increased to ~49% of the grafted population. The majority of the grafted cells were positive for GABA (~85%), and other MSN markers were also identified within the grafted population (including FOXP2, CALB1, TAC1, and DRD2). However, the grafts were smaller than observed in previous studies, and no functional improvements were observed.

2.5 Adil et al. (2018) cultured MSN progenitors in 3D biomaterials

In 2018, Adil and colleagues (Adil et al., 2018) used SHH and WNT inhibition (*via* DKK1) to explore the addition of a 3D biomaterial in their differentiation with the goal of improving the MSN differentiation. Cells were initially cultured in the 3D hydrogel for 26 days to generate striatal progenitors and were then differentiated in a 2D culture system, with the whole *in vitro* period lasting 60 days. *In vitro* characterization at 60 DIV revealed 78% of cells in the culture were MAP2 positive neurons. Of these neurons, 55% were DARPP32 positive, 61% were GABA+ and 70% were CTIP2 positive. In addition, 69% of cells were reported to have fired action potentials showing that they were electrophysiologically active. Of note, cells that were differentiated entirely in the 2D culture system, rather than the 3D-to-2D method, were not observed to be electrophysiologically active, suggesting that factors such as cell-to-cell contact may be important in differentiation of MSN-like cells. To assess these cells *in vivo*, cells were cultured in 3D hydrogel for 26 days, then in 2D for a further 10 days, and then 100,000 cells were transplanted into the striatum of 5-week-old transgenic R6/2 mice. Four months post-transplantation, authors reported survival of ~6% of transplanted cells within the graft areas, with grafts yielding DARPP32 expressing cells; expression of other striatal markers were not reported. Of note, there was no graft overgrowth observed which could in part be due to the cells being transplanted at a more mature stage, although it is difficult to determine this as proliferative markers were not assessed.

2.6 Wu et al. (2018) published a clinical grade protocol

In 2018, Wu and colleagues (Wu et al., 2018) published a clinical grade protocol that used dual SMAD inhibition in hESCs, with SHH and WNT manipulation for ventral patterning. They replaced the protein components (Noggin and SHH) with small molecules (LDN, SB, XAV and SAG), and used DAPT to accelerate neuronal maturation. At DIV21 immunofluorescence revealed that 87% of MAP2 and 85% of GABA positive cells expressed DARPP32, and of TUJ1 positive cells, 89% expressed GABA, 90% expressed CALBINDIN and 90% expressed DARPP32. While immunofluorescence indicated a high DARPP32 yield, quantitative polymerase chain reaction (QPCR) analysis showed substantial variation and in one experiment MSN progenitors were reported as expressing less DARPP32 than naïve hESCs. Cells also expressed PSD95 and SYNAPSIN1. Patch clamp recordings of cells on DIV20–24 detected spontaneous action potentials, and cells responded to the addition of GABA in the cultures with action potentials blocked following the addition of a GABA antagonist. *In vivo* assessment was performed by transplanting 100,000 cells into neonatal

P1severe combine immunodeficient mouse (SCID) mice and into adult QA-lesioned SCID mice. In the neonate-grafted mice at 12 weeks post-transplantation approximately 50% of the transplanted cells were positive for DARPP32, 95% for FOXP1, 95% for FOXG1, 40% for CALBINDIN, 80% for GABA, 10% for CHAT, 12% for SP and 8% for ENK. In the adult mice, at 16 weeks post-transplantation, 48.7% of transplanted cells were positive for DARPP32, 87% for GABA, 90% for TUJ1, 95% for FOXP1, 80% for CTIP2, 25% for SP and 10% for ENK.

2.7 Comella-Bolla et al. (2020) consistently produced neuronal populations with a high CTIP2 content

In a recent publication, Comella-Bolla et al. (Comella-Bolla et al., 2020) described a differentiation protocol that did not use SHH or Activin A to obtain ventral forebrain neurons. They combined dual-SMAD inhibition with WNT inhibition, and then regionalized the cells by continuing inhibition of the BMP and WNT pathways. This protocol induced expression of subpallial genes including GSX1, GSX2, ASCL1, DLX1, DLX2, DLX5, DLX6, and EBF1. As these cells matured, they became populated with high numbers of MAP2-expressing neurons (95–99% of the total culture), of which $\sim$40% were positive for GABA. At 37 DIV, $\sim$60% of cultures expressed CTIP2 and 6% of cells co-expressed CTIP2 and DARPP32. While this, compared with other studies presented here, appears to be among the lower reported quantifications of DARPP32 expressing cells, Comella-Bolla et al. note that differences in DARPP32 expression may vary between studies due to the differences in antibodies used. Specifically, different antibodies label different isoforms of the DARPP32 protein that undergoes different post-translational modifications depending on the state of the cell and the environmental signals. However, in their QPCR analysis they observed comparable levels of DARPP32 at DIV 0 and DIV 37, suggesting that this protocol did not increase DARPP32 expression above levels expressed by naïve pluripotent cultures. Additionally, they observed expression of the MGE marker LHX1 and striatal interneuron marker SST in their cultures, and between 20% and 50% of the mature neurons were positive for tyrosine hydroxylase (TH), which suggests that non-MSN phenotypes were produced in this differentiation. Single-cell Ca^{2+} imaging revealed neurons that showed action potentials and various patterns of firing activity, indicating functional maturity. For *in vivo* assessment of these cells, neonatal mice received 15,000 DIV 16 cells unilaterally and grafts were analyzed 1- and 3-months post-transplantation, with no evidence of tissue overgrowth. Grafted cells were reported to express CTIP2 (89% at 1 month and 82% at 3 months post-transplant) and DARPP32 (7% at 1 month and 5% at 3 months post-transplant). The authors concluded that the protocol will be a good platform for *in vitro* studies or drug screening and has potential for CRT.

2.8 Besusso et al. (2020) comprehensively studied early striatal gene expression in hPSC derived MSNs

In 2020, the lab group which published Delli Carri et al. (2013) modified their MSN differentiation strategy in a further advancement of their work (Besusso et al., 2020). Cell phenotype was determined using high content QPCR at different time points of the differentiation protocol and compared to pluripotent populations at 0 DIV. Gene panels based on developing brain regions were used to indicate the relative success of the differentiation. At 20 DIV, cultures expressed neuronal/ectodermal markers MAP2 and NEUROD1. These DIV 20 cultures also expressed pan-GE markers DLX1, DLX5, and DLX6, and LGE markers GAD1, GAD2, ISL1 and MEIS at this time. The authors also reported DIV 20 cultures as positive for off target genes, such as pallial marker EMX2, and MGE marker NKX2.1. Expression of these genes generally persisted through the maturation process to DIV 50, which coincided with a relative increase in the expression of most of these markers (MAP2, DLX1, DLX5, DLX6, GAD1, GAD2, ISL1, MEIS, NKX2.1). Several of the genes expressed at DIV 50 are understood to be preferentially expressed by immature cells (FOXG1, DLX1, DLX5, DLX6, and NKX2.1), which could be a sign that immature phenotypes remain in the culture at this time. Notably, some genes thought to be pivotal for MSN development were either not expressed, or expressed only at low levels (PAX6, ASCL1, DLX2, GSX2, CALB1, CALB2, CTIP2, EBF1, FOXP1, FOXP2, RARB).

However, data collected from immunocytochemical staining for several of these markers suggested they were indeed present in these cultures. The number of cells expressing PAX6 (which is expressed at high levels per cell in the pallium, but also at low levels in the dorsal LGE) decreased from ~40% in early cultures and then plateaued at around 20%. The subpallium markers GSX2 and ASCL1 were also observed, with GSX2 identified in ~10–20%

and ASCL1 in ~2–10% of the cell population across the culture. LGE markers CTIP2, EBF1, and ISL1 were also examined, and at DIV 50–30% of the cultures expressed CTIP2, ~20% expressed EBF1, and ~15% expressed ISL1. Further quantification of immunofluorescent analysis showed that the number of Ki67 positive cells decreased over time in culture, although there were still immuno-positive cells at 50 DIV. Finally, approximately 18% of the cultures co-expressed the MSN markers CTIP2 and DARPP32.

QA lesioned rats received intra-striatal infusions of 300,000 cells, (20 DIV) and at 1- or 2-months post-transplantation, all animals were taken for graft analysis. Within the graft there were high percentages of progenitor cells, with ~9% of graft-derived cells expressing Ki67 and ~68% of graft-derived cells expressing the neural progenitor marker Nestin, but there was no evidence of graft overgrowth. 45% of the grafted cells expressed CTIP2: of these only ~3% also co-expressed DARPP32, but the majority of DARPP32 positive cells co-expressed CTIP2. Although the numbers of CTIP2-expressing cells were greater following engraftment (having been transplanted at 20 DIV when CTIP2 expression was close to 0%), the proportion of co-labeled CTIP2/DARPP32-expressing cells was ~6 times lower in the *in vivo* analysis than in the *in vitro* analysis. Taken together, this suggests that, at this early time point post-transplantation, these grafts are still proliferative and although they express key markers indicating the presence of striatal cells, they have not yet developed into fully mature MSNs.

2.9 Summary

In this section, we have summarized key findings from papers that have (i) directly differentiated hPSCs toward striatal MSN phenotypes; (ii) transplanted MSN-precursor-like cell therapy products into HD models; and (iii) assessed some aspect of graft functionality following transplantation. These data suggest that hPSCs can be differentiated into MSN-like neurons *in vitro*, that these cells survive within the brain post-transplantation and that they are capable of maturation *in vivo* into mature neurons with features characteristic of striatal MSNs (Adil et al., 2018; Arber et al., 2015; Besusso et al., 2020; Comella-Bolla et al., 2020; Delli Carri et al., 2013; Jeon et al., 2012; Ma et al., 2012). Across these studies, determining the successful derivation of striatal MSNs has most commonly been achieved by examining the expression of CTIP2 and DARPP32 (which are expressed near-ubiquitously by mature striatal MSNs, with CTIP2 also expressed in MSN precursors).

With respect to protocols that yielded the greatest proportions of DARPP32 or CTIP2, Ma et al. reported ~75% of total cells *in vitro* expressed DARPP32, with evidence of MEIS2, GABA, and GAD co-expression (but no assessment of CTIP2 co-expression) (Ma et al., 2012). Cultures from Arber et al. resulted in 80% of cells expressing CTIP2 and 40% of the total culture also expressing DARPP32 (Arber et al., 2015). Post-transplantation, Ma et al. reported the highest yield of DARPP32 expressing cells (59%), with Arber et al. the second highest (49%), though in both cases CTIP2 expression was not reported, although additional MSN markers were assessed. The highest yield of CTIP2-expressing cells post-transplantation was reported by Comella-Bolla et al. (89% after 1 month and 82% after 3 months), though DARPP32 expression was substantially lower (7–5%) (Comella-Bolla et al., 2020); similarly, *in vitro* they reported the second highest yield of CTIP2-expressing cells (~60%) with low DARPP32 expression (6%), and in QPCR analysis DARPP32 expression was not seen to increase above levels observed in pluripotent cultures. While expression of DARPP32 and CTIP2 are key, a limited number of other markers have been used in the above studies, which add additional weight to the interpretation that these cells are in fact comparable to *bone fide* striatal MSNs.

There is still a need of further investigation to determine the success of the directed MSN differentiation. There is a plethora of other markers of striatal MSNs that could be employed (ideally in co-expression studies) to determine the subtype specificity of the cells being assessed. These could include markers that define patch or matrix MSNs, direct or indirect pathway associated MSNs, and those that indicate stage of maturation and positional/regional identity/specificity.

Is *in vitro* characterization of striatal MSN content of a cell therapy product prior to transplantation predictive of the subsequent *in vivo* yield of striatal MSNs? In the majority of studies discussed above, the level of DARPP32 (which was the only marker of striatal MSNs used across all the above studies both pre- and post-transplantation) observed *in vitro* prior to transplantation, was seen to decrease *in vivo*. The exception is Arber et al., in which slightly higher proportions (40% *in vitro,* rising to 49% *in vivo*) were generated. Overall, levels of DARPP32 expression in hPSC-derived MSNs post-transplantation were variable, even from studies using comparable protocols based on the same developmental principles. Thus, identification of *in vitro* markers that are predictive of graft outcome would be beneficial to the field. Moreover, it is critical to understand the extent to which any of these engrafted hPSC-derived MSNs can alleviate behavioral deficits, and this is examined in the next section.

3. Evidence for functional recovery following CRT and underlying mechanisms

The success of a CRT product can be characterized based on various aspects of functional efficacy including; the recovery of cognitive and motor deficits; restoring neurotransmission and normalizing electrophysiological activity within the network; and the integration and formation of synapses with relevant efferent and afferent brain regions, such as the globus pallidus or substantia nigra.

In this section, we consider the functional assessments of the grafts produced in the pre-clinical studies previously discussed in Section 2. These are discussed by the type of assessment performed, highlighting the key outputs reported in each study.

3.1 Behavioral recovery

3.1.1 Drug induced behavior

Apomorphine (APO)-induced rotations are commonly used to assess unilateral excitotoxic models. Animals with a unilateral QA lesion, show a contralateral rotation behavior caused by an imbalance of dopamine in the nigro-striatal pathway between the intact and lesioned side (Ungerstedt, 1971). This test is used not only to investigate the extent of the lesion and graft functional motor recovery, but also to provide information about graft composition (Lelos et al., 2016; Pundt, Kondoh, Conrad, & Low, 1996; Sanberg et al., 1997).

There are three studies where hPSC-derived MSNs were transplanted into HD rodent models and APO-induced rotations were performed ((Arber et al., 2015; Delli Carri et al., 2013; Jeon et al., 2012); see Table 1). Arber et al., assessed APO-induced rotations up to 16 weeks post-transplantation but found no recovery in hESC-derived MSN transplanted animals compared with lesion-only controls (Arber et al., 2015). In the other two studies, APO-induced rotation assessments revealed very early improvements, based on the reduction of the number of contralateral rotations post-transplantation (Delli Carri et al., 2013; Jeon et al., 2012). In Delli-Carri et al., animals that received hESC-derived MSN transplants rotated significantly less than lesion-only controls at 3- and 6-weeks, but not by 9 weeks post-transplantation. However, bilateral transplants were performed in this study and as APO-rotations test the lesioned or

Table 1 Summary of key manuscripts reporting human PSC derived MSN cell therapy products.

Study	Model	Cells	Transplant	Functional assessments
Ma et al. (2012)	Unilateral QA SCID mouse Unilateral transplant	Source: hESC Target: LGE progenitors *In vitro*: DIV 47 (ICC) β-TUB 93%; D32 75% GABA 83.7%	Cells TX: 100,000 Graft: 4 months, 58.6% of grafted cells were D32+/GABA+	Rotarod: Improved at *16 weeks* Open field: Improved centre ratio and crossing at *8–16 weeks*. Improved total distance at *16 weeks* Treadscan: Improved at *16 weeks*
Jeon et al. (2012)	Unilateral QA rat Unilateral transplant	Source: hESC, hiPSC & HD-hiPSC (72CAG) Target: forebrain GABAergic NPCs *In vitro:* DIV 30 (ICC) MAP2 89%; D32 27%; GABA 75%	Cells TX: 200,000 Graft: 12 weeks some D32+ cells; % not reported	Adjusting stepping test: hiPSC improved at *4–12 weeks*. hESC and HD-hiPSC improved at *6–12 weeks* Staircase: hESC and hiPSC improved at *2–12 weeks*, HD-hiPSC, improved at *4–12 weeks* APO: hESC and HD-hiPSC improved at *6–12 weeks*, compared to the sham group. hiPSC improved at *2–8 weeks* only
Delli Carri et al. (2013)	Unilateral QA rat Unilateral transplant	Source: hiPSC/hESC Target: Ventral telencephalic prec *In vitro:* DIV 80 (ICC) MAP2 51%; D32 10.2%; CTIP2 30.6%; GABA 40%	Cells TX: 500,000 Graft: 9 weeks, HuNu+/D32+ & other neuronal/MSN markers identified in graft; % not reported	APO: Improved at *3–6 weeks*
Jeon et al. (2014)	YAC128 Bilateral transplant *TX at 12 mo*	Source: HD-hiPSC (72CAG) Target: forebrain GABAergic NPCs *In vitro*: reference Jeon 2012 (see above)	Cells TX: 100,000 Graft: 12 weeks some D32+ cells; % not reported	Rotarod: Improved at 3 *weeks*
Arber et al. (2015)	Unilateral QA rat Unilateral transplant	Source: hESC Target: LGE progenitors *In vitro:* DIV 43 (ICC) D32 40%; CTIP2 60%; $ARPP21^+$; $CALB1^+$; $PENK^+$; $TAC1^+$; $GAD1^+$	Cells TX: 400,000 Graft: 16 weeks, 86% & 49% HuNu+ cells were GABA+ & D32+ (respectively). Further MSN markers identified	APO: No improvements observed
Adil et al. (2018)	R6/2 mouse Bilateral transplant *TX at 5 weeks*	Source: hESC Target: MSN precursor *In vitro:* DIV 60 MAP2 78%; D32 43%; CTIP2 54.6%; GABA 48%	Cells TX: 100,000 per hemisphere Graft: 12–15 weeks HNA+/D32+ cells in graft; % not reported	Rotarod: Improved at *5 weeks* Clasping: Improved at *2–20 weeks* Fluorogold injection in the substantia nigra (SN) 10 weeks after transplant showed retrograde labelling

Continued

Table 1 Summary of key manuscripts reporting human PSC derived MSN cell therapy products.—cont'd

Study	Model	Cells	Transplant	Functional assessments
Wu et al. (2018)	Unilateral QA SCID mouse Unilateral transplant	Source: hESC Target: LGE progenitors *In vitro*: DIV 21 D32 87% of MAP2; D32 85% of GABA; D32 90% of TUJ1; GABA 89% of TUJ1; CALBINDIN—90% of TUJ1; $PSD95^{+}$, $SYNAPSIN1^{+}$	Cells TX: 100,000 Graft: 16 weeks hN+ cells express 48.7% D32, 90% GABA, 90% TUJ1, 80% CTIP2, 30% SP and 10% ENK	Rotarod: Improved at 8–16 *weeks* Open field: Improved total distance at *16 weeks*
Comella-Bolla et al., 2020	B6CBA mouse *TX at P2*	Source: hiPSC/hESC Target: Ventral telencephalic *In vitro*: DIV 37 MAP2 95–99%; D32/CTIP2 6%; CTIP2 60%; GABA 40% of MAP2; GSX1, GSX2, ASCL1, DLX1, DLX2, DLX5, DLX6, EBF1	Cells TX: 15,000 cells Graft: 82% Ctip2, 5% D32	Transmission electron microscopy (TEM) showed synaptic connections from grafted cells with GP neurons. Inhibitory and excitatory connections were identified
Besusso et al., 2020	Unilateral QA rat Unilateral transplant	Source: hESC Target: MSN precursor *In vitro:* DIV 50 CTIP2—30% EBF1—20% ISL1—15% D32/CTIP2—18%	Cells TX: 300,000 Graft: 2 months 45.3% Ctip2+ of total graft; 2.8% D32+/Ctip2+	Rotarod: No improvements observed Vibrissae-Evoked hand placing: Improved at *1 month*, not at 2 months Adjusting step test: Improved at *1–2 months* Monosynaptic tracing using modified rabies virus: Traced cells were only observed in the striatum. Starter and traced cell numbers were increased from 1 to 2 months. Electrophysiology: at 2 month post-graft in acute brain slices. Grafted cells have a complex dendritic arborization, generate action potentials and show spontaneous excitatory potentials. PET: Raclopride showed no changes in expression of D2/D3 receptors compared with sham controls

Abbreviations: QA, quinolinic acid lesion; TX, transplanted; mo, months; wk, weeks; HNA, human nuclear antigen; NSC, neural stem cell; NPC, neural progenitor cell; APO, apomorphine-induced rotations; D32, DARPP-32; DIV, days in vitro; ICC, immunocytochemistry; P, postnatal day.

transplanted hemisphere against the intact, non-transplanted hemisphere, the results may be confounded.

In Jeon et al., the authors transplanted progenitors derived from three cell lines (a HD mutant hiPSC line, a non-mutant hiPSC line, and a non-mutant hESC). At 2 weeks post-transplantation they observed improvements in animals that received hiPSC-derived cells (the control, non-mutant line) compared to those animals that received either sham grafts or HD mutant hiPSC derived grafts. At both 6- and 8-weeks post-transplantation, all animals that received cell transplants showed significant recovery compared to sham-grafted controls. This was maintained through to the final time-point at 12 weeks for the HD-mutant hiPSC and hESC lines. However, from 10 weeks post-transplant, the animals with non-mutant hiPSC derived grafts returned to rotation levels nearing those of the sham transplants. This latter observation emphasizes the importance of longer-term studies to ascertain whether changes in functional levels are maintained and robust.

3.1.2 Tests of motor and cognitive function

Motor tasks are a common choice for functional assessment as they are simple, effective and reliable. Motor function is directly impaired by striatal atrophy, which makes this form of behavior ideal for measuring disease progression and for identifying functional improvements post-transplantation (Dunnett & Brooks, 2018). Cognitive impairments are also relevant to conditions of striatal degeneration; however, although the ability of rodent fetal grafts to improve cognitive processing has been reported, the impact of hPSC-derived grafts on cognitive function has yet to be explored (Brasted et al., 2000; Brasted, Watts, Robbins, & Dunnett, 1999; Brasted, Watts, Torres, et al., 1999).

There are five published studies where hPSC-derived MSNs have been transplanted into HD models and motor tasks have been used for functional readouts ((Adil et al., 2018; Besusso et al., 2020; Jeon et al., 2012; Jeon et al., 2014; Ma et al., 2012; Wu et al., 2018); see Table 1). Ma et al. (Ma et al., 2012) transplanted hESC-derived MSN progenitors into a QA lesion mouse model and carried out the rotarod task at monthly intervals post-transplantation. At 16 weeks post-transplantation, they report a significant increase in the latency to fall from the rotarod for mice that received hESC-derived MSN progenitor transplants compared with controls. They also observed that these transplanted mice showed improved performance on the open field test. Additionally, gait analysis on Treadscan revealed significantly greater stride length compared to control mice. These data suggest that some locomotor deficits may improve following transplantation of LGE-like progenitors.

Wu et al. (Wu et al., 2018) transplanted hESC-derived MSNs into adult QA-lesioned mice and assessed motor behaviors using open field and rotarod. Grafted animals showed a significant increase in total distance moved in the open field test at 16 weeks post-transplantation compared to sham controls. On the rotarod task, transplanted mice performed significantly better than sham controls from 8 weeks post-transplantation, with a greater latency to fall which was maintained to the end of the study at 16-weeks post-transplantation.

In Besusso et al. (2020), no significant differences were seen in rotarod performance between QA-lesioned rats transplanted with hESC-derived MSN precursors and sham controls, however, significant improvements were observed in vibrissae-evoked hand placing and adjusted steps tests by 1-month post-transplant compared with sham controls, with improvement maintained to 2 months post-transplantation for the stepping, but not for the vibrissae test (Besusso et al., 2020). Thus, these studies demonstrate that improvement in motor function in the QA-lesion models is feasible post-transplantation of hPSC-derived MSNs,

Jeon et al. (Jeon et al., 2012) transplanted neural progenitors derived from three cell lines (a HD-mutant hiPSC line, a non-mutant hiPSC line, and a non-mutant hESC) into a unilateral QA lesion rat model. Functional assessments were performed every 2 weeks from 2 until 12 weeks post-transplant. At 2 weeks post-transplant, the non-mutant hESC and non-mutant hiPSC transplant groups improved on the staircase task compared to the sham transplanted animals and the HD-mutant hiPSC cell transplant group. Two weeks later, the HD-mutant hiPSC transplant group also showed significant improvement on the staircase task compared to the sham transplants. These improvements were maintained to the 12-week time point. In the stepping task, the non-mutant hiPSC transplant group showed significant improvements from 4 weeks and all cell transplant groups improved significantly from 6 weeks post-transplant, again with improvement maintained to 12 weeks post-transplant. While all transplanted groups achieved functional improvement compared to the sham transplant group, it is challenging to understand the scale of these improvements as unlesioned animals were not used in this study, nor were raw ipsilateral data reported (Jeon et al., 2012). In a later study from this group (Jeon et al., 2014), HD-mutant hiPSCs derived neural precursors were transplanted into the YAC128 HD mouse model at 12 months of age. An improvement in rotarod activity from 3 to 12 weeks post-transplantation was observed, compared to untreated mutant controls. Notably, in both studies, the improvement exhibited by the transplanted groups was variable, highlighting the need to conduct functional assessments over multiple sessions and over a prolonged period post-transplantation to fully observe the impact of a transplant.

Adil et al. (Adil et al., 2018) performed bilateral transplantations of hESC-derived MSN precursor cells embedded in a 3D hydrogel into R6/2 HD transgenic mice at 5-weeks of age. Two weeks post-transplantation weekly motor assessments were initiated comparing R6/2 mice that had received transplants, untreated R6/2 mice and untreated wild-type mice. Rotarod data were presented as normalized latency to fall and showed that transplanted mice performed significantly better than non-transplanted R6/2 mice from 10 weeks of age (5 weeks post-transplant). This improvement over untreated mutant controls was maintained until 16–17 weeks of age (i.e., 11–12 weeks post-transplant), whereupon the untreated mutant animals had all succumbed to the disease. Additionally, treated mutants exhibited a $\sim$19% increase in lifespan and continued motor benefits up to week 20. Onset of clasping was delayed and the phenotype less marked in transplanted R6/2 mice compared with untreated R6/2 mice. This data, together with the 2014 study from Jeon et al., shows the capacity of hPSC-derived precursor cells to alleviate deficits associated with an aggressive degenerative HD mouse model.

There has been no report of cognitive assessment to date in animals following the transplantation of hPSC-derived MSNs or human fetal striatal tissue, although, rat fetal allograft studies have provided proof of principle that intrastriatal transplants in models of HD can produce recovery on cognitive tasks (Brasted, Watts, Robbins, & Dunnett, 1999; Brasted, Watts, Torres, et al., 1999; Dobrossy & Dunnett, 2007). Cognitive deficits are an important part of the HD phenotype, and thus there is a need in future studies to assess the capacity of hPSC-derived MSNs to improve impaired cognitive function following transplantation,

3.2 Neurochemical, electrophysiological and neuroanatomical assessments

Neurochemical and electrophysiological recovery following CRT relate to the restoration of neurotransmitter expression and the expression of appropriate electrophysiological activity of the transplanted cells in the brain. These measures can be assessed using techniques such as positron emission tomography (PET) scans, electron microscopy and electrophysiology. An additional level of assessment of graft functionality is neuroanatomical integration, which can be assessed in tracing studies, for example, fluorogold tracing and monosynaptic tracing based on the modified rabies virus. Table 1 highlights the studies that have included one or more of these assessments *in vivo* following transplantation of PSC-derived MSNs. These studies are discussed in more detail below.

To date there has been only one reported study where hPSC-derived MSN transplants were assessed using PET scans (Besusso et al., 2020). In this study, PET quantification of [^{11}C]Raclopride (a specific D_2/D_3-dopamine receptor antagonist) was used to detect D_2/D_3-dopamine receptor availability at 2 months post-transplantation. There was no significant difference in the uptake of [^{11}C]Raclopride between transplanted animals and sham controls, suggesting that there was no change in D_2/D_3-dopamine receptor expressing cells in the striatum. Besusso et al. also examined the electrical potential of their transplanted hESC-MSN product 2 months following engraftment by performing patch-clamp recordings. Evidence of functional maturity was reported, as determined by the ability of some grafted cells to generate single and repetitive action potentials, and the presence of large voltage-gated Na^+ and K^+ currents. In addition, they reported that grafted cells could receive functional synaptic inputs by recording spontaneous postsynaptic potentials, suggesting some level of integration of the grafted cells with surrounding host tissue.

Besusso et al. was one of two studies to assess neuroanatomical integration. They used monosynaptic tracing based on the modified rabies virus to show that host striatal neurons can make synaptic contacts with grafted cells, although other areas of the basal ganglia were not found to form synaptic connections with the grafted cells. While they did not assess the capacity of grafted cells to integrate into the host striatal neurons and form synaptic connections, histological findings suggested that grafted cells do project into the host brain, since hNCAM positive fibers were observed in various areas of the basal ganglia, including the globus pallidus and subthalamic nucleus. Additionally, they report fiber projections to the ventral medial thalamic nucleus and the substantia nigra pars reticulata. The second study that assessed neuroanatomical integration of hPSC-derived grafts was Adil et al. (2018). They used the retrograde tracer Fluorogold to demonstrate that graft-derived cells projected to, and connected with, cells of the host substantia nigra from bilateral transplant in R6/2 mice (Adil et al., 2018).

Electron microscopy can also provide evidence of circuit reconstruction by demonstrating synaptic contact between host and graft-derived neurons. Comella-Bolla et al. (2020) used GFP-immunogold labeling in conjunction with electron microscopy to demonstrate synaptic connections between cells of the mouse host brain and grafted hPSC-derived cells (Comella-Bolla et al., 2020). Specifically, grafted cells received excitatory synaptic inputs from host neurons and formed inhibitory synaptic outputs onto target host cells. This provides evidence that hPSC-derived MSNs can innervate within the normal striatum and can play a functional role in the host striatum.

3.3 Summary

The reviewed studies from this section highlight the need to include more assessments of functional recovery in CRT preclinical studies, in order to fully characterize the ability of hPSC-derived MSN grafts to mature, integrate and improve behavioral deficits in HD. To date, only a limited number of studies (Adil et al., 2018; Besusso et al., 2020; Delli Carri et al., 2013; Jeon et al., 2014, 2012; Ma et al., 2012; Wu et al., 2018) have assessed motor function, and no study has measured functional recovery in cognitive tasks post-graft. There are also only limited numbers of studies (three such papers) investigating neurochemical, electrophysiological or neuroanatomical aspects of recovery.

Overall, the studies discussed above have provided some evidence that grafted cells can function *in vivo*: graft-derived cells have the ability to generate single and repetitive action potentials (Besusso et al., from patch clamp recordings); they can make synaptic connections with host striatal cells within specific areas of the basal ganglia circuitry (e.g., SN in Adil et al. using fluorogold); and host striatal cells can connect with transplanted cells (Besusso et al. monosynaptic tracing using rabies virus). Furthermore, hPSC-derived MSN grafts are able to bring about some level of functional recovery in different behavioral tasks, with Jeon et al. (2014) and Adil et al. (2018) reporting improvements on the rotarod task in transplanted transgenic animals, and Ma et al. (2012) and Wu et al. (2018) reporting motor improvements in the QA lesion model. Improvements on the APO-induced rotation test were also reported in Jeon et al. (2012) and Delli Carri et al. (2013) in the QA-lesion model. However, almost no studies have used non-invasive imaging to characterize the graft or the host response to the graft, with the exception of Besusso et al. (2020).

Thus, the data available to date suggest that, in some circumstances, hPSC-derived grafts can mature in the brain into MSN-like cells and can confer functional benefit. However, limited data are available for each cell therapy product (with each study comprising differences in starting cell line and differentiation protocol), making it pertinent to collect more systematic and thorough data to better understand the benefits and limitations of each potential therapeutic option. Additionally, differences in the models and species used (genetically-modified vs lesion models, mice vs rats), makes direct comparison between the approaches challenging. Looking forward, the addition of non-invasive imaging to future studies of this type, in combination with functional and neuroanatomical assessments, will enable us to obtain a more complete and translational vision of the graft and its functionality.

4. Challenges

Up to now, this chapter has considered what has been achieved to date from CRT in HD with respect to functional recovery and circuit reconstruction. This section will focus on challenges that we see as critical for progressing this field to the clinic—both those that have arisen from the above discussed studies and those that have not yet been tackled.

4.1 Creating an efficacious cell therapy product

4.1.1 Homogeneity vs heterogeneity of the cell product: Should the cell product for transplantation in HD comprise a pure striatal MSN population or is a more diverse range of cells required for a successful transplant?

Whether or not we are transplanting the most appropriate cells for repair in HD requires systematic evaluation and has been reviewed previously (Precious & Kelly, 2017). The focal loss of MSNs in the striatum makes them an attractive donor cell target for CRT, but MSNs cannot be defined as a single phenotype. Indeed, MSNs may be categorized in various ways: by striatal subcompartments (i.e., striatal patch (striosome) or matrix neurons); by the pathways in which they function (i.e., direct vs indirect pathways); by their biochemical composition (i.e., expression of different receptors, neurotransmitters and neuropeptides); and by additional variations in gene expression (Bocchi et al., 2021; Gokce et al., 2016). Fully defining these differences and understanding how they impact on the success of a cell therapy product remains an important challenge for the advancement of hPSC-MSNs to clinical application. This work has already begun and recently the RNA expression profile of human fetal striatal tissue (specifically LGE in this case) was explored at the single cell level, providing a "molecular atlas" of this developmental structure and the MSN progenitors within (at 7-, 9-, 11- and 20-weeks post-conception, (Bocchi et al., 2021)). Further work is required to similarly define hPSC-MSNs, and comparisons of hPSC-derived MSN subtypes to those from human fetal striatal tissues would be very valuable for the field.

In addition to furthering our understanding of the optimum MSN content, the ideal cell therapy product may contain cells other than MSNs. The majority of early preclinical studies exploring CRT for HD transplanted the whole ganglionic eminence (WGE), which can be further subdivided into the lateral side (LGE, from which MSN progenitors arise), medial side (MGE), and caudal tail (CGE). MSNs do not arise from either the MGE nor CGE, but both substructures contribute interneurons to the striatum, and thus may be relevant for MSN development. Watts, Brasted, and Dunnett (2000) conducted a direct comparison of striatal transplants of cells from different GE dissections (LGE, MGE, and the collective WGE) (Watts et al., 2000) into QA lesioned rats and observed that of these three transplanted products, WGE gave rise to more DARPP32 positive MSNs than LGE or MGE did alone. This suggests that the MGE derived cells could support the LGE progeny following engraftment, either by providing important developmental signals or neurotrophic support. Interestingly, different levels of functional recovery across different behavioral tasks were observed with each of these two tissues, which could further indicate that cells from both LGE and MGE support different aspects of functional recovery.

Taking these studies into consideration, the limited data showing functional improvements with hPSC-derived MSN products may in part be because we are not yet transplanting an optimal cell therapy product. Considering this, it may be fruitful to differentiate a range of desirable cell types (including striatal interneurons and glial progenitors) in order to improve cell integration and function. Identifying the most suitable population of cells for transplantation will undoubtedly be a key element to bringing about reliable efficacy.

4.1.2 *Authenticity of the cell product: Is a lack of authenticity a problem? Do we require an authentic cell type to produce a successful CRT product?*

Optimizing a range of cell types for transplantation may lead to improved functionality of grafts; however, an alternative explanation for the inconsistent recovery observed with hPSC-derived MSN cell products could be that these cells are not sufficiently "authentic." Determining the authenticity of hPSC-derived MSNs is particularly challenging as what constitutes an authentic MSN is not yet fully defined, but to be considered authentic, the cell should have the full complement of characteristics required by endogenous striatal MSNs. Currently, as detailed above, cell products for transplantation are assessed for MSN fate specification using a small number of markers that do not characterize all aspects of an authentic MSN; for example, co-expression of CTIP2 and DARPP32, does not necessarily mean a cell comprises all critical traits to ensure an authentic MSN phenotype. By the nature of protocol optimization, it is likely that we are biasing cell populations to preferentially express the markers we use to measure success. Thus, in doing so, we may be ignoring other critical genes and features of striatal MSNs.

Overcoming such issues has been challenging as the preferred methods used for assessment (such as immunocytochemistry, QPCR, electrophysiology) are commonly limited in their measures and outputs, produce low resolution data, and can be prone to bias. In the case of QPCR, data are not straightforward to interpret as expression is usually reported as a relative fold change compared to a control. If the control is unsuitable or unreliable, readouts could be misguided, resulting in incorrect conclusions. Furthermore, as the method is a bulk analysis tool (that pools all cells in a sample together), heterogeneity of the control population or hPSC-derived MSN product will also produce only a shallow data set that is hard to interpret without further information such as histological analysis. However, histological analysis is also limited. For example, in their post-transplantation graft analysis, Ma et al. (Ma et al., 2012) reported that 58.6% of their grafted cells expressed DARPP32, and therefore they produced perhaps the highest MSN content. However, they also noted that the intensity of DARPP32 staining in their hPSC-derived product was lower than that observed in the endogenous MSN population (though this was not quantified), suggesting that DARPP32 was indeed expressed by their transplanted cells, but at lower levels than the authentic MSNs of the host mouse brain. Technologies such as single cell sequencing methods can help address such difficulties, but currently these methods are substantially more expensive and less accessible, which limits their use in systematic differentiation optimization studies. It is also important to note that the interpretation of much data relies on comparison to a good control, so future work should aim to include comparisons to authentic MSNs where possible, such as *bona fide* human MSNs harvested from human brain (Bartley, Lelos, Gray, & Rosser, 2021).

As mentioned above, this work has already begun: RNA sequencing (both bulk and single cell) has been used to assess the gene expression of adult mouse MSNs (Gokce et al., 2016) and human LGE MSN progenitors (Bocchi et al., 2021). Data from these studies have offered some insight into how MSN-related genes are expressed throughout specific periods of development, how they define fate specification, and how they continue to define MSNs following maturation and integration. However, it is also necessary to determine which features of an authentic MSN are necessary for a successful cell therapy product, and which aspects (if any) are not required. Rather than simply pursue authenticity, we need to maintain a goal that the ideal cell therapy product is one that restores function and overcomes disease mediated deficits.

4.1.3 Methods for generating a cell product for HD: Why is there variability in the output of PSC-derived MSN directed differentiations? Is directed differentiation of PSCs the optimal method? How does the method of generating MSNs influence the final cell product?

We have discussed the progress made in developing protocols for directed differentiation of hPSCs toward producing MSNs. Generally, these protocols manipulate three main pathways (SHH, WNT and TGFβ) to promote an LGE-like fate based on developmental principles. SHH is released by the notochord which is positioned along the ventral midline of the neural tube. It orientates progenitors within the neural tube in a gradient fashion and increased exposure of SHH (or its agonists) to neural forebrain progenitors has been shown to increase the expression of ventral telencephalic genes (Nicoleau et al., 2013; Rallu et al., 2002). Conversely, engagement of the canonical WNT pathway causes forebrain progenitors to commit to a more dorsal fate, and thus inhibiting this pathway (commonly using DKK1 or XAV939) is used to prevent dorsalization (Backman et al., 2005; Nicoleau et al., 2013). The third pathway of note, the TGFβ pathway, was suggested to be involved in striatal neuron differentiation following the finding that activin receptors and Smad2 are expressed in the developing LGE (Maira, Long, Lee, Rubenstein, & Stifani, 2010). Addition of Activin A, a TGFβ family protein, has been demonstrated to rapidly induce an LGE-like phenotype, without increasing more ventral genes (which are more commonly associated with the MGE) (Arber et al., 2015). While each of these pathways is relevant for generating the LGE phenotype, no one protocol has emerged as consistently capable of efficient generation of MSN-like phenotypes.

Replication of these protocols has been shown to be challenging, with vast discrepancies in the neuronal and MSN yield reported between labs (Le Cann et al., 2021). Furthermore, Le Cann et al. showed differing results for a range of cell lines using the same protocol. The reasons for these differences are unknown, however they could be related to parameters of the starting cell source, for example, phase of the cell cycle, genetic variations, epigenetic variations, clonal differences, or passage number when initiating a differentiation (passage number is reported in two of the studies discussed in Section 2 (Aubry et al., 2008; Ma et al., 2012)). Of the papers discussed in Section 2, all publications report the names of the cell lines used, whether hESC or hiPSC (of note, Le Cann et al. did not use any of the same cell lines as Stanslowsky et al. and Fjodorova et al. (Fjodorova & Li, 2018; Stanslowsky et al., 2016), the two published protocols which they compared). Wu et al. (2018) report comparisons of the yield of DARPP32 and GABA positive neurons in three different ESC lines (Wu et al., 2018). However, there is no indication in the other papers that the cell lines used respond differently to the directed differentiation protocol applied (unless the difference in cell lines forms part of the experimental design, i.e., normal hiPSC v HD hiPSC), and thus no indication of differences in the phenotypic outputs. Systematic comparisons (including the testing, validation, and optimisation of protocols across different cell lines) have been reported for the generation of midbrain dopaminergic neurons for PD (Nolbrant, Heuer, Parmar, & Kirkeby, 2017), and similar comparisons in hPSC-MSN differentiations could give further insight into the importance of the above parameters.

An alternative strategy for deriving striatal cells is direct reprogramming. Direct reprogramming may result in a more homogeneous cell phenotype than is currently achievable with directed differentiation methods, which in turn may prove beneficial if this is what is required for functional efficacy. Victor et al. (2014) reprogrammed human adult fibroblasts *in vitro* to MSNs using a lenti-viral vector co-expressing two important neuronal microRNAs (miR-9/9* and miR-124) with four striatal transcription factors (BCL11B, DLX1, DLX2 and MYT1L) (Victor et al., 2014). Reprogrammed cells, 5 weeks post-transduction, expressed neuronal markers MAP2, β-III-tubulin and NeuN (90%, 87% and 84% of total cells, respectively). In addition, 90% of MAP2 positive cells expressed GABA and 70% of MAP2 positive cells expressed DARPP32. To assess these cells *in vivo*, reprogrammed cells tagged with eGFP, were transplanted at 2 weeks post-transduction (20,000 cells injected) into the striatum of P0–P1 immunodeficient mice.

Histological analysis of transplanted cells showed expression of MAP2 (93%), DARPP32 (91% of eGFP), FOXP1 (91% of eGFP) and GABA (76% of eGFP). At 16-weeks post-transplantation, grafted cells were shown to be electrophysiologically active with membrane properties similar to those seen in host MSNs. Furthermore, eGFP positive projections were observed between the striatum and the substantia nigra.

Another approach to direct reprogramming is *in vivo* direct conversion. Wu et al. (2020) used an AAV vector with two transcription factors (NeuroD1 and Dlx2) and an mCherry reporter, which they delivered to the brain to directly reprogram striatal astrocytes into GABAergic neurons in two transgenic models of HD; R6/2 mice (at 2 months of age) and YAC128 mice (at 15 months of age) (Wu et al., 2020). In the R6/2 mice, 1 month after delivery, 78.6% of infected astrocytes expressed neuronal marker NeuN, with 15% expressing astrocyte marker S100β. Assessment of MSN subtypes revealed expression of DARPP32 (56.6.% of converted cells); GAD67 (82.4%); GABA (88.7%) and parvalbumin (8.4%). Converted neurons displayed electrophysiological properties similar to host neurons, reduced mhtt aggregates and were observed to project to principal areas of the basal ganglia circuit such as the globus pallidus and substantia nigra. While similar success was observed in the aged YAC128 mic, the conversion was less efficient as numbers were generally lower. Functional assessment in the R6/2 mice showed that treated animals improved in different behavioral tasks, such as stride length, increased total distance traveled in open field and reduction in clasping phenotype, with no change in the grip strength test.

These two studies of direct reprogramming offer alternative methods to replace lost MSNs in the degenerated HD brain. It would be interesting to look further into the MSN subtypes generated from these direct reprogramming methods, to explore their capacity for transplantation into various HD models and carry out systematic functional assessments to determine the extent of functional recovery. More studies are needed to increase our understanding of both the requirements and the implications of generating MSNs in this way, in order to understand what stage cells should be reprogrammed to along the MSN-development/maturation pathway; which subtypes are being generated; whether reprogrammed cells miss key features because critical developmental steps were bypassed; and whether they suffer from the same (or other) issues as observed with directed differentiation cell products.

Overall, we conclude that a number of elements of the *in vitro* differentiation process remain poorly understood and some aspects are not systematically controlled or compared. It is not known whether cells need to progress through all phases of cell development in order to generate authentic MSNs or whether bypassing developmental stages *via* direct reprogramming may also result in a desirable cell type that can be functionally efficacious upon transplantation. Cells harvested from regions of the fetal brain are already lineage-committed, having undergone endogenous fate specification. Considering this natural trajectory taken by *bone fide* striatal MSNs, directed differentiation protocols for PSCs have applied appropriate developmental knowledge. More will be learnt about cells in specific striatal sub-regions and at different gestational ages from in-depth studies of fetal brain tissues using more recent advanced technologies such as RNA sequencing (Bocchi et al., 2021).

4.1.4 Alternative neural cell types as CRT products: Does the cell product for transplantation need to be MSN precursors?

It is worth mentioning here that several studies have reported successful functional recovery in HD models following transplantation of hPSC-derived neural precursor cells (NPCs), despite not being patterned to an MSN-like phenotype. In 2007, Song et al. co-cultured hESCs with PA6 stromal feeder cells to obtain neuronal progenitors that were transplanted into unilateral QA lesioned rats (20,000 cells grafted) (Song et al., 2007). Animals were perfused 3 weeks post-transplantation and histological processing of the brains showed grafted cells positive for Nestin, TUJ1 (βIII-tubulin) and DCX, corresponding to an early neuronal phenotype, but negative for the more mature markers MAP2 and NeuN. Animals demonstrated improvement in APO rotations compared to sham-transplanted rats from 1 week post-transplantation and remained improved during the 3 weeks of study. Importantly, expression of striatal markers was not reported.

Reidling et al. (2018) used hESCs to obtain NSCs (positive for CD24, SOX1, SOX2, Nestin and Pax6), and transplanted them bilaterally into 5-week-old R6/2, and 2-month-old Q140, HD mice (100,000 cells per hemisphere) (Reidling et al., 2018). Four weeks post-transplantation into R6/2 mice, the grafted cells showed expression of MAP2 and βIII-tubulin, but no specific MSN or striatal markers were reported. In the Q140s, approximately 25% of cells survived 6 months post-transplant, 84% were GFAP positive and 16% were DCX positive. Electrophysiological recordings demonstrated that transplanted hNSCs had properties of immature neurons (small membrane capacitance, and higher membrane resistance), but despite this they displayed spontaneous excitatory and inhibitory postsynaptic currents, although at a lower frequency than endogenous host MSNs. This may indicate integration with surrounding host and/or grafted cells, a notion supported by electron microscopy findings that showed symmetrical and asymmetrical contacts of the host with transplanted cells. In terms of functional recovery, grafted R6/2 mice showed increased latency to fall at 1- and 3-weeks post-transplantation compared to non-grafted R6/2 mice on the rotarod test. Transplanted R6/2 mice also performed better in other motor tasks such as the pole test and the forelimb grip test, where improvements were observed after 4 weeks. Grafted Q140 mice showed improvements in the pole test at 3- and 5- months post-grafting. In terms of cognitive deficits, the Q140 mice demonstrated improvements in a novel object recognition task after 5 months. Interestingly, a reduction in mhtt aggregates in the Q140 mice was also reported, which may have contributed to this functional response.

Park et al., 2021 differentiated a hiPSC line into NPCs which expressed NESTIN, SOX2 and MUSASHI, and later, (6 weeks after neural differentiation) expressed 26% NeuN, 3% O4 (oligodendrocyte marker) and 3.7% GFAP (Park et al., 2021). YAC128 mice received bilateral striatal transplants at 6 months old (400,000 cells/hemisphere), and at 5 months post-transplantation approximately 33% of grafted cells survived and expressed 26% MAP2, 2.6% O4 and 66% GFAP. Grafted cells expressed hDARPP32, with some co-expression with CTIP2 or synaptophysin. At 1-month post-transplant mice showed improvements on the rotarod and grip strength tests; at 3-months improvements in the novel object recognition task, and at 4-months improvements in the simple swim test.

Recently, the same authors looked at knockdown of SUPT4H1 (a gene required in HD for specific expression of the trinucleotide and hexanucleotide repeat expansions) in an HD-hiPSC line (Park et al., 2022). Cells were spontaneously differentiated into a neural progenitor phenotype and transplanted into 6-month-old YAC128 mice. Three months post-transplantation, approximately 30% of SUPT4H1-grafted cells survived with a proportion expressing DARPP32 and MAP2. Transplanted YAC128 mice showed improvements on the rotarod at 1- and 3-months post-transplantation and improvements in the grip strength task at 3-months post-transplantation.

In an alternative approach, Benraiss et al. (2016), transplanted human glial progenitor cells (hGPCs) isolated from human fetal brain into neonatal R6/2 mice (bilateral transplants of 50,000 cells/hemisphere) in order to understand the therapeutic potential of glia in HD (Benraiss et al., 2016). Transplanted R6/2 mice survived longer (mean increase of 12 days) and showed delayed onset of motor deterioration compared to untreated R6/2 controls. Specifically, at 12 weeks of age rotarod performance was significantly better in transplanted R6/2 mice. Further additional motor assessments (using a SmartCube and NeuroCube, Psychogenetics Inc. USA) demonstrated significant phenotypic differences between treated and untreated R6/2 mice (including locomotion, rearing, stride length, age-dependent gait, body movement variability), and transplanted R6/2 mice performed better in the T-maze compared to R6/2 sham treated mice. Interestingly, authors noted that there was substantial replacement of host mouse astroglia with normal htt-expressing human counterparts, and the graft-derived cells integrated into the host brain as astrocytes or GPCs and not neurons.

These studies highlight that functional recovery following transplantation in HD may be achieved even without an MSN-based cell product. Of note, in the studies from Reidling et al. (2018) and Park et al. (2021, 2022) cell survival post-transplantation was in the range of 25–33%, and of the surviving cells in these studies, 85% (Reidling et al., 2018) and 66% (Park et al., 2021) were identified as astrocytes. Although the role of glial cells in HD is not entirely understood, the study by Benraiss demonstrated the potential of glial progenitor cell transplants in HD. Taken together, these studies might suggest that functional recovery may occur through multiple mechanisms (i.e., not just circuitry repair, but also through trophic or other effects) and that a combination transplants of GPCs with NPCs/MSNs could have a beneficial impact on functional recovery in HD models.

4.2 Effective study of recovery

4.2.1 Capturing successful behavioral recovery in animal models: Do we have the necessary tools to determine whether a graft is successful? What aspects of behavior can cell therapies affect?

This chapter has highlighted studies that aimed to assess functional recovery in HD models following transplantation of hPSC-derived MSNs. However, to date, evidence of graft-mediated recovery remains modest (see Section 3.1) and is based only on simple motor or drug-induced assays. But we begin by asking whether we have the appropriate tools (preclinical rodent models, behavioral assays) to robustly measure functional recovery long-term.

Both lesion and genetically modified rodents are available to test therapeutics, however, interestingly graft survival has been shown to be model- and species-dependent. There are two main models of HD, QA lesion model and genetically modified rodents (Lelos & Dunnett, 2018). Intrastriatal infusion of the excitotoxin QA has been widely used to model

HD; selectively ablating the striatal MSNs and sparing other cell types such as striatal interneurons (Sanberg, Calderon, Giordano, Tew, & Norman, 1989). The unilateral striatal QA lesion model has been demonstrated to produce motor (Dobrossy & Dunnett, 2005; Fricker et al., 1997; Kalonia, Kumar, & Kumar, 2011; Lelos et al., 2016; Pundt et al., 1996; Sanberg et al., 1997) and cognitive deficits (Dunnett, Nathwani, & Brasted, 1999; Lelos, Harrison, & Dunnett, 2011, 2012; Shear et al., 1998). Thus, the QA lesion model has the advantage of modeling striatal atrophy and some functional deficits with relevance to HD, but it lacks the cellular pathology associated with the huntingtin mutation. Of note, the QA-lesion model is not only used in the majority of the hPSC-derived MSN transplants discussed in this chapter, but also in all the published human fetal striatal transplant studies.

Since the identification of the huntingtin gene, a multitude of genetic models of HD have been developed. These models reproduce various aspects of the disease, including neuropathological features and behavioral impairments (e.g., reviewed in (Brooks, Jones, & Dunnett, 2012; Hickey & Chesselet, 2011; Menalled & Chesselet, 2002; Pouladi, Morton, & Hayden, 2013)). Ascertaining the extent of deficits on behavioral tasks of a given model requires longitudinal, systematic studies following the time-course over a variety of tasks over their lifespan (Brooks, Higgs, Janghra, Jones, & Dunnett, 2012; Brooks, Higgs, Jones, & Dunnett, 2012; Brooks, Janghra, et al., 2012; Menalled et al., 2012; Rising et al., 2011; Yhnell, Dunnett, & Brooks, 2016a, 2016b). The most widely characterized rodent genetic HD models are in the mouse, but intrastriatal grafts in the mouse host have been reported to be less successful than intrastriatal grafts in the rat host (Roberton et al., 2013). Moreover, grafts have been reported to survive less well in the mutant HD mouse brain than in the wild-type mouse brain (Adil et al., 2018). More recently, genetically modified rat HD models have been generated (Plank et al., 2018) but to date these have not been used as transplant recipients. An advantage of using a mutant rat model of HD would be the ability to assess behavioral phenotypes induced by disease pathology, while the 51 CAG repeat construct expressed in the F344tgHD model (Plank et al., 2018) also has the advantage of being more similar in length to the mutation expressed in humans.

There are inherent limitations of our preclinical models, which include short life-span and the need to modulate the immune system for long-term survival of xenografts. Thus, in this context, we are only able to maintain cell survival for up to ~6 months post-transplantation in an immune-competent rodent treated with systemic immunosuppressants (and immunodeficient rodents can harbor grafts for longer, but systematic behavioral analysis is not feasible). As a result, we are not able to characterize behavioral recovery long-term to determine the ultimate magnitude of recovery induced by a graft. Similarly, we are not able to characterize the long-term stability or robustness of any recovery.

An additional consideration is when to deliver the therapeutic intervention with respect to disease stage. As an example, from the studies reviewed in Sections 2 and 3, Jeon used YAC128 mice (observed to have deficits on the rotarod at 4 months of age (Brooks, Higgs, Janghra, et al., 2012)), transplanted post-disease onset at 12 months of age and observed improvements 3-weeks post-graft in the rotarod task. On the other hand, Adil used R6/2 mice (observed to have deficits at 8.5 weeks of age, with normal rotarod behavior at 4.5 weeks (Hickey, Gallant, Gross, Levine, & Chesselet, 2005)) and transplanted early in the disease course, at 5 weeks of age, and reporting improvements 10-weeks post-transplantation. In these two experiments, mice were transplanted at different times in the disease process. Although both demonstrated improvements on the task, this raises the question as to whether more systematic studies are needed to determine the optimal time for administration. In PD, it has been reported that transplantation into the aged brain results in less innervation and poorer recovery than in young brain (Grinand & Takahashi, 2022), but less is known about the impact of age or disease progression in the HD brain. Striatal atrophy is detectable in HD patients many years prior to the onset of motor symptoms (Aylward et al., 2004), however, genetic models of HD present with relatively little overt atrophy (Brooks, Jones, & Dunnett, 2012). When considering CRT in HD, it has been reported that infusion of cells into a striatum with little or no striatal atrophy produces smaller grafts than transplantation into an atrophied striatum (Watts & Dunnett, 1998). Conversely, if the striatum has suffered severe atrophy, there will not be enough remaining structure in which transplanted donor cells can be injected, populate and integrate. Thus, understanding the relationship between age, extent of disease and the capacity for neural repair remains important in determining the optimal time for therapeutic intervention, and current data hint that early intervention may be optimal for graft survival or innervation.

Another pertinent question in the field is: to what extent can MSN grafts alleviate functional deficits? As discussed in Section 3.1, only a limited number of studies have reported improvements in motor function. Additionally, it should be recognized that the prominent motor feature of chorea is not adequately modeled in rodents, making it unfeasible to assess the impact of cell therapeutics on this key motor feature.

While evidence of some improvement in motor dysfunction is encouraging for the field, no studies have reported the use of cognitive or psychiatric assays after transplantation of hPSC-derived MSNs. The lack of cognitive and psychiatric assessments after engraftement of hPSC-derived MSN cell therapy products is paralleled by the lack of data from primary human fetal striatal tissue pre-clinical transplant studies. Indeed, exploring cognitive and psychiatric behavior after transplantation of both cell types (human fetal striatal tissue and hPSC-derived MSNs) in HD models will be interesting and critical to the field, to determine the extent of functional recovery possible for these aspects of behavior. While there are various cognitive and psychiatric tests that could be employed, including both automated operant tasks (Yhnell & Heuer, 2018) and non-automated tasks (e.g., tail-suspension test, (Sanchis et al., 2019)), cognitive testing often requires specialist equipment and extensive behavioral training, making them more challenging to employ. Thus, given the prevalence of these symptoms in patients, it is pertinent for future studies to consider the impact of any novel therapeutic on cognitive and neuropsychiatric behaviors.

While it is challenging to take all aspects of patient presentation into account using animal models (e.g., age, extent of disease, expanded repeat length, profile of motor, cognitive and psychiatric impairments), it is important to understand how aspects of the models map onto the human condition, and the limitations that exist in the models. Better models will help the translation of cell therapeutics to the clinic and should be used to identify the key parameters for effective graft survival and function.

4.2.2 Mechanism of recovery: How long do we expect it to take for graft induced recovery to occur? How does graft-derived functional recovery happen?

The mechanism(s) underpinning the efficacy of CRT has not been definitively established. Key concepts include graft-mediated trophic release to support and protect host neurons from further damage and/or reconstruction of critical elements of host neural circuitry. Data from clinical trials using fetal derived tissues suggest the first signs of recovery were observed in HD patients, at the earliest, at 6 months post-transplantation (Bachoud-Levi & Perrier, 2014). However, in the reviewed studies from Section 3, the earliest recovery timepoint in spontaneous behavior is 2 weeks post-transplant in the clasping task in R6/2 model (Adil et al., 2018), in the staircase task and in APO-induced rotations in QA lesion model (Jeon et al., 2012). Observations of behavioral improvements at such short time points post-transplantation raises questions about the mechanism underlying this recovery. In particular, it raises the possibility that the mechanisms supporting short-term improvements in animal models (possibly trophic effects) are different to those bringing about more long-term functional recovery (i.e., the former may be driven by trophic effects, while the latter may be driven by circuitry reconstruction).

The concept of graft-mediated trophic release is based on the idea that neurotrophic factors, such as glial cell derived neurotrophic factor (GDNF) or BDNF, can protect and support host cells under threat from neurodegenerative processes (Allen, Watson, Shoemark, Barua, & Patel, 2013). This is consistent with reports that some functional benefit, such as behavioral recovery in motor tasks and reduction of striatal atrophy, can be observed in transplants where long-term cell survival is very low (e.g., such as those using mesenchymal stem cells as the transplant material) (Im et al., 2010; Lin et al., 2011; Pollock et al., 2016). Indeed, in some circumstances, the transplant could be used as a vector to deliver trophic molecules or neuroprotective agents (Dey et al., 2010; Pollock et al., 2016). As discussed in Section 4.1.4, the study from Benraiss and colleagues (Benraiss et al., 2016) reported improvements in motor and cognitive deficits after transplanting hGPCs. Based on these results, glial cells may have an important role supporting the remaining striatal population, restoring K^+ homeostasis, Ca^{2+} signaling or BDNF expression (Siracusa, Fusco, & Cuzzocrea, 2019). This highlights the idea of a CRT product combining neuronal and glial progenitors. However, in cases where substantial degeneration has already occurred, simply supporting endogenous neurons may prove insufficient.

Rather than simply supporting endogenous neurons, a different approach is likely to involve both replacement of the cells lost to the disease and neural repair. Circuit reconstruction incorporates the idea that once cells are transplanted (i) they will survive long-term and will not be rejected by the host; (ii) they will integrate within the host brain and mature into the desired phenotype; (iii) they will extend projections to form new neural networks; (iv) and these neural networks will reconnect the appropriate circuit and support multiple aspects of functional recovery. The extent of reconstruction is not likely to perfectly recapitulate the original anatomy, however, it is anticipated that the new cells would have sufficient plasticity to adapt and form relevant synaptic connections with the host. It has been hypothesized that circuit reconstruction by itself is not enough to recovery the functionality of the damaged area (Brasted, Watts, Robbins, & Dunnett, 1999; Coffey, Lund, & Rawlins, 1989; Dobrossy & Dunnett, 2003); indeed, functional recovery has been shown to be affected by the host experiences and training (Dobrossy & Dunnett, 2001, 2003, 2004). Specifically, preclinical studies suggest that environmental enrichment of animal housing (social interaction, activity and play) or training can enhance the degree of innervation and extent of synaptic contacts formed between the graft and host brain, and ultimately improve the extent of graft-mediated recovery (Brasted et al., 2000; Dobrossy & Dunnett, 2001, 2005).

Assessment of circuit reconstruction is key to develop a better understanding of the mechanisms underpinning graft-mediated recovery. Although only limited data on graft-host integration has been reported after grafting of hPSC-derived MSNs thus far, assessment of integration can be achieved using several neuroanatomical technologies, such as neuronal tracers (conventional retrograde tracers (Horse radish peroxidase (HRP), Cholera toxin B or Hydoxystilbamidine (FluoroGold™)); conventional anterograde tracers (*Phaseolus vulgaris*-luecoagglutinin (PHA-L), dextra-amines (DAs)); and viral tracers (Recombinant vectors derived from human immunodeficiency virus (HIV), herpes simplex virus type 1 (HSV-1), adeno-associated virus (AAV) or rabies)) (Saleeba, Dempsey, Le, Goodchild, & McMullan, 2019). Tracing studies can be supported by immunohistochemical (IHC) characterization of fiber outgrowth (hNCAM, STEM121) (Besusso et al., 2020) or neuronal activity using immediate early genes (c-fos, EGR-1) (Labandeira-Garcia & Guerra, 1994; Mandel, Wictorin, Cenci, & Bjorklund, 1992); as well as electrophysiological recordings on brain slices (Besusso et al., 2020), electron microscopy (Comella-Bolla et al., 2020) or PET scans assessing dopamine function (as ^{18}F-dopa, ^{11}C-raclopride) or oxygen and glucose levels ($^{15}O_2$ and 18F-2-fluoro-2-deoxy-D-glucose (^{18}FDG)-PET) (Brooks, 2004). Although there are numerous techniques to assess circuit reconstruction, there is very limited evidence of it in studies reporting transplants of hPSC-derived MSNs, as has been discussed in Section 3. Other new technologies, such as optogenetics and Designer Receptors Exclusively Activated by Designer Drugs (DREADDs), have not been reported in HD CRT but have been used in other models, such as PD or stroke (Chen et al., 2016; Steinbeck et al., 2015; Tonnesen et al., 2011; Tornero et al., 2017). Ideally, combining neuroanatomical technologies with behavioral output will allow scientists to report a more robust proof of functional efficacy by simultaneously assessing different aspects of graft-mediated recovery.

To date, the most success in terms of achieving circuit reconstruction in HD has been following homotopic transplantation of donor striatal cells harvested from rodent or human primary fetal tissue. Numerous studies have demonstrated that grafted fetal striatal progenitors survive and differentiate in the host brain. Studies using xenograft transplantation of human fetal tissue (WGE) have reported connectivity with the host brain using IHC (TH immunostaining, tau immunocytochemistry and NF_{70}) (Armstrong, Watts, Svendsen, Dunnett, & Rosser, 2000; Grasbon-Frodl, Nakao, Lindvall, & Brundin, 1996; Kelly et al., 2007). Rodent allograft transplantation has also provided further evidence of striatal circuit reconstruction. Fetal striatal tissues (WGE) from both mouse and rat have been successfully engrafted in HD rodent models and are capable of integration assessed by neuronal tracing (Kelly et al., 2007; Klein, Lane, & Dunnett, 2013; Wictorin et al., 1990), and electrophysiology (Mazzocchi-Jones et al., 2009). Rodent WGE allografts have also been directly compared with hWGE xenografts, where human xenografts produced larger grafts (Hurelbrink & Barker, 2005; Lelos et al., 2016) and exhibited more extensive migration (Hurelbrink & Barker, 2005) showing how helpful these comparisons could be for clinical translation. However, the work in the rodent stem cells' field is much less than in human stem cells, which can be a disadvantage when it comes to understanding the basic mechanisms of integration and survival of cells.

The mechanism(s) of functional recovery after graft transplantation are not yet elucidated and greater understanding of the capacity for synaptic integration and electrophysiological activity is required. This knowledge will help us understand the recovery process after transplantation. Indeed, early recovery in a specific task is not necessarily is an indicator of sustained improvements long-term, and recovery in any individual assay needs to be interpreted with caution and if possible, confirmed by the use of multiple assessments to allow robust interpretation of the data, as well as exploration of the mechanism.

4.2.3 The impact of graft development and health: How do graft characteristics correlate with the capacity for functional recovery? For how long can grafts survive? Will grafts succumb to neurodegenerative disease?

The size and health of a graft are also important elements that need to be further refined for cell therapies in HD, with particular consideration given to their effect on functional recovery. These are influenced by various factors including the content of the cell product, the number of cells transplanted, the number of deposits, the location of transplant, the local neural environment, and the disease model in use.

While Aubry and colleagues did not examine functional recovery, they did explore the effect that stage of cell maturity had on final graft content and overgrowth following transplantation (Aubry et al., 2008). Their data showed that early hPSC derived neural crest progenitors were too immature and not yet lineage restricted, as grafts using these cells generated teratomas, and transplantation of mature late-stage cells exhibited poor survival, suggesting the mature MSNs were not able to survive the transplantation process. From this they determined that the ideal time point for transplantation was after progenitors had committed to an LGE/ventral telencephalic fate, but not yet undergone terminal differentiation. They then conducted a long-term transplantation study, in which they grafted hPSC derived MSN progenitors at this 'ideal' developmental stage. While the subsequent grafts yielded neuronal populations enriched for DARPP32, the authors reported declining health in grafted animals and found substantial overgrowth of the graft. As highlighted by this example, the issues of teratoma formation and graft overgrowth must be overcome as we progress cell therapies to clinical application.

Later works (such as those reviewed in Section 2) have further optimized neural differentiation methods to minimize the possibility that a cell product contains pluripotent or overly proliferative populations, but the data demonstrating lack of uncontrolled growth will need to be robust for clinical applications. The removal of pluripotent cells to prevent teratoma formation can be achieved using selective elimination techniques such as cell sorting or treatment with pluripotent specific cytotoxins (e.g., SCD-1 inhibition, (Ben-David et al., 2013)). However, overcoming the issue of neural overgrowth perhaps poses a greater challenge. The obvious solution is to ensure cell products are terminally differentiated and therefore have limited proliferative potential. Yet, Aubry and colleagues demonstrated that mature populations show reduced viability following transplantation, and recent work in the PD field has shown that mature populations are less able to integrate with the host brain and elicit functional recovery (Hills et al., 2022). Similarly, in rat allograft studies, transplanting more mature cells is associated with reduced functional efficacy (Fricker et al., 1997; Fricker-Gates, White, Gates, & Dunnett, 2004; Watts et al., 2000). Therefore, further work is required to determine the optimal level of maturity and proliferation through systematic study of the ideal time-point for transplantation and refinement of quality control (QC) measures prior to transplantation. Furthermore, additional measures may need to be in place to ensure that cell products can be made safe for clinical application. Notably, there is ongoing work establishing the use of cell lines with "suicide genes", that can act as a failsafe measure once CRT approaches clinical application (de Luzy et al., 2021).

Following transplantation there is inevitable loss of some transplanted cells due to several variables including: undergoing the physical or enzymatic dissociation process used to generate a cell suspension; the type of suspension generated (tissue/cluster, quasi-single, or single cell); the lack of exogenous trophic support prior to and following transplantation; the process of being drawn up and then expelled through a narrow-gauge needle for the transplant procedure; and the prolonged period settling in the host brain, which may be hypoxic until angiogenesis can occur and is subject to the innate immune response creating a hostile environment that can increase the risk of graft rejection (Praet et al., 2015). Combined, these issues make it challenging to determine the optimal number of cells to be transplanted and the number of deposits to be made. Even after successful grafting, it remains unknown how long grafts can remain healthy and functional. Additionally, the implications of transplanting into diseased brains (i.e., those with expanded CAG repeats) and the long-term impact of the internal environment of the diseased host brain on the donor transplanted cells, needs to be determined. It has been reported that grafts survive less well in the HD mutant brain compared to the wild-type brain (Adil et al., 2018). Additionally, it has been suggested that long-term transplants of human fetal tissue into HD patients have expressed mhtt aggregates, although this would need to be confirmed through co-localization with graft-specific markers (Cicchetti et al., 2014). This latter effect remains to be studied in hPSC-derived transplants and the functional consequences of this pathology remain unknown. Therefore, optimizing for each of these factors represents a considerable challenge for the field and each new transplantation study is a valuable resource that can be used to inform on these issues.

4.3 Wider challenges

There are further challenges which must be overcome to progress CRT for HD to the clinic. These are not specifically related to hPSC-based products, and instead represent issues more directly related to clinical translation. For example, the transition to a GMP protocol, the scale up of manufacturing, the methods of product delivery, methods of overcoming the immune response, patient safety, and designing and running appropriate clinical trials that accurately measure the efficacy of such a treatment. For example, for translation to the clinic, it may be that larger brain models, e.g., mini-pigs, sheep, non-human primates, will provide a more representative scenario of transplantation into HD patients and should therefore be considered for future CRT studies. The advantages of these models lie in their larger brain size, which is ideal for assessing neurosurgical techniques and establishing appropriate cell dosing regimens. These issues have been discussed elsewhere and are beyond the intended scope of this chapter, however, they constitute critical challenges for the wider field (Rosser et al., 2022).

5. Conclusions and authors perspectives

In this chapter, we have reviewed studies that have used hPSC-derived MSNs for transplantation in HD models and assessed functional recovery. Substantial work has been conducted to determine methods of generating MSN-like products and there is evidence of functional recovery across motor behavioral tasks, as well as neurochemical restoration and neuroanatomical reconstruction.

However, it is also true that much remains to be achieved using such cell therapy products. We have outlined many of the challenges that remain for the field to overcome, in doing so, we asked ourselves:

"What do we want to see in future papers published in the field of CRT for HD?"

(1) Consistent and complete characterization of the cell product being transplanted

Each CRT study begins with the cell product that is being transplanted. Therefore, to fully understand and extrapolate from the results following transplantation, the reader must have a detailed understanding of the cell product. Previously, most groups have characterized their cell product by reporting expression of a small selection of genes or other characteristics. With no agreed methods of characterization, it is often hard to compare these data between studies. Ideally, cell products would be characterized in more depth and using methods that readily allow for direct comparisons. The increased availability of single cell sequencing techniques may address this issue, especially as it becomes more common to share raw data from such studies which will allow direct comparisons and more advanced assessment and quantification of a cell product.

(2) A fully defined disease model

As discussed above, there are several disease models of HD that have various benefits and weaknesses. These models are often not equivalent, even between studies. It is therefore essential to have appropriate controls in each study to fully interpret any observed graft functionality. Additionally, consideration must be given to the impact of the age of the rodent, the duration of disease and the nature of the model, to allow more precise interpretation of the functional efficacy data.

(3) Functional assessment across a wide range of well controlled paradigms

Functional assessment in relevant papers often only entails one or two behavioral paradigms. While these can collectively advance the field, we could potentially learn more by performing a wider battery of assessments that encompass behaviors more translational and reminiscent of patient symptoms (to include motor, cognitive and psychiatric symptoms), which will in turn require well-controlled studies with well-powered cohorts of animals.

(4) A well-founded theory of functional recovery

Achieving functional recovery is a central goal of CRT for HD but understanding the mechanisms underpinning such recovery is also deeply important. Complementary to behavioral tasks, additional assessments of connectivity, anatomical and morphological characterization of grafts, and study of graft mediated responses are therefore required.

(5) Long-term characterization

If CRT for HD is to be a viable clinical option, we must endeavor to study recovery over longer periods of time and to observe how function changes during this time. Most studies explore recovery for only a few weeks, while this is partially due to limitations of rodent lifespans, it is possible to study such grafted rodents over longer periods (e.g., ~6 months). Furthermore, once established in rodent models, it will be important to explore cell therapy products in longer-lived and larger brain animal models.

(6) Continued use of human fetal tissues

Human fetal striatal tissue still has a pivotal role to play in guiding us to better understand the characteristics of an authentic MSN. Therefore, it is useful to include these tissues for control transplants in cell therapy studies, and more can be learned from further use of these tissues (Bartley et al., 2021).

(7) Disclosure of negative results

Finally, it is important to disclose negative results, as often understanding where our cell products are failing will allow us to determine what further issues must be tackled.

References

Adil, M. M., Gaj, T., Rao, A. T., Kulkarni, R. U., Fuentes, C. M., Ramadoss, G. N., et al. (2018). hPSC-derived striatal cells generated using a scalable 3D hydrogel promote recovery in a Huntington disease mouse model. *Stem Cell Reports*, *10*(5), 1481–1491. https://doi.org/10.1016/j.stemcr.2018.03.007.

Allen, S. J., Watson, J. J., Shoemark, D. K., Barua, N. U., & Patel, N. K. (2013). GDNF, NGF and BDNF as therapeutic options for neurodegeneration. *Pharmacology & Therapeutics*, *138*(2), 155–175. https://doi.org/10.1016/j.pharmthera.2013.01.004.

Anderson, K. D., & Reiner, A. (1991). Immunohistochemical localization of DARPP-32 in striatal projection neurons and striatal interneurons: Implications for the localization of D1-like dopamine receptors on different types of striatal neurons. *Brain Research*, *568*(1–2), 235–243. https://doi.org/10.1016/0006-8993(91)91403-n.

Arber, C., Precious, S. V., Cambray, S., Risner-Janiczek, J. R., Kelly, C., Noakes, Z., et al. (2015). Activin A directs striatal projection neuron differentiation of human pluripotent stem cells. *Development*, *142*(7), 1375–1386. https://doi.org/10.1242/dev.117093.

Armstrong, R. J., Watts, C., Svendsen, C. N., Dunnett, S. B., & Rosser, A. E. (2000). Survival, neuronal differentiation, and fiber outgrowth of propagated human neural precursor grafts in an animal model of Huntington's disease. *Cell Transplantation*, *9*(1), 55–64. https://doi.org/10.1177/096368970000900108.

Aubry, L., Bugi, A., Lefort, N., Rousseau, F., Peschanski, M., & Perrier, A. L. (2008). Striatal progenitors derived from human ES cells mature into DARPP32 neurons in vitro and in quinolinic acid-lesioned rats. *Proceedings of the National Academy of Sciences of the United States of America*, *105*(43), 16707–16712. https://doi.org/10.1073/pnas.0808488105.

Aylward, E. H., Sparks, B. F., Field, K. M., Yallapragada, V., Shpritz, B. D., Rosenblatt, A., et al. (2004). Onset and rate of striatal atrophy in preclinical Huntington disease. *Neurology*, *63*(1), 66–72. https://doi.org/10.1212/01.wnl.0000132965.14653.d1.

Bachoud-Levi, A. C., Gaura, V., Brugieres, P., Lefaucheur, J. P., Boisse, M. F., Maison, P., et al. (2006). Effect of fetal neural transplants in patients with Huntington's disease 6 years after surgery: A long-term follow-up study. *Lancet Neurology*, *5*(4), 303–309. https://doi.org/10.1016/S1474-4422(06)70381-7.

Bachoud-Levi, A. C., & Perrier, A. L. (2014). Regenerative medicine in Huntington's disease: Current status on fetal grafts and prospects for the use of pluripotent stem cell. *Revue Neurologique (Paris)*, *170*(12), 749–762. https://doi.org/10.1016/j.neurol.2014.10.007.

Backman, M., Machon, O., Mygland, L., van den Bout, C. J., Zhong, W., Taketo, M. M., et al. (2005). Effects of canonical Wnt signaling on dorso-ventral specification of the mouse telencephalon. *Developmental Biology*, *279*(1), 155–168. https://doi.org/10.1016/j.ydbio.2004.12.010.

Bartley, O. J. M., Lelos, M. J., Gray, W. P., & Rosser, A. E. (2021). Do foetal transplant studies continue to be justified in Huntington's disease? *Neuronal. Signals*, *5*(4). https://doi.org/10.1042/NS20210019.

Ben-David, U., Gan, Q. F., Golan-Lev, T., Arora, P., Yanuka, O., Oren, Y. S., et al. (2013). Selective elimination of human pluripotent stem cells by an oleate synthesis inhibitor discovered in a high-throughput screen. *Cell Stem Cell*, *12*(2), 167–179. https://doi.org/10.1016/j.stem.2012.11.015.

Benraiss, A., Wang, S., Herrlinger, S., Li, X., Chandler-Militello, D., Mauceri, J., et al. (2016). Human glia can both induce and rescue aspects of disease phenotype in Huntington disease. *Nature Communications*, 7, 11758. https://doi.org/10.1038/ncomms11758.

Besusso, D., Schellino, R., Boido, M., Belloli, S., Parolisi, R., Conforti, P., et al. (2020). Stem cell-derived human striatal progenitors innervate striatal targets and alleviate sensorimotor deficit in a rat model of Huntington disease. *Stem Cell Reports*, *14*(5), 876–891. https://doi.org/10.1016/j.stemcr.2020.03.018.

Bocchi, V. D., Conforti, P., Vezzoli, E., Besusso, D., Cappadona, C., Lischetti, T., et al. (2021). The coding and long noncoding single-cell atlas of the developing human fetal striatum. *Science*, *372*(6542). https://doi.org/10.1126/science.abf5759.

Boer, G. J. (1994). Ethical guidelines for the use of human embryonic or fetal tissue for experimental and clinical neurotransplantation and research. Network of European CNS Transplantation and Restoration (NECTAR). *Journal of Neurology*, *242*(1), 1–13. https://doi.org/10.1007/BF00920568.

Brasted, P. J., Watts, C., Robbins, T. W., & Dunnett, S. B. (1999). Associative plasticity in striatal transplants. *Proceedings of the National Academy of Sciences of the United States of America*, *96*(18), 10524–10529. https://doi.org/10.1073/pnas.96.18.10524.

Brasted, P. J., Watts, C., Torres, E. M., Robbins, T. W., & Dunnett, S. B. (1999). Behavioural recovery following striatal transplantation: Effects of postoperative training and P-zone volume. *Experimental Brain Research*, *128*(4), 535–538. https://doi.org/10.1007/s002210050877.

Brasted, P. J., Watts, C., Torres, E. M., Robbins, T. W., & Dunnett, S. B. (2000). Behavioral recovery after transplantation into a rat model of Huntington's disease: Dependence on anatomical connectivity and extensive postoperative training. *Behavioral Neuroscience*, *114*(2), 431–436. https://www.ncbi.nlm.nih.gov/pubmed/10832803.

Brooks, D. J. (2004). Positron emission tomography imaging of transplant function. *NeuroRx*, *1*(4), 482–491. https://doi.org/10.1602/neurorx.1.4.482.

Brooks, S., Higgs, G., Janghra, N., Jones, L., & Dunnett, S. B. (2012). Longitudinal analysis of the behavioural phenotype in YAC128 (C57BL/6J) Huntington's disease transgenic mice. *Brain Research Bulletin*, *88*(2–3), 113–120. https://doi.org/10.1016/j.brainresbull.2010.05.005.

Brooks, S., Higgs, G., Jones, L., & Dunnett, S. B. (2012). Longitudinal analysis of the behavioural phenotype in Hdh(CAG)150 Huntington's disease knock-in mice. *Brain Research Bulletin*, *88*(2–3), 182–188. https://doi.org/10.1016/j.brainresbull.2010.05.004.

Brooks, S. P., Janghra, N., Workman, V. L., Bayram-Weston, Z., Jones, L., & Dunnett, S. B. (2012). Longitudinal analysis of the behavioural phenotype in R6/1 (C57BL/6J) Huntington's disease transgenic mice. *Brain Research Bulletin*, *88*(2–3), 94–103. https://doi.org/10.1016/j.brainresbull.2011.01.010.

Brooks, S. P., Jones, L., & Dunnett, S. B. (2012). Comparative analysis of pathology and behavioural phenotypes in mouse models of Huntington's disease. *Brain Research Bulletin*, *88*(2–3), 81–93. https://doi.org/10.1016/j.brainresbull.2011.10.002.

Chambers, S. M., Fasano, C. A., Papapetrou, E. P., Tomishima, M., Sadelain, M., & Studer, L. (2009). Highly efficient neural conversion of human ES and iPS cells by dual inhibition of SMAD signaling. *Nature Biotechnology*, *27*(3), 275–280. https://doi.org/10.1038/nbt.1529.

Chen, Y., Xiong, M., Dong, Y., Haberman, A., Cao, J., Liu, H., et al. (2016). Chemical control of grafted human PSC-derived neurons in a mouse model of Parkinson's disease. *Cell Stem Cell*, *18*(6), 817–826. https://doi.org/10.1016/j.stem.2016.03.014.

Cicchetti, F., Lacroix, S., Cisbani, G., Vallieres, N., Saint-Pierre, M., St-Amour, I., et al. (2014). Mutant huntingtin is present in neuronal grafts in Huntington disease patients. *Annals of Neurology*, *76*(1), 31–42. https://doi.org/10.1002/ana.24174.

Coffey, P. J., Lund, R. D., & Rawlins, J. N. (1989). Retinal transplant-mediated learning in a conditioned suppression task in rats. *Proceedings of the National Academy of Sciences of the United States of America, 86*(18), 7248–7249. https://doi.org/10.1073/pnas.86.18.7248.
Comella-Bolla, A., Orlandi, J. G., Miguez, A., Straccia, M., Garcia-Bravo, M., Bombau, G., et al. (2020). Human pluripotent stem cell-derived neurons are functionally mature in vitro and integrate into the mouse striatum following transplantation. *Molecular Neurobiology, 57*(6), 2766–2798. https://doi.org/10.1007/s12035-020-01907-4.
de Luzy, I. R., Law, K. C. L., Moriarty, N., Hunt, C. P. J., Durnall, J. C., Thompson, L. H., et al. (2021). Human stem cells harboring a suicide gene improve the safety and standardisation of neural transplants in parkinsonian rats. *Nature Communications, 12*(1), 3275. https://doi.org/10.1038/s41467-021-23125-9.
Delli Carri, A., Onorati, M., Lelos, M. J., Castiglioni, V., Faedo, A., Menon, R., et al. (2013). Developmentally coordinated extrinsic signals drive human pluripotent stem cell differentiation toward authentic DARPP-32+ medium-sized spiny neurons. *Development, 140*(2), 301–312. https://doi.org/10.1242/dev.084608.
Dey, N. D., Bombard, M. C., Roland, B. P., Davidson, S., Lu, M., Rossignol, J., et al. (2010). Genetically engineered mesenchymal stem cells reduce behavioral deficits in the YAC 128 mouse model of Huntington's disease. *Behavioural Brain Research, 214*(2), 193–200. https://doi.org/10.1016/j.bbr.2010.05.023.
Dobrossy, M. D., & Dunnett, S. B. (2001). The influence of environment and experience on neural grafts. *Nature Reviews. Neuroscience, 2*(12), 871–879. https://doi.org/10.1038/35104055.
Dobrossy, M. D., & Dunnett, S. B. (2003). Motor training effects on recovery of function after striatal lesions and striatal grafts. *Experimental Neurology, 184*(1), 274–284. https://doi.org/10.1016/s0014-4886(03)00028-1.
Dobrossy, M. D., & Dunnett, S. B. (2004). Environmental enrichment affects striatal graft morphology and functional recovery. *The European Journal of Neuroscience, 19*(1), 159–168. https://doi.org/10.1111/j.1460-9568.2004.03105.x.
Dobrossy, M. D., & Dunnett, S. B. (2005). Training specificity, graft development and graft-mediated functional recovery in a rodent model of Huntington's disease. *Neuroscience, 132*(3), 543–552. https://doi.org/10.1016/j.neuroscience.2005.01.016.
Dobrossy, M. D., & Dunnett, S. B. (2007). The corridor task: Striatal lesion effects and graft-mediated recovery in a model of Huntington's disease. *Behavioural Brain Research, 179*(2), 326–330. https://doi.org/10.1016/j.bbr.2007.02.030.
Dunnett, S. B., & Brooks, S. P. (2018). Motor assessment in Huntington's disease mice. *Methods in Molecular Biology, 1780*, 121–141. https://doi.org/10.1007/978-1-4939-7825-0_7.
Dunnett, S. B., Nathwani, F., & Brasted, P. J. (1999). Medial prefrontal and neostriatal lesions disrupt performance in an operant delayed alternation task in rats. *Behavioural Brain Research, 106*(1–2), 13–28. https://doi.org/10.1016/s0166-4328(99)00076-5.
Fjodorova, M., & Li, M. (2018). Robust induction of DARPP32-expressing GABAergic striatal neurons from human pluripotent stem cells. *Methods in Molecular Biology, 1780*, 585–605. https://doi.org/10.1007/978-1-4939-7825-0_27.
Fricker, R. A., Torres, E. M., Hume, S. P., Myers, R., Opacka-Juffrey, J., Ashworth, S., et al. (1997). The effects of donor stage on the survival and function of embryonic striatal grafts in the adult rat brain. II. Correlation between positron emission tomography and reaching behaviour. *Neuroscience, 79*(3), 711–721. https://doi.org/10.1016/s0306-4522(96)00657-4.
Fricker-Gates, R. A., White, A., Gates, M. A., & Dunnett, S. B. (2004). Striatal neurons in striatal grafts are derived from both post-mitotic cells and dividing progenitors. *The European Journal of Neuroscience, 19*(3), 513–520. https://doi.org/10.1111/j.1460-9568.2004.03149.x.

Gaura, V., Bachoud-Levi, A. C., Ribeiro, M. J., Nguyen, J. P., Frouin, V., Baudic, S., et al. (2004). Striatal neural grafting improves cortical metabolism in Huntington's disease patients. *Brain, 127*(Pt 1), 65–72. https://doi.org/10.1093/brain/awh003.

Gokce, O., Stanley, G. M., Treutlein, B., Neff, N. F., Camp, J. G., Malenka, R. C., et al. (2016). Cellular taxonomy of the mouse striatum as revealed by single-cell RNA-Seq. *Cell Reports, 16*(4), 1126–1137. https://doi.org/10.1016/j.celrep.2016.06.059.

Grasbon-Frodl, E. M., Nakao, N., Lindvall, O., & Brundin, P. (1996). Phenotypic development of the human embryonic striatal primordium: A study of cultured and grafted neurons from the lateral and medial ganglionic eminences. *Neuroscience, 73*(1), 171–183. https://doi.org/10.1016/0306-4522(96)00008-5.

Grinand, L., & Takahashi, J. (2022). Automated measurement of fluorescence signals reveals a significant increase of the graft-derived neurite extension in neonates compared to aged rats. *Regenerative Therapy, 19*, 97–106. https://doi.org/10.1016/j.reth.2022.01.002.

Hickey, M. A., & Chesselet, M.-F. (2011). Behavioral assessment of genetic mouse models of Huntington's disease. In E. L. Lane, & S. B. Dunnett (Eds.), *II. Animal models of movement disorders* (pp. 3–19). Humana Press. https://doi.org/10.1007/978-1-61779-301-1_1.

Hickey, M. A., Gallant, K., Gross, G. G., Levine, M. S., & Chesselet, M. F. (2005). Early behavioral deficits in R6/2 mice suitable for use in preclinical drug testing. *Neurobiology of Disease, 20*(1), 1–11. https://doi.org/10.1016/j.nbd.2005.01.024.

Hills, R., Mossman, J. A., Bratt-Leal, A. M., Tran, H., Williams, R. M., Stouffer, D. G., et al. (2022). Functional recovery from human induced pluripotent stem cell-derived dopamine neuron grafts is dependent on neurite outgrowth. *bioRxiv*. https://doi.org/10.1101/2022.04.19.488213.

Hou, Y., Dan, X., Babbar, M., Wei, Y., Hasselbalch, S. G., Croteau, D. L., et al. (2019). Ageing as a risk factor for neurodegenerative disease. *Nature Reviews. Neurology, 15*(10), 565–581. https://doi.org/10.1038/s41582-019-0244-7.

Hurelbrink, C. B., & Barker, R. A. (2005). Migration of cells from primary transplants of allo- and xenografted foetal striatal tissue in the adult rat brain. *The European Journal of Neuroscience, 21*(6), 1503–1510. https://doi.org/10.1111/j.1460-9568.2005.03963.x.

Hurelbrink, C. B., Tyers, P., Armstrong, R. J., Dunnett, S. B., Barker, R. A., & Rosser, A. E. (2003). Long-term hibernation of human fetal striatal tissue does not adversely affect its differentiation in vitro or graft survival: Implications for clinical trials in Huntington's disease. *Cell Transplantation, 12*(7), 687–695. https://doi.org/10.3727/000000003108747307.

Im, W., Lee, S. T., Park, J. E., Oh, H. J., Shim, J., Lim, J., et al. (2010). Transplantation of patient-derived adipose stem cells in YAC128 Huntington's disease transgenic mice. *PLoS Currents, 2*. https://doi.org/10.1371/currents.RRN1183.

Jeon, I., Choi, C., Lee, N., Im, W., Kim, M., Oh, S. H., et al. (2014). In vivo roles of a patient-derived induced pluripotent stem cell line (HD72-iPSC) in the YAC128 model of Huntington's disease. *International Journal of Stem Cells*, 7(1), 43–47. https://doi.org/10.15283/ijsc.2014.7.1.43.

Jeon, I., Lee, N., Li, J. Y., Park, I. H., Park, K. S., Moon, J., et al. (2012). Neuronal properties, in vivo effects, and pathology of a Huntington's disease patient-derived induced pluripotent stem cells. *Stem Cells, 30*(9), 2054–2062. https://doi.org/10.1002/stem.1135.

Kalonia, H., Kumar, P., & Kumar, A. (2011). Licofelone attenuates quinolinic acid induced Huntington like symptoms: Possible behavioral, biochemical and cellular alterations. *Progress in Neuro-Psychopharmacology & Biological Psychiatry, 35*(2), 607–615. https://doi.org/10.1016/j.pnpbp.2011.01.003.

Kelly, C. M., Precious, S. V., Penketh, R., Amso, N., Dunnett, S. B., & Rosser, A. E. (2007). Striatal graft projections are influenced by donor cell type and not the immunogenic background. *Brain, 130*(Pt 5), 1317–1329. https://doi.org/10.1093/brain/awm053.

Klein, A., Lane, E. L., & Dunnett, S. B. (2013). Brain repair in a unilateral rat model of Huntington's disease: New insights into impairment and restoration of forelimb movement patterns. *Cell Transplantation*, *22*(10), 1735–1751. https://doi.org/10.3727/096368912X657918.

Labandeira-Garcia, J. L., & Guerra, M. J. (1994). Cortical stimulation induces fos expression in intrastriatal striatal grafts. *Brain Research*, *652*(1), 87–97. https://doi.org/10.1016/0006-8993(94)90321-2.

Le Cann, K., Foerster, A., Rosseler, C., Erickson, A., Hautvast, P., Giesselmann, S., et al. (2021). The difficulty to model Huntington's disease in vitro using striatal medium spiny neurons differentiated from human induced pluripotent stem cells. *Scientific Reports*, *11*(1). https://doi.org/10.1038/s41598-021-85656-x. https://doi.org/ARTN 6934.

Lelos, M. J., & Dunnett, S. B. (2018). Generating excitotoxic lesion models of Huntington's disease. *Methods in Molecular Biology*, *1780*, 209–220. https://doi.org/10.1007/978-1-4939-7825-0_11.

Lelos, M. J., Harrison, D. J., & Dunnett, S. B. (2011). Impaired sensitivity to Pavlovian stimulus-outcome learning after excitotoxic lesion of the ventrolateral neostriatum. *Behavioural Brain Research*, *225*(2), 522–528. https://doi.org/10.1016/j.bbr.2011.08.017.

Lelos, M. J., Harrison, D. J., & Dunnett, S. B. (2012). Intrastriatal excitotoxic lesion or dopamine depletion of the neostriatum differentially impairs response execution in extrapersonal space. *The European Journal of Neuroscience*, *36*(10), 3420–3428. https://doi.org/10.1111/j.1460-9568.2012.08256.x.

Lelos, M. J., Roberton, V. H., Vinh, N. N., Harrison, C., Eriksen, P., Torres, E. M., et al. (2016). Direct comparison of rat- and human-derived ganglionic eminence tissue grafts on motor function. *Cell Transplantation*, *25*(4), 665–675. https://doi.org/10.3727/096368915X690297.

Lin, Y. T., Chern, Y., Shen, C. K., Wen, H. L., Chang, Y. C., Li, H., et al. (2011). Human mesenchymal stem cells prolong survival and ameliorate motor deficit through trophic support in Huntington's disease mouse models. *PLoS One*, *6*(8), e22924. https://doi.org/10.1371/journal.pone.0022924.

Lindvall, O., & Bjorklund, A. (2004). Cell replacement therapy: Helping the brain to repair itself. *NeuroRx*, *1*(4), 379–381. https://doi.org/10.1602/neurorx.1.4.379.

Ma, L., Hu, B., Liu, Y., Vermilyea, S. C., Liu, H., Gao, L., et al. (2012). Human embryonic stem cell-derived GABA neurons correct locomotion deficits in quinolinic acid-lesioned mice. *Cell Stem Cell*, *10*(4), 455–464. https://doi.org/10.1016/j.stem.2012.01.021.

Maira, M., Long, J. E., Lee, A. Y., Rubenstein, J. L., & Stifani, S. (2010). Role for TGF-beta superfamily signaling in telencephalic GABAergic neuron development. *Journal of Neurodevelopmental Disorders*, *2*(1), 48–60. https://doi.org/10.1007/s11689-009-9035-6.

Mandel, R. J., Wictorin, K., Cenci, M. A., & Bjorklund, A. (1992). Fos expression in intrastriatal striatal grafts: Regulation by host dopaminergic afferents. *Brain Research*, *583*(1–2), 207–215. https://doi.org/10.1016/s0006-8993(10)80026-5.

Mazzocchi-Jones, D., Dobrossy, M., & Dunnett, S. B. (2009). Embryonic striatal grafts restore bi-directional synaptic plasticity in a rodent model of Huntington's disease. *The European Journal of Neuroscience*, *30*(11), 2134–2142. https://doi.org/10.1111/j.1460-9568.2009.07006.x.

Menalled, L. B., & Chesselet, M. F. (2002). Mouse models of Huntington's disease. *Trends in Pharmacological Sciences*, *23*(1), 32–39. https://doi.org/10.1016/s0165-6147(00)01884-8.

Menalled, L. B., Kudwa, A. E., Miller, S., Fitzpatrick, J., Watson-Johnson, J., Keating, N., et al. (2012). Comprehensive behavioral and molecular characterization of a new knock-in mouse model of Huntington's disease: zQ175. *PLoS One*, 7(12), e49838. https://doi.org/10.1371/journal.pone.0049838.

Nicoleau, C., Varela, C., Bonnefond, C., Maury, Y., Bugi, A., Aubry, L., et al. (2013). Embryonic stem cells neural differentiation qualifies the role of Wnt/beta-catenin signals in human telencephalic specification and regionalization. *Stem Cells, 31*(9), 1763–1774. https://doi.org/10.1002/stem.1462.

Nolbrant, S., Heuer, A., Parmar, M., & Kirkeby, A. (2017). Generation of high-purity human ventral midbrain dopaminergic progenitors for in vitro maturation and intracerebral transplantation. *Nature Protocols, 12*(9), 1962–1979. https://doi.org/10.1038/nprot.2017.078.

Ouimet, C. C., Langley-Gullion, K. C., & Greengard, P. (1998). Quantitative immunocytochemistry of DARPP-32-expressing neurons in the rat caudatoputamen. *Brain Research, 808*(1), 8–12. https://doi.org/10.1016/s0006-8993(98)00724-0.

Park, H. J., Han, A., Kim, J. Y., Choi, J., Bae, H. S., Cho, G. B., et al. (2022). SUPT4H1-edited stem cell therapy rescues neuronal dysfunction in a mouse model for Huntington's disease. *NPJ Regenerative Medicine,* 7(1), 8. https://doi.org/10.1038/s41536-021-00198-0.

Park, H. J., Jeon, J., Choi, J., Kim, J. Y., Kim, H. S., Huh, J. Y., et al. (2021). Human iPSC-derived neural precursor cells differentiate into multiple cell types to delay disease progression following transplantation into YAC128 Huntington's disease mouse model. *Cell Proliferation, 54*(8), e13082. https://doi.org/10.1111/cpr.13082.

Pauly, M. C., Piroth, T., Dobrossy, M., & Nikkhah, G. (2012). Restoration of the striatal circuitry: From developmental aspects toward clinical applications. *Frontiers in Cellular Neuroscience, 6,* 16. https://doi.org/10.3389/fncel.2012.00016.

Plank, A. C., Canneva, F., Raber, K. A., Urbach, Y. K., Dobner, J., Puchades, M., et al. (2018). Early alterations in operant performance and prominent huntingtin aggregation in a congenic F344 rat line of the classical CAGn51trunc model of Huntington disease. *Frontiers in Neuroscience, 12,* 11. https://doi.org/10.3389/fnins.2018.00011.

Pollock, K., Dahlenburg, H., Nelson, H., Fink, K. D., Cary, W., Hendrix, K., et al. (2016). Human mesenchymal stem cells genetically engineered to overexpress brain-derived neurotrophic factor improve outcomes in Huntington's disease mouse models. *Molecular Therapy, 24*(5), 965–977. https://doi.org/10.1038/mt.2016.12.

Pouladi, M. A., Morton, A. J., & Hayden, M. R. (2013). Choosing an animal model for the study of Huntington's disease. *Nature Reviews. Neuroscience, 14*(10), 708–721. https://doi.org/10.1038/nrn3570.

Praet, J., Santermans, E., Daans, J., Le Blon, D., Hoornaert, C., Goossens, H., et al. (2015). Early inflammatory responses following cell grafting in the CNS trigger activation of the subventricular zone: A proposed model of sequential cellular events. *Cell Transplantation, 24*(8), 1481–1492. https://doi.org/10.3727/096368914X682800.

Precious, S. V., & Kelly, C. (2017). Transplantation in HD: Are we transplanting the right cells? *IntechOpen.* https://doi.org/10.5772/66490.

Prince, M., Bryce, R., Albanese, E., Wimo, A., Ribeiro, W., & Ferri, C. P. (2013). The global prevalence of dementia: A systematic review and metaanalysis. *Alzheimers Dement, 9*(1), 63–75 e62. https://doi.org/10.1016/j.jalz.2012.11.007.

Pundt, L. L., Kondoh, T., Conrad, J. A., & Low, W. C. (1996). Transplantation of human fetal striatum into a rodent model of Huntington's disease ameliorates locomotor deficits. *Neuroscience Research, 24*(4), 415–420. https://doi.org/10.1016/0168-0102(95)01009-2.

Rallu, M., Machold, R., Gaiano, N., Corbin, J. G., McMahon, A. P., & Fishell, G. (2002). Dorsoventral patterning is established in the telencephalon of mutants lacking both Gli3 and Hedgehog signaling. *Development, 129*(21), 4963–4974. https://doi.org/10.1242/dev.129.21.4963.

Reidling, J. C., Relano-Gines, A., Holley, S. M., Ochaba, J., Moore, C., Fury, B., et al. (2018). Human neural stem cell transplantation rescues functional deficits in R6/2 and Q140 Huntington's disease mice. *Stem Cell Reports, 10*(1), 58–72. https://doi.org/10.1016/j.stemcr.2017.11.005.

Rising, A. C., Xu, J., Carlson, A., Napoli, V. V., Denovan-Wright, E. M., & Mandel, R. J. (2011). Longitudinal behavioral, cross-sectional transcriptional and histopathological characterization of a knock-in mouse model of Huntington's disease with 140 CAG repeats. *Experimental Neurology*, *228*(2), 173–182. https://doi.org/10.1016/j.expneurol.2010.12.017.

Roberton, V. H., Evans, A. E., Harrison, D. J., Precious, S. V., Dunnett, S. B., Kelly, C. M., et al. (2013). Is the adult mouse striatum a hostile host for neural transplant survival? *Neuroreport*, *24*(18), 1010–1015. https://doi.org/10.1097/WNR.0000000000000066.

Rosser, A. E., & Bachoud-Levi, A. C. (2012). Clinical trials of neural transplantation in Huntington's disease. *Progress in Brain Research*, *200*, 345–371. https://doi.org/10.1016/B978-0-444-59575-1.00016-8.

Rosser, A. E., Busse, M. E., Gray, W. P., Badin, R. A., Perrier, A. L., Wheelock, V., et al. (2022). Translating cell therapies for neurodegenerative diseases: Huntington's disease as a model disorder. *Brain*, *145*(5), 1584–1597. https://doi.org/10.1093/brain/awac086.

Saleeba, C., Dempsey, B., Le, S., Goodchild, A., & McMullan, S. (2019). A Student's guide to neural circuit tracing. *Frontiers in Neuroscience*, *13*, 897. https://doi.org/10.3389/fnins.2019.00897.

Sanberg, P. R., Borlongan, C. V., Koutouzis, T. K., Norgren, R. B., Jr., Cahill, D. W., & Freeman, T. B. (1997). Human fetal striatal transplantation in an excitotoxic lesioned model of Huntington's disease. *Annals of the New York Academy of Sciences*, *831*, 452–460. https://doi.org/10.1111/j.1749-6632.1997.tb52217.x.

Sanberg, P. R., Calderon, S. F., Giordano, M., Tew, J. M., & Norman, A. B. (1989). The quinolinic acid model of Huntington's disease: Locomotor abnormalities. *Experimental Neurology*, *105*(1), 45–53. https://doi.org/10.1016/0014-4886(89)90170-2.

Sanchis, A., Garcia-Gimeno, M. A., Canada-Martinez, A. J., Sequedo, M. D., Millan, J. M., Sanz, P., et al. (2019). Metformin treatment reduces motor and neuropsychiatric phenotypes in the zQ175 mouse model of Huntington disease. *Experimental & Molecular Medicine*, *51*(6), 1–16. https://doi.org/10.1038/s12276-019-0264-9.

Shear, D. A., Dong, J., Haik-Creguer, K. L., Bazzett, T. J., Albin, R. L., & Dunbar, G. L. (1998). Chronic administration of quinolinic acid in the rat striatum causes spatial learning deficits in a radial arm water maze task. *Experimental Neurology*, *150*(2), 305–311. https://doi.org/10.1006/exnr.1998.6767.

Siracusa, R., Fusco, R., & Cuzzocrea, S. (2019). Astrocytes: Role and functions in brain pathologies. *Frontiers in Pharmacology*, *10*, 1114. https://doi.org/10.3389/fphar.2019.01114.

Song, J., Lee, S. T., Kang, W., Park, J. E., Chu, K., Lee, S. E., et al. (2007). Human embryonic stem cell-derived neural precursor transplants attenuate apomorphine-induced rotational behavior in rats with unilateral quinolinic acid lesions. *Neuroscience Letters*, *423*(1), 58–61. https://doi.org/10.1016/j.neulet.2007.05.066.

Stanslowsky, N., Reinhardt, P., Glass, H., Kalmbach, N., Naujock, M., Hensel, N., et al. (2016). Neuronal dysfunction in iPSC-derived medium spiny neurons from chorea-acanthocytosis patients is reversed by Src kinase inhibition and F-actin stabilization. *Journal of Neuroscience*, *36*(47), 12027–12043. https://doi.org/10.1523/Jneurosci.0456-16.2016.

Steinbeck, J. A., Choi, S. J., Mrejeru, A., Ganat, Y., Deisseroth, K., Sulzer, D., et al. (2015). Optogenetics enables functional analysis of human embryonic stem cell-derived grafts in a Parkinson's disease model. *Nature Biotechnology*, *33*(2), 204–209. https://doi.org/10.1038/nbt.3124.

Tabrizi, S. J., Reilmann, R., Roos, R. A., Durr, A., Leavitt, B., Owen, G., et al. (2012). Potential endpoints for clinical trials in premanifest and early Huntington's disease in the TRACK-HD study: Analysis of 24 month observational data. *Lancet Neurology*, *11*(1), 42–53. https://doi.org/10.1016/S1474-4422(11)70263-0.

Tonnesen, J., Parish, C. L., Sorensen, A. T., Andersson, A., Lundberg, C., Deisseroth, K., et al. (2011). Functional integration of grafted neural stem cell-derived dopaminergic neurons monitored by optogenetics in an in vitro Parkinson model. *PLoS One*, *6*(3), e17560. https://doi.org/10.1371/journal.pone.0017560.

Tornero, D., Tsupykov, O., Granmo, M., Rodriguez, C., Gronning-Hansen, M., Thelin, J., et al. (2017). Synaptic inputs from stroke-injured brain to grafted human stem cell-derived neurons activated by sensory stimuli. *Brain*, *140*(3), 692–706. https://doi.org/10.1093/brain/aww347.

Ungerstedt, U. (1971). Postsynaptic supersensitivity after 6-hydroxy-dopamine induced degeneration of the nigro-striatal dopamine system. *Acta Physiologica Scandinavica. Supplementum*, *367*, 69–93. https://doi.org/10.1111/j.1365-201x.1971.tb11000.x.

Vasic, V., Barth, K., & Schmidt, M. H. H. (2019). Neurodegeneration and neuro-regeneration-Alzheimer's disease and stem cell therapy. *International Journal of Molecular Sciences*, *20*(17). https://doi.org/10.3390/ijms20174272.

Victor, M. B., Richner, M., Hermanstyne, T. O., Ransdell, J. L., Sobieski, C., Deng, P. Y., et al. (2014). Generation of human striatal neurons by microRNA-dependent direct conversion of fibroblasts. *Neuron*, *84*(2), 311–323. https://doi.org/10.1016/j.neuron.2014.10.016.

Vonsattel, J. P., Myers, R. H., Stevens, T. J., Ferrante, R. J., Bird, E. D., & Richardson, E. P., Jr. (1985). Neuropathological classification of Huntington's disease. *Journal of Neuropathology and Experimental Neurology*, *44*(6), 559–577. https://doi.org/10.1097/00005072-198511000-00003.

Watts, C., Brasted, P. J., & Dunnett, S. B. (2000). Embryonic donor age and dissection influences striatal graft development and functional integration in a rodent model of Huntington's disease. *Experimental Neurology*, *163*(1), 85–97. https://doi.org/10.1006/exnr.1999.7341.

Watts, C., & Dunnett, S. B. (1998). Effects of severity of host striatal damage on the morphological development of intrastriatal transplants in a rodent model of Huntington's disease: Implications for timing of surgical intervention. *Journal of Neurosurgery*, *89*(2), 267–274. https://doi.org/10.3171/jns.1998.89.2.0267.

Wictorin, K., Clarke, D. J., Bolam, J. P., Brundin, P., Gustavii, B., Lindvall, O., et al. (1990). Extensive efferent projections of intra-striatally transplanted striatal neurons as revealed by a species-specific neurofilament marker and anterograde axonal tracing. *Progress in Brain Research*, *82*, 391–399. https://doi.org/10.1016/s0079-6123(08)62627-8.

Wu, Z., Parry, M., Hou, X. Y., Liu, M. H., Wang, H., Cain, R., et al. (2020). Gene therapy conversion of striatal astrocytes into GABAergic neurons in mouse models of Huntington's disease. *Nature Communications*, *11*(1). https://doi.org/10.1038/s41467-020-14855-3. https://doi.org/ARTN 1105.

Wu, M., Zhang, D., Bi, C., Mi, T., Zhu, W., Xia, L., et al. (2018). A chemical recipe for generation of clinical-grade striatal neurons from hESCs. *Stem Cell Reports*, *11*(3), 635–650. https://doi.org/10.1016/j.stemcr.2018.08.005.

Yhnell, E., Dunnett, S. B., & Brooks, S. P. (2016a). A longitudinal motor characterisation of the HdhQ111 mouse model of Huntington's disease. *The Journal of Huntington's Disease*, *5*(2), 149–161. https://doi.org/10.3233/JHD-160191.

Yhnell, E., Dunnett, S. B., & Brooks, S. P. (2016b). A longitudinal operant assessment of cognitive and Behavioural changes in the HdhQ111 mouse model of Huntington's disease. *PLoS One*, *11*(10), e0164072. https://doi.org/10.1371/journal.pone.0164072.

Yhnell, E., & Heuer, A. (2018). Automated operant assessments of Huntington's disease mouse models. *Methods in Molecular Biology*, *1780*, 143–162. https://doi.org/10.1007/978-1-4939-7825-0_8.

CHAPTER TWO

Challenges of cell therapies for retinal diseases

Monville Christelle[a,b,*], Morizur Lise[a,b,c], and Karim Ben M'Barek[a,b,c]
[a]INSERM U861, I-Stem, AFM, Institute for Stem cell Therapy and Exploration of Monogenic diseases, Corbeil-Essonnes, France
[b]Paris Saclay Evry U861, Corbeil-Essonnes, France
[c]CECS, Corbeil-Essonnes, France
*Corresponding author: e-mail address: cmonville@istem.fr

Contents

Abstract

Because retinal cells could not regenerate in mammals, retinal degeneration is an irreversible phenomenon caused by various disease conditions including age-related macular degeneration (AMD) and retinitis pigmentosa (RP). During the course of these diseases, photoreceptors (PRs) are susceptible to degenerate due to their malfunctions or to a primary dysfunction of the retinal pigment epithelium (RPE). These diseases affect millions of individuals worldwide. Many therapies currently under development are designed to treat retinal pathologies and among those, cell-based replacement strategies hold significant promise for treatment of retinal degenerative diseases, especially when atrophy of retinal tissue occurs. This review will focus on the use of hPSCs to slow-down or treat retinal disorders and discuss current challenges to achieve successful engraftment of cell products.

International Review of Neurobiology, Volume 166
ISSN 0074-7742
https://doi.org/10.1016/bs.irn.2022.09.004

1. Introduction

In the past decade, personalized medicine approaches have seen major changes with breakthrough discoveries such as human induced pluripotent stem cells (hiPSCs) (Takahashi et al., 2007) and genome editing technologies like CRISPR-Cas9 (clustered regularly interspaced short palindromic repeats-associated (Cas)), (For a review Rezazade Bazaz & Dehghani, 2022).

The eye is a complex organ but also a very good target for cellular replacement therapies because of its small size, ease in visual function monitoring and surgical accessibility. It is formed by highly specialized tissues that are derived from different cell lineages during development. Located in the posterior pole of the eye, the retina is the functional part that converts light inputs into electrical signals, which are ultimately transmitted to the brain. Photoreceptors (PR), cones and rods, are the primary cells involved in light detection. Rods are sensitive to dim light whereas cones detect specific wavelengths allowing color vision. Electrical signals generated by PR are further processed by other neurons (bipolar, amacrine, horizontal cells) and transmitted to the brain through ganglion cells. During retinal development, multipotent retinal progenitors emerging from the neuroectoderm give rise to all types of neurons, retinal pigmented epithelial (RPE) cells and Müller glia cells in a sequential manner (Bassett & Wallace, 2012; Wallace, 2011). RPE cells form a monolayer of non-dividing polygonal cells that lies beneath the neurosensory retina and sustain numerous crucial functions insuring PR health (Strauss, 1995, 2016; Yang, Zhou, & Li, 2021).

Eye health and vision have widespread and profound implications in many aspects of life, health, sustainable development, and economy (Burton et al., 2021). A number of retinal degenerative diseases such as age-related macular degeneration (AMD) and retinitis pigmentosa (RP) leads to the death of PR, which ultimately results in vision impairment and in some cases in blindness. These diseases affect millions of individuals worldwide (Burton et al., 2021; Flaxman et al., 2017; Verbakel et al., 2018; Wong et al., 2014). Their global economic impact is estimated around US$411 billion annually (Burton et al., 2021). Many therapies currently under development are designed to treat retinal pathologies (Cehajic-Kapetanovic, Singh, Zrenner, & MacLaren, 2022; Ludwig & Gamm, 2021). Among those, cell-based replacement strategies hold significant promise for treatment of retinal degenerative diseases, especially when atrophy of retinal tissue occurs (Ben M'Barek, Habeler, Regent, & Monville, 2019).

This review will focus on the use of hPSCs to slow-down or treat retinal disorders and discuss current challenges to achieve successful engraftment of cell products. Challenges associated with large-scale manufacturing will also be discussed.

2. Retinal disorders: Focus on AMD and RP

In 2020, an estimated 596 million people had distance vision impairment worldwide, of whom 43 million were blind (Burton et al., 2021). By 2050, population aging, growth, and urbanization might lead to an estimated 895 million people with distance vision impairment, of whom 61 million will be blind (Burton et al., 2021). Major causes of vision impairment worldwide are uncorrected refractive errors and cataracts for which effective treatments are available. On the other hand, retinal degenerative diseases are the leading cause of untreatable sight-loss in the industrialized world.

2.1 Age-related macular degeneration

AMD represents the third leading cause of moderate or severe vision loss worldwide (Burton et al., 2021; Flaxman et al., 2017). Projections for 2040 are alarming with an estimated number of people affected with all stages of the disease of 288 million (Wong et al., 2014).

Initial stage of AMD (early AMD) is usually asymptomatic (Mitchell, Liew, Gopinath, & Wong, 2018). However, ophthalmic examinations reveal the presence of medium drusen in the macula (diameter between 63 and 125 μm). Progression to intermediate AMD engenders changes in retinal pigmentation and the presence of large drusen (larger than 125 μm) (Mitchell et al., 2018). As disease progresses, distortion of central vision starts to appear, leading ultimately to scotomas in the central visual field. Two types of evolution are possible: a wet form characterized by the presence of macular choroidal neovascularizations (corresponding to 10–20% of cases) or a dry form defined by a macular atrophy (corresponding to 80–90% of cases) (Al-Khersan, Hussain, Ciulla, & Dugel, 2019; Mitchell et al., 2018).

Vision loss in wet AMD can be very fast and severe. Indeed, new vessels susceptible to leak and generate hemorrhages develop in the macular area. Thus, upon leakage, the macula can be severely damaged (Handa et al., 2019). Dry AMD is more progressive with the degeneration of RPE cells located in the macula. The primary death of RPE cells causes the secondary degeneration of PR located into this macula.

Although the leading risk is age, genetic and environmental factors also play a role in AMD development. In European population, prevalence of all stages of AMD rises from 3.5% between 55–59 years to 27.3% after 85-year old (Colijn et al., 2017). An important avoidable factor that favors the development of late AMD is smoking (Mitchell, Wang, Smith, & Leeder, 2002). Alcohol, diet, hypertension or sunlight exposition are also environmental risks factors. Complement factor H (*CFH*) and age-related maculopathy susceptibility 2/high-temperature requirement factor A1 (*ARMS2/HTRA1*) are the most important gene polymorphisms associated with increased risk for AMD (Handa et al., 2019).

Exact disease mechanisms at the origin of AMD are poorly understood. Modifications of the microenvironment caused by inflammation processes, oxidative stress, metabolic dysfunctions, the complement system, lipids and extracellular matrix may be involved (Brown et al., 2018; Handa et al., 2019; Hussain, Lee, & Marshall, 2020). In addition, accumulation of proteins and lipids into Bruch's membrane can stimulate inflammation and alter circulation of nutrients and liquids. Such modifications may damage RPE cells.

Regarding therapeutic options, there is no treatment available for the dry form of AMD. Anti-VEGF therapy delivered at regular intervals is the treatment of choice for wet AMD. Such treatment prevents vision worsening and may induce a moderate improvement in visual acuity. Unfortunately, some patients do not respond to therapy (Al-Khersan et al., 2019). Treatment compliance is suggested as an explanation of incomplete response. This is probably due to the high frequency of injections required (every month or every 2 months) and to its cost. Diet supplementation with zinc and antioxidants may also reduce moderately the risk of disease progression to advanced AMD stages (Age-Related Eye Disease Study Research Group, 2001).

2.2 Retinitis pigmentosa

1.5 million of patients worldwide are affected with RP (Verbakel et al., 2018). RP is a heterogeneous group of inherited retinal dystrophies for which transmission is autosomal-recessive in 50–60% of cases and linked to chromosome X in 5–15% of patients (Hartong, Berson, & Dryja, 2006; Pfeiffer, Marc, & Jones, 2020). RP can be associated to other extra-ocular symptoms (syndromic forms) in 20–30% of cases (Hartong et al., 2006).

An ophthalmic eye fundus examination reveals the classical triad of RP features: (i) bone spicule pigmentation corresponding to melanin deposits from migrating RPE cells, (ii) attenuation of retinal vessels that could be secondary to PR loss and (iii) a waxy pallor of the optic nerve (likely due to the presence of glial cells) (Verbakel et al., 2018). The median age of symptom appearance is 29-year old (Iftikhar et al., 2019). RP patients typically experiment night blindness that goes unnoticed at early stages. The visual field is then progressively reduced (tunnel vision). Large portions of the center of vision are lost in advanced stages leading to legal blindness by 40-year old (Hartong et al., 2006). Photopsia is also commonly experienced by patients at different stages of the disease and its occurrence is modulated by stress (Bittner, Diener-West, & Dagnelie, 2009; Bittner, Haythornthwaite, Diener-West, & Dagnelie, 2012).

A combination of specific gene mutation, genetic background and environmental factors modulate age of onset, clinical manifestations as well as disease progression. To date, more than 75 genes involved in RP have been described (https://sph.uth.edu/retnet/sum-dis.htm) but a number of cases remains without known causal genes (30% of non-syndromic RP and 50% of autosomal-dominant RP) (Hartong et al., 2006; Verbakel et al., 2018). Additional studies are necessary to identify other genes underlying RP.

The primary cellular defect is localized into rods or, for about 5% of cases, into the RPE (Ben M'Barek, Habeler, & Monville, 2018). At advanced stages of the disease, the ganglion cell layer and inner nuclear layer remain relatively well preserved while the PR layer is completely depleted. However, after decades of degeneration, rewiring, glial hypertrophy and global cell death occur following the complete PR loss (Pfeiffer et al., 2020). This important aspect of the pathology suggests that the retinal circuitry does not degenerate immediately when PR are dead. RP retinas may thus be amenable to PR cell transplantation during a specific time window (before retinal remodeling) if donor cells are able to integrate into the preexisting circuitry (Morizur, Herardot, Monville, & Ben M'Barek, 2020).

Few therapeutic options are available to treat RP (Scholl et al., 2016; Tezel & Ruff, 2021). Luxturna™ (voretigene neparvovec-rzyl) is the only gene therapy treatment approved to date for patients with *RPE65* gene mutations (2% of RP cases). This therapy was shown to be safe and induced an improvement in navigational abilities and light sensitivity (Maguire et al., 2021, 2019). When cataract is present (40–50% of RP patients), its removal by surgery improves visual outcomes (Auffarth, Tetz, Krastel, Blankenagel, & Völcker, 1997; Bayyoud, Bartz-Schmidt, & Yoeruek, 2013). In addition, specific devices can help improve night vision (i.e., specific goggles) (Hartong, Jorritsma, Neve, Melis-Dankers, & Kooijman, 2004; Ikeda et al., 2019). Very recently, optogenetic tools were used to restore visual function at the late stages of RP (Sahel et al., 2021, NCT03326336).

3. Human pluripotent stem cells for retinal cell therapy

The retina is particularly well suited for the benefits of stem cell therapy because of the relatively easy anatomic access and physiological monitoring (Ben M'Barek et al., 2019; Morizur et al., 2020; Singh et al., 2020). The interdependence nature of PR and RPE cells in the retina and the non-regenerative capacities of these cells are at the core of retinal degenerative mechanisms. Primary dysfunction in either population often causes secondary damage in the other (Jones, Marc, & Pfeiffer, 1995; Ludwig & Gamm, 2021; Sparrow, Hicks, & Hamel, 2010).

Depending on the retinal degenerative disease and its specific stage of progression, different cell types are lost. Therefore, different cell therapy products should be considered. In the case of AMD, RPE cells are the primary cell population affected followed by PR. In RP where the defective gene is involved in PR function, PR degenerate first. However, in RPE gene associated RP, RPE cells are dysfunctional and PR degenerate because of this. Therefore, RPE cells or PR or both are required for replacement according to these different situations (Ben M'Barek et al., 2019; Gagliardi, Ben M'Barek, & Goureau, 2019; Morizur et al., 2020).

Another major aspect of tissue restoration is to determine precisely the site of injection as the entirety of the retina cannot be covered by the grafted material. Even if RP affects first peripheral retina, the macula becomes ultimately damaged. Thus, in AMD and RP, the macula is usually considered as the targeted site of implantation to maximize the visual recovery as visual acuity mostly depends on this area.

Different cell sources for retinal cell therapy were investigated including fetal or adult retinal tissues which were expected to replace lost cells or non-retinal cell types which provide neuroprotection to prevent complete PR loss (Ben M'Barek et al., 2019; Morizur et al., 2020). However, these cell sources are limited because of their supply chain, donor to donor variability and scale-up possibilities (limited cell amplification). Due to their self-renewal potency and pluripotency, hPSCs have aroused interest of the scientific community in the last decade. hPSCs can generate every cell type of the human retina *in vitro*. Current efforts are directed to the development of robust differentiation protocols from these cells, which could be used for disease modeling or cell therapy (Ben M'Barek, Regent, &

Monville, 2015; Morizur et al., 2020; Takahashi et al., 2007; Thomson et al., 1998). Differentiation processes are tightly regulated by exogenous signals coming from adjacent tissues and cell-intrinsic factors *in vivo* (Ben M'Barek et al., 2019; Weisz et al., 1999). Levering on developmental studies in animal models, the overall strategy is to expose hPSCs to the corresponding signaling molecules in a time and dose dependent manner along the differentiation process to generate the retinal cell of interest (Morizur et al., 2020).

3.1 RPE cell therapy

The first demonstration of an RPE cell therapy was performed in monkeys in the early 80s (Gouras, Flood, Kjedbye, Bilek, & Eggers, 1985; Gouras, Flood, & Kjeldbye, 1984) and later confirmed in rats (Lopez et al., 1989). Historically, the first protocol was based on the spontaneous differentiation of hPSCs into RPE cells. Indeed, upon FGF2 withdrawal from the culture medium, hPSCs differentiate into RPE cells following several weeks or months (Ben M'Barek et al., 2017; Buchholz et al., 2009; Ferguson, Balaiya, Mynampati, Sambhav, & Chalam, 2015; Klimanskaya et al., 2004). However, such "spontaneous" protocol is highly operator-dependent, time-consuming (2–3 months) and does not allow for the production of large cell banks. Indeed, only few clusters of pigmented RPE cells are collected per culture dish (about 5% of the cells). To overcome these limitations, studies have focused on directing RPE differentiation by adding signaling molecules that mimic *in vivo* developmental cues known to be important in RPE development and specification at critical time points (Ben M'Barek et al., 2015; Buchholz et al., 2013; Regent et al., 2019).

As a step toward clinical use in regenerative medicine, differentiation protocols were optimized to obtain RPE cells: (i) in conditions using a good manufacturing practice (GMP)-compliant production process and (ii) in large numbers to treat the millions of patients affected by RPE-associated retinal degeneration. Recent studies have started to address these issues by developing xeno-free/feeder-free protocols (Choudhary et al., 2017; Reichman et al., 2017). The development of directed differentiation protocols has also opened new avenues for the implementation of automated systems to upscale the production process and increase its robustness (Regent et al., 2019).

hPSCs derived from embryos (human embryonic stem cells or hESCs) or following adult cell reprogramming (hiPSCs) both have the potency to differentiate into RPE cells that are equally effective in protecting the retina from degeneration following transplantation into the subretinal space

(SRS) of a RP rodent model (Riera et al., 2016). Pioneer studies used hESCs and led to the first clinical trials. RPE cells derived from hiPSCs have only recently come to clinical trials (Table 1).

The mode of delivery of RPE cell therapy weighs on therapeutic outcomes. Two modes of cell formulation for implantation are proposed, either as a cell suspension or as an organized epithelial monolayer (Ben M'Barek et al., 2019). Both formulations are currently evaluated in human clinical trials (Table 1). Cells in suspension offer the advantage of a simplified logistic as vials of cells are sent frozen to hospitals where they are stored and thawed when desired (off-the-shell strategy). Moreover, little manipulation is required before transplantation into the SRS and a simple syringe is used for eye injection. The Royal College of Surgeons (RCS) rat is a widely studied, classic model of recessively inherited retinal degeneration in which the RPE fails to phagocytose shed outer segments, and photoreceptor cells subsequently die (D'Cruz et al., 2000; Thomas et al., 2018). The dystrophic RCS rats are characterized by RPE dysfunction due to the deletion in the Mer tyrosine kinase (MerTK) receptor that abolishes internalization of PR outer segments by RPE cells (Nandrot & Dufour, 2010). The phagocytosis of shed photoreceptor outer segments (OS) is a prerequisite for maintaining normal retinal physiology for which RPE plays a major role. Accumulation of debris in the subretinal space can lead to severe photoreceptor degeneration and rapid loss of vision. In RCS rats, although retinal thickness remains close to the level of normal eyes at 1 month of age (Ambati & Fowler, 2012; McGill et al., 2012), the outer nuclear layer (ONL) is reduced to a single layer at the age of 3 months, with the debris zone occupying the former outer segment area. Finally, at 6 months, the inner nuclear layer is in direct contact with the RPE cell layer, with very few photoreceptors surviving (DiLoreto, del Cerro, Cox, & del Cerro, 1998; Feng, Yasumura, Matthes, LaVail, & Vollrath, 2002; Goldman & O'Brien, 1978; Mullen & LaVail, 1976; Sauvé, Girman, Wang, Lawrence, & Lund, 2001). When hPSC-derived RPE cells are delivered as a cell suspension into the SRS of the RCS rat model, PR degeneration is slowed down and consequently visual functions is preserved (Ben M'Barek et al., 2017; Carr et al., 2009; Idelson et al., 2009; Lu et al., 2009; Lund et al., 2006). The first hPSC-RPE clinical trials based on this strategy targeted AMD and Stargardt's disease. The safety profiles were reassuring as no adverse event related to the use of hPSCs was observed (Schwartz et al., 2012, 2015). Moreover, some preliminary signs of efficacy sustained for up to 4 years in some patients (Schwartz, Tan, Hosseini &

Table 1 A brief description of the clinical trials of cell therapy of RP and AMD which target RPE cells.

Targeted disease	Sponsor/company	Formulation	Stage of development	References
Allogenic hESC-RPE/Cell suspension				
AMD, SMD	Astellas Institute for Regenerative Medicine	Cell suspension injection (5, 10, 15 or 20×10^4 cells/SR transplantation)	Phase I, II	NCT01344993 (Schwartz et al., 2012, Schwartz, Tan, Hosseini, & Nagiel, 2016)
SMD	Moorefields Eye Hospital NHS Foundation Trust London, United Kingdom	Cell suspension injection (5, 10, 15 or 20×10^4 cells/SR transplantation)	Phase I, II	NCT01469832 (Mehat et al., 2018)
Dry AMD	CHA Bundang Medical Center Seongnam-si, Republic of Korea/CHABiotech CO., Ltd.	Cell suspension injection (5, 10, 15 or 20×10^4 cells/SR transplantation)	Phase I/II	NCT03305029 NCT01674829 (Sung et al., 2021)
Dry AMD, GA	Cell Cure Neurosciences Ltd./Hoffmann-La Roche	cell suspension injection (5-20×10^4 cells/SR transplantation)	Phase I/II	NCT02286089
Dry AMD, SMD	Hospital of Zhengzhou, Beijing Tongren Hospital, China	Cell suspension injection (1×10^6 cells/SR transplantation)	Phase I/II	NCT03046407 NCT02755428

Continued

Table 1 A brief description of the clinical trials of cell therapy of RP and AMD which target RPE cells.—cont'd

Targeted disease	Sponsor/company	Formulation	Stage of development	References
RP	Beijing Tongren Hospital, Capital Medical University Beijing, China	Cell suspension injection (15×10^4 cells/SR transplantation)	Phase I/II	NCT03944239 (Li et al., 2021)
Dry AMD, SMD	Southwest Hospital Chongqing, Chongqing, China	Cell suspension injection (15×10^4 cells/SR transplantation)	Phase I/II	NCT02749734
Dry AMD, wet AMD, SMD	Federal University of São Paulo	as cell suspension (10^5 cells/SR transplantation) *vs.* as a sheet on a polymeric substrate	Phase I/II	NCT02903576
Allogenic hESC-RPE/Epithelium				
Dry AMD, GA	Regenerative Patch Technologies, LLC	Epithelium on a parylene membrane	Phase I/II	NCT02590692 (Kashani et al., 2018, 2022, 2021, 2020)
Acute wet AMD	Pfizer/London Project to Cure Blindness	Epithelium on polyester membrane	Phase I	NCT01691261 (da Cruz et al., 2018)
RP with genetic defects in LRAT, MERTK genes	CHNO des Quinze-Vingts, France/CECS IStem	Epithelium on amniotic membrane	Phase I/II	NCT03963154

Allogenic hiPSC-RPE/Cell suspension				
RP	Kobe City Eye Hospital	Cell suspension injection (15×10^4 cells/SR transplantation)	Phase I	JPRN-jRCTa050200027
Wet AMD	Kobe City Eye Hospital	Cell suspension injection (15×10^4 cells/SR transplantation)	Phase I	JPRN-UMIN000026003
Autologous hiPSC-RPE/Epithelium				
AMD	Kobe City Eye Hospital	Epithelium without substrate	Phase I/II	JPRN-UMIN000011929 (Mandai, Watanabe, et al., 2017)
AMD	National Eye Institute	Epithelium on biodegradable scaffold	Phase I/II	NCT04339764

AMD: Age-related macular degeneration, hESC: human embryonic stem cells, LCA: Leber Congenital Amaurosis, RP: Retinitis Pigmentosa, GA: Geographic Atrophy, SR: Subretinal, RPE: Retinal Pigment Epithelium, hiPSC: human induced pluripotent stem cells, LRAT: Lecithin Retinol Acyltransferase, MERTK: MER Proto-Oncogene, Tyrosine Kinase.

Nagiel, et al., 2016). A total of 38 patients were treated without major serious adverse events such as cell dispersion out of the eye or teratoma formation (Schwartz et al., 2016). Since then, several teams have reported safety results with the same approach (Li et al., 2021; Liu et al., 2017; Mehat et al., 2018; Song et al., 2015; Sung et al., 2021). However, it is still not clear if RPE cells grafted as a cell suspension can form a confluent, polarized monolayer in the SRS in the transplantation zone (Diniz et al., 2013; Hu et al., 2012).

The second formulation approach is to deliver an already formed hPSC-RPE epithelium. Indeed, formation of tight junctions between neighboring RPE cells is required for the cells to be fully polarized with the formation of specific apical microvilli needed for the interaction with PR and their crucial functions (Singh, Park, et al., 2020; Strauss, 2005). This strategy significantly complicates the transport (living cells), storage (limited stability) and subretinal delivery (need of specific implantation devices). As a living monolayer of cells, the graft should be kept for a limited period at a controlled temperature (Ben M'Barek et al., 2020; Hori et al., 2019). An hiPSC-RPE patch without any additional substrate was tested in a patient with wet form of AMD (Mandai, Watanabe, et al., 2017). A supporting scaffold made of polymers or of biological composition may nonetheless be needed during epithelial reformation of hPSC-derived RPE cells to ensure that cells can be removed from the culture dish (Ben M'Barek, Habeler, & Monville, 2018; Ben M'Barek, Habeler, Plancheron, et al., 2018; Diniz et al., 2013; Kashani et al., 2018; Sharma et al., 2019). For implantation, specific devices should be set up to deliver the hPSC-RPE sheet into the SRS (Ben M'Barek et al., 2020, 2017; da Cruz et al., 2018; Fernandes et al., 2017; Stanzel et al., 2014). Compared to cell suspension injections, the viability of hPSC-derived RPE cells grafted in the SRS of Nude rat and functional outcomes were improved when delivered as a cell sheet (Ben M'Barek et al., 2017; Diniz et al., 2013; Hu et al., 2012; Kashani et al., 2020; Sharma et al., 2019). Our group demonstrated that hPSC-RPE cells injected as a RPE sheet was more effective to preserve PR cell layer and visual functions in retinal dystrophic rats compared to the same cells injected as a cell suspension. Preliminary results of three human clinical trials based on the sheet formulation to treat AMD were published recently (da Cruz et al., 2018; Kashani et al., 2018, 2021, 2022; Mandai, Watanabe, et al., 2017). While these studies were not designed to evaluate efficacy, first data are encouraging. A very recent study was able to assess RPE cell survival and showed that subretinally-implanted, HLA-mismatched donor

RPE cells survived after a 2-year follow up in an AMD patient, expressed functional markers, and did not elicit clinically detectable intraocular inflammation or serologic immune responses even without long-term immunosuppression (Kashani et al., 2022).

3.2 PR cell therapy

Over the last decade, substantial progress has been made in the development of cell therapy approaches with the successful transplantation of PR precursors derived from PSCs—either as a retinal sheet or as a cell suspension—in rodent and non-human primate models (Aboualizadeh et al., 2020; Gagliardi et al., 2018, 2019; Ludwig & Gamm, 2021; Tu et al., 2019). Large body of work from several teams reported on the successful transplantation of post-mitotic PR precursors isolated from mouse neonatal retina (Pearson et al., 2012; Santos-Ferreira et al., 2015; Singh et al., 2013, 2020). Transplantation of PR derived from mouse PSCs into the SRS of different blind mice models showed integration into the degenerating retina, differentiation into mature PRs, and formation of synaptic connections, with some data hinting at a possible partial recovery of visual function (Decembrini, Koch, Radtke, Moulin, & Arsenijevic, 2014; Gonzalez-Cordero et al., 2013; Kruczek et al., 2017; Mandai, Fujii, et al., 2017; Santos-Ferreira, Völkner, et al., 2016; Tucker et al., 2011). However, studies published from late 2016 indicated that the observed functional recovery was achieved through the transfer of cytoplasmic material from transplanted mouse PR to remaining host PR rather than through real integration into the recipient outer nuclear layer (ONL) (Decembrini et al., 2017; Ortin-Martinez et al., 2017; Pearson et al., 2016; Santos-Ferreira, Llonch, et al., 2016; Singh et al., 2016). It is unclear whether human PR may also be able to transfer cytoplasmic material to host cells (Zhu, Cifuentes, Reynolds, & Lamba, 2017). In healthy, wild-type retinas, material transfer appears to be the dominant process and recent studies show that it occurs through the formation of direct contact between endogenous PR subtypes (Heisterkamp et al., 2022) and between grafted and endogenous PR by open-end nanotubes-like processes (Kalargyrou et al., 2021; Ortin-Martinez et al., 2021). These processes permit the direct transfer of cytoplasmic and membrane-bound material *in vitro* and *in vivo* and gain of function in acceptor cells (Kalargyrou et al., 2021; Ortin-Martinez et al., 2021). In models of retinal degeneration that have significant PR loss, precursor incorporation may be the dominant process (Waldron et al., 2018). Recent studies have reported successful transplantation of cone-precursors derived from hPSCs in blind mouse (Ribeiro et al., 2021). Two clinical trials are testing the delivery of human retinal progenitor cells (hRPCs) with a subretinal injection (*Re*-Neuron, NCT02464436) or with intravitreal injection (jCyte, NCT03073733) as treatments for RP. The first clinical trial using hiPSC-derived neuroretinal sheets has been initiated in Japan (RIKEN, JRCT ID jRCTa050200027; Maeda, Sugita, Kurimoto, & Takahashi, 2021).

4. Transposing bench-scale practices to industrialization

The work of the last decade has demonstrated in both preclinical models and clinical trials that the use of hPSC derivatives is a safe and viable therapeutic option for the treatment of retinal degenerative diseases. However, a number of important questions must be addressed before this strategy becomes part of the arsenal of clinical tools, among which the challenge of transplant immune rejection, the development of more complex retinal grafts through tissue engineering, automation, cryopreservation and the use of artificial intelligence (AI) (Fig. 1).

4.1 Graf rejection: Immunological considerations

The success of cell replacement therapies does not rely solely on the preparation, quality and maturation of the grafted cell product but also on our

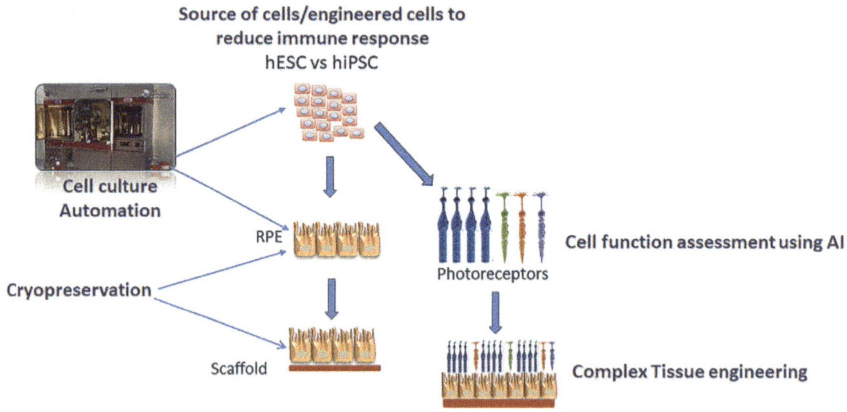

Fig. 1 Transposing bench-scale practices to industrialization. hESC: human embryonic stem cells, hiPSC: human induced pluripotent stem cells, RPE: Retinal pigment epithelium.

understanding of host immunological responses following transplantation. The eye as a whole represents a unique immunologic environment initially thought protected from immune rejection. The "ocular immune privilege" is not what was hoped but features of SRS create a relatively safe target for transplantation of retinal cells derived from hPSCs (Zhou & Caspi, 2010). Moreover, this issue is difficult to address as retinal cells derived from hPSC are usually grafted in a xenogeneic context and therefore under heavy immune suppression (Barnea-Cramer et al., 2016; Ben M'Barek et al., 2020, 2017; Gagliardi et al., 2018; Iraha et al., 2018; Shirai et al., 2016; West et al., 2010; Zhu et al., 2017). In addition, the diseased ocular microenvironment may further increase the risk of immune rejection. The evaluation of cell rejection is notably difficult due to the absence of parameter that indicate rejection of a retinal transplant (Petrash, Palestine, & Canto-Soler, 2021; Sugita, Mandai, Kamao, & Takahashi, 2021). For example, hyper-pigmentation or absence of pigmentation does not strictly correlate to graft survival (Mehat et al., 2018; Petrash et al., 2021).

The literature demonstrates two strategies to prevent immune rejection of RPE grafts in clinical trials: (1) pharmacologic immune modulation and (2) reducing the immunogenicity of the graft. So far, except in the cases of autologous or MHC-matched grafts, every study that has demonstrated evidence of temporary survival of grafted RPE cells have applied pharmacologic immune modulation. However, the best immunosuppression strategy is still debated in ongoing and planned human clinical trials, with current strategies ranging from local to high systemic immunosuppression regimen (da Cruz et al., 2018; Kashani et al., 2021; Schwartz et al., 2012, 2015, 2016). Recent data showed for the first time that non-matched transplanted RPE cells could survive in one patient who received a short course of immunosuppression consisting of 0.075 mg/kg/day tacrolimus from day −8 to day 42 (Kashani et al., 2022).

To reduce graft immunogenicity, the ideal cell product is autologous RPE grafts. hiPSC technology is used to obtain RPE cells derived from the patient's cells (Kamao et al., 2014; Mandai, Watanabe, et al., 2017; Takagi et al., 2019). But high risk of genetic instability (Huang et al., 2019) and high cost (Bravery, 2015; Jacquet et al., 2013) suggest that such strategy is not yet achievable for wide-scale clinical practice. An alternative would be to establish hPSC banks from "super-donors" homozygous for conserved human leucocyte antigen (HLA) haplotypes that are representative of a specific population to enable HLA matching and therefore minimize the risk of allograft rejection (Petrus-Reurer et al., 2021; Sugita et al., 2021; Taylor, Bolton, & Bradley, 2011). Current estimates predict that a bank of hiPSCs from 100 to 150 highly selected donors would meet the needs of 50–90% of the recipient population in UK (Gourraud, Gilson, Girard, & Peschanski, 2012; Lee et al., 2018; Nakajima, Tokunaga, & Nakatsuji, 2007; Pappas et al., 2015; Taylor, Peacock, Chaudhry, Bradley, & Bolton, 2012; Turner et al., 2013). Another promising strategy is the engineering of allogenic hPSCs to be hypoimmunogenic to both adaptive and innate immune responses using gene editing methods (Deuse et al., 2019; Gornalusse et al., 2017; Petrus-Reurer, Kumar, et al., 2020; Petrus-Reurer, Winblad, et al., 2020; Riolobos et al., 2013). Thus, the generation of hypoimmunogenic hPSCs holds the promise of universal cell grafts from a unique cell bank.

4.2 Automation, cryopreservation and final formulation

Although the differentiation efficiency of hPSCs into retinal cells has improved in recent years, it remains a long and laborious process requiring meticulous manipulations. Automation will not only allow the scale up of bioproduction processes but also increase the robustness of differentiation protocols (Regent et al., 2019). It could thus enable larger and more reliable cell production for clinical applications. Robots as well as appropriate programing have been developed for the amplification of hPSCs (Jenkins & Farid, 2015; Konagaya, Ando, Yamauchi, Suemori, & Iwata, 2015; Paull, Sevilla, Zhou, Kim, & Napolitano, 2015; Terstegge et al., 2007; Thomas et al., 2009). More recently, our group was able to fully automate the differentiation of hPSCs into RPE cells and their subsequent amplification (Regent et al., 2019).

RPE sheets currently grafted in clinical trials are maintained in culture until the time of administration, which drastically limits shelf life, restricts transport to treatment sites, and requires the coordination of surgical procedures with product manufacturing. Cryopreservation of RPE-seeded implants would significantly increase product availability by extending shelf-life, increasing the flexibility of manufacturing and clinical schedules, and enabling wide-spread on-demand distribution. The capacity to cryopreserve these implants and administer an adherent cellular therapy immediately post-thaw has been identified as a crucial milestone for commercialization of such therapies (Sharma, Bose, Maminishkis, & Bharti, 2020). However, previous attempts to cryopreserve or vitrify adherent monolayers of different mammalian cells have resulted in highly variable (35–89%) post-thaw survival rates (Bailey et al., 2015; Okumura et al., 2019; Pasch, Schiefer,

Heschel, & Rau, 1999). No clinical application of a cryopreserved, adherent RPE cell-based therapy has yet been reported. One recent study has demonstrated successful cryopreservation of a substrate-adherent, hESC-derived RPE cell monolayer on an implantable scaffold (Pennington et al., 2021), opening new avenues for RPE translation to clinic.

The final formulation of cell therapy products has a major impact on correct integration and polarization of the grafted material in the host cellular environment (Ben M'Barek et al., 2019; Ben M'Barek, Habeler, & Monville, 2018). Attempts at delivering single cell suspensions of hPSC-derived PR and RPE cells have led to poor integration rate and limited visual recovery compared to their transplantation as cell sheets (Ben M'Barek et al., 2020, 2017; Iraha et al., 2018; Mandai, Watanabe, et al., 2017; Shirai et al., 2016). Importantly, PR transplanted as cell sheets from retinal organoids formed synaptic connections and developed polarized segments (Iraha et al., 2018; Mandai, Fujii, et al., 2017; Shirai et al., 2016). However, the formation of rosettes was also observed, within which outer segments of PR were separated from the support of endogenous RPE cells (Iraha et al., 2018; Shirai et al., 2016). Levering on the clinical success of RPE sheet transplants grown on biocompatible substrates, the addition of supportive scaffolds is currently explored to deliver PR in a more physiologically structured way. This may improve implantation and facilitate proper apical-basal polarization of the graft (Jung et al., 2018; Thompson et al., 2019; Worthington et al., 2017; Worthington, Do, Smith, Tucker, & Salem, 2019). In addition, the design of bio-engineered tissues composed of a combination of layers of organized RPE cells and PR might be a suitable graft for advanced stage AMD and RP where RPE cells are defective and PR lost.

4.3 Cell function prediction using artificial intelligence

Cell therapy products undergo extensive quality controls that are complex, destructive and time-consuming (reverse transcription polymerase chain reaction, protein quantifications, immunofluorescence assays). Predicting RPE cell functions in stem cell implants using non-invasive tests such as light microscopy, for example, is a critical task in order to reduce product cost, increase quality and move toward potential marketing. Such cell function predictions can be carried out using Artificial Intelligence (AI)-based models (Padi et al., 2020; Schaub et al., 2020). Moreover, several recent studies have demonstrated the utility of non-invasive imaging for comprehensive studying integration and therapeutic efficacy of transplanted cells (Aboualizadeh et al., 2020; Liu et al., 2021). Although in its infancy, AI holds great promises to improve quality control procedures.

5. Conclusion

Regenerative medicine based on stem cells is likely to make important contributions to the cure of human blindness. This is the main focus of the public creating both fears and hopes. Cell therapy for retinal dystrophies is at the forefront of the clinical investigation compared to other diseases in which such treatment approach could be applied. Indeed, with the ease to access and the available imaging techniques as well as the sophisticated functional assays, retinal diseases serve as a proof of concept of regenerative medicine application. In the coming years, results from many clinical trials will help determine the best formulation strategy for each category of patients as well as the level of immune suppression required. Extensive work on the industrialization of manufacturing processes and development of robust quality controls are still needed to upscale the production processes and treat the millions of patients affected by retinal degeneration. This will in combination with other therapeutic strategies open the future for marketing new products for patients in need.

Funding Statement

This work was supported by grants from NeurATRIS: A Translational Research Infrastructure for Biotherapies in Neurosciences (Investissements d'Avenir—ANR-11-INBS-0011) and INGESTEM: the National Infrastructure Engineering for Pluripotent and differentiated Stem cells (Investissements d'Avenir—ANR-11-INBS-000). I-Stem is part of the Biotherapies Institute for Rare Diseases supported by the Association Française contre les Myopathies-Téléthon.

References

Aboualizadeh, E., Phillips, M. J., McGregor, J. E., DiLoreto, D. A., Strazzeri, J. M., Dhakal, K. R., et al. (2020). Imaging transplanted photoreceptors in living nonhuman primates with single-cell resolution. *Stem Cell Reports*, *15*, 482–497. https://doi.org/10.1016/j.stemcr.2020.06.019.

Age-Related Eye Disease Study Research Group. (2001). A randomized, placebo-controlled, clinical trial of high-dose supplementation with vitamins C and E and beta carotene for age-related cataract and vision loss: AREDS report no. 9. *Archives of Ophthalmology*, *119*, 1439–1452. https://doi.org/10.1001/archopht.119.10.1439.

Al-Khersan, H., Hussain, R. M., Ciulla, T. A., & Dugel, P. U. (2019). Innovative therapies for neovascular age-related macular degeneration. *Expert Opinion on Pharmacotherapy*, *20*, 1879–1891. https://doi.org/10.1080/14656566.2019.1636031.

Ambati, J., & Fowler, B. J. (2012). Mechanisms of age-related macular degeneration. *Neuron, 75*, 26–39. https://doi.org/10.1016/j.neuron.2012.06.018.

Auffarth, G. U., Tetz, M. R., Krastel, H., Blankenagel, A., & Völcker, H. E. (1997). Complicated cataracts in various forms of retinitis pigmentosa. Type and incidence. *Der Ophthalmologe, 94*, 642–646. https://doi.org/10.1007/s003470050175.

Bailey, T. L., Wang, M., Solocinski, J., Nathan, B. P., Chakraborty, N., & Menze, M. A. (2015). Protective effects of osmolytes in cryopreserving adherent neuroblastoma (Neuro-2a) cells. *Cryobiology, 71*, 472–480. https://doi.org/10.1016/j.cryobiol.2015.08.015.

Barnea-Cramer, A. O., Wang, W., Lu, S.-J., Singh, M. S., Luo, C., Huo, H., et al. (2016). Function of human pluripotent stem cell-derived photoreceptor progenitors in blind mice. *Scientific Reports, 6*, 29784. https://doi.org/10.1038/srep29784.

Bassett, E. A., & Wallace, V. A. (2012). Cell fate determination in the vertebrate retina. *Trends in Neurosciences, 35*, 565–573. https://doi.org/10.1016/j.tins.2012.05.004.

Bayyoud, T., Bartz-Schmidt, K. U., & Yoeruek, E. (2013). Long-term clinical results after cataract surgery with and without capsular tension ring in patients with retinitis pigmentosa: A retrospective study. *BMJ Open, 3*, e002616. https://doi.org/10.1136/bmjopen-2013-002616.

Ben M'Barek, K., Habeler, W., Plancheron, A., Jarraya, M., Regent, F., Terray, A., et al. (2017). Human ESC-derived retinal epithelial cell sheets potentiate rescue of photoreceptor cell loss in rats with retinal degeneration. *Science Translational Medicine, 9*, eaai7471. https://doi.org/10.1126/scitranslmed.aai7471.

Ben M'Barek, K., Bertin, S., Brazhnikova, E., Jaillard, C., Habeler, W., Plancheron, A., et al. (2020). Clinical-grade production and safe delivery of human ESC derived RPE sheets in primates and rodents. *Biomaterials, 230*, 119603. https://doi.org/10.1016/j.biomaterials.2019.119603.

Ben M'Barek, K., Habeler, W., & Monville, C. (2018). Stem cell-based RPE therapy for retinal diseases: Engineering 3D tissues amenable for regenerative medicine. *Advances in Experimental Medicine and Biology, 1074*, 625–632. https://doi.org/10.1007/978-3-319-75402-4_76.

Ben M'Barek, K., Habeler, W., Regent, F., & Monville, C. (2019). Developing cell-based therapies for RPE-associated degenerative eye diseases. *Advances in Experimental Medicine and Biology, 1186*, 55–97. https://doi.org/10.1007/978-3-030-28471-8_3.

Ben M'Barek, K., Regent, F., & Monville, C. (2015). Use of human pluripotent stem cells to study and treat retinopathies. *World Journal of Stem Cells, 7*, 596–604. https://doi.org/10.4252/wjsc.v7.i3.596.

Ben M'Barek, K., Habeler, W., Plancheron, A., Jarraya, M., Goureau, O., & Monville, C. (2018). Engineering transplantation-suitable retinal pigment epithelium tissue derived from human embryonic stem cells. *Journal of Visualized Experiments*, (139), 58216. https://doi.org/10.3791/58216.

Bittner, A. K., Diener-West, M., & Dagnelie, G. (2009). A survey of photopsias in self-reported retinitis pigmentosa: Location of photopsias is related to disease severity. *Retina, 29*, 1513–1521. https://doi.org/10.1097/IAE.0b013e3181af0d57.

Bittner, A. K., Haythornthwaite, J. A., Diener-West, M., & Dagnelie, G. (2012). Photopsias are related in part to perceived stress and positive mood in retinitis pigmentosa. *Eye (London, England), 26*, 101–108. https://doi.org/10.1038/eye.2011.247.

Bravery, C. A. (2015). Do human leukocyte antigen-typed cellular therapeutics based on induced pluripotent stem cells make commercial sense? *Stem Cells and Development, 24*, 1–10. https://doi.org/10.1089/scd.2014.0136.

Brown, C. N., Green, B. D., Thompson, R. B., den Hollander, A. I., Lengyel, I., & Eye-Risk Consortium. (2018). Metabolomics and age-related macular degeneration. *Metabolites, 9*, E4. https://doi.org/10.3390/metabo9010004.

Buchholz, D. E., Hikita, S. T., Rowland, T. J., Friedrich, A. M., Hinman, C. R., Johnson, L. V., et al. (2009). Derivation of functional retinal pigmented epithelium from induced pluripotent stem cells. *Stem Cells*, *27*, 2427–2434. https://doi.org/10.1002/stem.189.

Buchholz, D. E., Pennington, B. O., Croze, R. H., Hinman, C. R., Coffey, P. J., & Clegg, D. O. (2013). Rapid and efficient directed differentiation of human pluripotent stem cells into retinal pigmented epithelium. *Stem Cells Translational Medicine*, *2*, 384–393. https://doi.org/10.5966/sctm.2012-0163.

Burton, M. J., Ramke, J., Marques, A. P., Bourne, R. R. A., Congdon, N., Jones, I., et al. (2021). The lancet global health commission on global eye health: Vision beyond 2020. *The Lancet. Global Health*, *9*(4), e489–e551. https://doi.org/10.1016/S2214-109X(20)30488-5.

Carr, A.-J., Vugler, A. A., Hikita, S. T., Lawrence, J. M., Gias, C., Chen, L. L., et al. (2009). Protective effects of human iPS-derived retinal pigment epithelium cell transplantation in the retinal dystrophic rat. *PLoS One*, *4*, e8152. https://doi.org/10.1371/journal.pone.0008152.

Cehajic-Kapetanovic, J., Singh, M. S., Zrenner, E., & MacLaren, R. E. (2022). Bioengineering strategies for restoring vision. *Nature Biomedical Engineering*, Epub ahead of print. https://doi.org/10.1038/s41551-021-00836-4.

Choudhary, P., Booth, H., Gutteridge, A., Surmacz, B., Louca, I., Steer, J., et al. (2017). Directing differentiation of pluripotent stem cells toward retinal pigment epithelium lineage. *Stem Cells Translational Medicine*, *6*, 490–501. https://doi.org/10.5966/sctm.2016-0088.

Colijn, J. M., Buitendijk, G. H. S., Prokofyeva, E., Alves, D., Cachulo, M. L., Khawaja, A. P., et al., EYE-RISK consortium, & European Eye Epidemiology (E3) consortium. (2017). Prevalence of age-related macular degeneration in Europe: The past and the future. *Ophthalmology*, *124*(12), 1753–1763. https://doi.org/10.1016/j.ophtha.2017.05.035.

D'Cruz, P. M., Yasumura, D., Weir, J., Matthes, M. T., Abderrahim, H., LaVail, M. M., et al. (2000). Mutation of the receptor tyrosine kinase gene Mertk in the retinal dystrophic RCS rat. *Human Molecular Genetics*, *9*, 645–651. https://doi.org/10.1093/hmg/9.4.645.

da Cruz, L., Fynes, K., Georgiadis, O., Kerby, J., Luo, Y. H., Ahmado, A., et al. (2018). Phase 1 clinical study of an embryonic stem cell-derived retinal pigment epithelium patch in age-related macular degeneration. *Nature Biotechnology*, *36*, 328–337. https://doi.org/10.1038/nbt.4114.

Decembrini, S., Martin, C., Sennlaub, F., Chemtob, S., Biel, M., Samardzija, M., et al. (2017). Cone genesis tracing by the Chrnb 4-EGFP mouse line: Evidences of cellular material fusion after cone precursor transplantation. *Molecular Therapy*, *25*, 634–653. https://doi.org/10.1016/j.ymthe.2016.12.015.

Decembrini, S., Koch, U., Radtke, F., Moulin, A., & Arsenijevic, Y. (2014). Derivation of traceable and transplantable photoreceptors from mouse embryonic stem cells. *Stem Cell Reports*, *2*, 853–865. https://doi.org/10.1016/j.stemcr.2014.04.010.

Deuse, T., Hu, X., Gravina, A., Wang, D., Tediashvili, G., De, C., et al. (2019). Hypoimmunogenic derivatives of induced pluripotent stem cells evade immune rejection in fully immunocompetent allogeneic recipients. *Nature Biotechnology*, *37*, 252–258. https://doi.org/10.1038/s41587-019-0016-3.

DiLoreto, D. A., del Cerro, C., Cox, C., & del Cerro, M. (1998). Changes in visually guided behavior of Royal College of Surgeons rats as a function of age: A histologic, morphometric, and functional study. *Investigative Ophthalmology & Visual Science*, *39*, 1058–1063.

Diniz, B., Thomas, P., Thomas, B., Ribeiro, R., Hu, Y., Brant, R., et al. (2013). Subretinal implantation of retinal pigment epithelial cells derived from human embryonic stem

cells: improved survival when implanted as a monolayer. *Investigative Ophthalmology & Visual Science, 54*, 5087–5096. https://doi.org/10.1167/iovs.12-11239.
Feng, W., Yasumura, D., Matthes, M. T., LaVail, M. M., & Vollrath, D. (2002). Mertk triggers uptake of photoreceptor outer segments during phagocytosis by cultured retinal pigment epithelial cells. *The Journal of Biological Chemistry, 277*, 17016–17022. https://doi.org/10.1074/jbc.M107876200.
Ferguson, L. R., Balaiya, S., Mynampati, B. K., Sambhav, K., & Chalam, K. V. (2015). Deprivation of bFGF promotes spontaneous differentiation of human embryonic stem cells into retinal pigment epithelial cells. *Journal of Stem Cells, 10*, 159–170.
Fernandes, R. A. B., Stefanini, F. R., Falabella, P., Koss, M. J., Wells, T., Diniz, B., et al. (2017). Development of a new tissue injector for subretinal transplantation of human embryonic stem cell derived retinal pigmented epithelium. *International Journal of Retina and Vitreous, 3*, 41. https://doi.org/10.1186/s40942-017-0095-6.
Flaxman, S. R., Bourne, R. R. A., Resnikoff, S., Ackland, P., Braithwaite, T., Cicinelli, M. V., et al. (2017). Global causes of blindness and distance vision impairment 1990-2020: A systematic review and meta-analysis. *The Lancet Global Health, 5*, e1221–e1234. https://doi.org/10.1016/S2214-109X(17)30393-5.
Gagliardi, G., Ben M'Barek, K., Chaffiol, A., Slembrouck-Brec, A., Conart, J.-B., Nanteau, C., et al. (2018). Characterization and transplantation of CD73-positive photoreceptors isolated from human iPSC-derived retinal organoids. *Stem Cell Reports, 11*, 665–680. https://doi.org/10.1016/j.stemcr.2018.07.005.
Gagliardi, G., Ben M'Barek, K., & Goureau, O. (2019). Photoreceptor cell replacement in macular degeneration and retinitis pigmentosa: A pluripotent stem cell-based approach. *Progress in Retinal and Eye Research, 71*, 1–25. https://doi.org/10.1016/j.preteyeres.2019.03.001.
Goldman, A. I., & O'Brien, P. J. (1978). Phagocytosis in the retinal pigment epithelium of the RCS rat. *Science, 201*, 1023–1025. https://doi.org/10.1126/science.567376.
Gonzalez-Cordero, A., West, E. L., Pearson, R. A., Duran, Y., Carvalho, L. S., Chu, C. J., et al. (2013). Photoreceptor precursors derived from three-dimensional embryonic stem cell cultures integrate and mature within adult degenerate retina. *Nature Biotechnology, 31*, 741–747. https://doi.org/10.1038/nbt.2643.
Gornalusse, G. G., Hirata, R. K., Funk, S. E., Riolobos, L., Lopes, V. S., Manske, G., et al. (2017). HLA-E-expressing pluripotent stem cells escape allogeneic responses and lysis by NK cells. *Nature Biotechnology, 35*, 765–772. https://doi.org/10.1038/nbt.3860.
Gouras, P., Flood, M. T., & Kjeldbye, H. (1984). Transplantation of cultured human retinal cells to monkey retina. *An. Acad. Brasil. Ciênc, 56*, 431–443.
Gouras, P., Flood, M. T., Kjedbye, H., Bilek, M. K., & Eggers, H. (1985). Transplantation of cultured human retinal epithelium to Bruch's membrane of the owl monkey's eye. *Current Eye Research, 4*, 253–265. https://doi.org/10.3109/02713688509000857.
Gourraud, P.-A., Gilson, L., Girard, M., & Peschanski, M. (2012). The role of human leukocyte antigen matching in the development of multiethnic "haplobank" of induced pluripotent stem cell lines. *Stem Cells, 30*, 180–186. https://doi.org/10.1002/stem.772.
Handa, J. T., Bowes Rickman, C., Dick, A. D., Gorin, M. B., Miller, J. W., Toth, C. A., et al. (2019). A systems biology approach towards understanding and treating non-neovascular age-related macular degeneration. *Nature Communications, 10*, 3347. https://doi.org/10.1038/s41467-019-11262-1.
Hartong, D. T., Berson, E. L., & Dryja, T. P. (2006). Retinitis pigmentosa. *Lancet, 368*, 1795–1809. https://doi.org/10.1016/S0140-6736(06)69740-7.
Hartong, D. T., Jorritsma, F. F., Neve, J. J., Melis-Dankers, B. J. M., & Kooijman, A. C. (2004). Improved mobility and independence of night-blind people using night-vision goggles. *Investigative Ophthalmology & Visual Science, 45*, 1725–1731. https://doi.org/10.1167/iovs.03-1061.

Heisterkamp, P., Borsch, O., Lezama, N. D., Gasparini, S., Fathima, A., Carvalho, L. S., et al. (2022). Evidence for endogenous exchange of cytoplasmic material between a subset of cone and rod photoreceptors within the adult mammalian retina via direct cell-cell connections. *Experimental Eye Research*, *219*, 109033. https://doi.org/10.1016/j.exer.2022.109033.

Hori, K., Kuwabara, J., Tanaka, Y., Nishida, M., Koide, N., & Takahashi, M. (2019). A simple and static preservation system for shipping retinal pigment epithelium cell sheets. *Journal of Tissue Engineering and Regenerative Medicine*, *13*, 459–468. https://doi.org/10.1002/term.2805.

Hu, Y., Liu, L., Lu, B., Zhu, D., Ribeiro, R., Diniz, B., et al. (2012). A novel approach for subretinal implantation of ultrathin substrates containing stem cell-derived retinal pigment epithelium monolayer. *Ophthalmic Research*, *48*, 186–191. https://doi.org/10.1159/000338749.

Huang, C.-Y., Liu, C.-L., Ting, C.-Y., Chiu, Y.-T., Cheng, Y.-C., Nicholson, M. W., et al. (2019). Human iPSC banking: Barriers and opportunities. *Journal of Biomedical Science*, *26*, 87. https://doi.org/10.1186/s12929-019-0578-x.

Hussain, A. A., Lee, Y., & Marshall, J. (2020). Understanding the complexity of the matrix metalloproteinase system and its relevance to age-related diseases: Age-related macular degeneration and Alzheimer's disease. *Progress in Retinal and Eye Research*, *74*, 100775. https://doi.org/10.1016/j.preteyeres.2019.100775.

Idelson, M., Alper, R., Obolensky, A., Ben-Shushan, E., Hemo, I., Yachimovich-Cohen, N., et al. (2009). Directed differentiation of human embryonic stem cells into functional retinal pigment epithelium cells. *Cell Stem Cell*, *5*, 396–408. https://doi.org/10.1016/j.stem.2009.07.002.

Iftikhar, M., Lemus, M., Usmani, B., Campochiaro, P. A., Sahel, J. A., Scholl, H. P. N., et al. (2019). Classification of disease severity in retinitis pigmentosa. *The British Journal of Ophthalmology*, *103*, 1595–1599. https://doi.org/10.1136/bjophthalmol-2018-313669.

Ikeda, Y., Nakatake, S., Funatsu, J., Fujiwara, K., Tachibana, T., Murakami, Y., et al. (2019). Night-vision aid using see-through display for patients with retinitis pigmentosa. *Japanese Journal of Ophthalmology*, *63*, 181–185. https://doi.org/10.1007/s10384-018-00644-5.

Iraha, S., Tu, H.-Y., Yamasaki, S., Kagawa, T., Goto, M., Takahashi, R., et al. (2018). Establishment of immunodeficient retinal degeneration model mice and functional maturation of human ESC-derived retinal sheets after transplantation. *Stem Cell Reports*, *10*, 1059–1074. https://doi.org/10.1016/j.stemcr.2018.01.032.

Jacquet, L., Stephenson, E., Collins, R., Patel, H., Trussler, J., Al-Bedaery, R., et al. (2013). Strategy for the creation of clinical grade hESC line banks that HLA-match a target population. *EMBO Molecular Medicine*, *5*, 10–17. https://doi.org/10.1002/emmm.201201973.

Jenkins, M. J., & Farid, S. S. (2015). Human pluripotent stem cell-derived products: Advances towards robust, scalable and cost-effective manufacturing strategies. *Biotechnology Journal*, *10*(1), 83–95. https://doi.org/10.1002/biot.201400348.

Jones, B. W., Marc, R. E., & Pfeiffer, R. L. (1995). Retinal degeneration, remodeling and plasticity. In H. Kolb, E. Fernandez, & R. Nelson (Eds.), *Webvision: The Organization of the Retina and Visual System*. Salt Lake City (UT): University of Utah Health Sciences Center.

Jung, Y. H., Phillips, M. J., Lee, J., Xie, R., Ludwig, A. L., Chen, G., et al. (2018). 3D microstructured scaffolds to support photoreceptor polarization and maturation. *Advanced Materials*, *30*, e1803550. https://doi.org/10.1002/adma.201803550.

Kalargyrou, A. A., Basche, M., Hare, A., West, E. L., Smith, A. J., Ali, R. R., et al. (2021). Nanotube-like processes facilitate material transfer between photoreceptors. *EMBO Reports*, *22*(11), e53732. https://doi.org/10.15252/embr.202153732.

Kamao, H., Mandai, M., Wakamiya, S., Ishida, J., Goto, K., Ono, T., et al. (2014). Objective evaluation of the degree of pigmentation in human induced pluripotent stem cell-derived RPE. *Investigative Ophthalmology & Visual Science*, *55*, 8309–8318. https://doi.org/10.1167/iovs.14-14694.

Kashani, A. H., Lebkowski, J. S., Rahhal, F. M., Avery, R. L., Salehi-Had, H., Dang, W., et al. (2018). A bioengineered retinal pigment epithelial monolayer for advanced, dry age-related macular degeneration. *Science Translational Medicine*, *10*, eaao4097. https://doi.org/10.1126/scitranslmed.aao4097.

Kashani, A. H., Uang, J., Mert, M., Rahhal, F., Chan, C., Avery, R. L., et al. (2020). Surgical method for implantation of a biosynthetic retinal pigment epithelium monolayer for geographic atrophy: Experience from a phase 1/2a study. *Ophthalmology. Retina.*, *4*, 264–273. https://doi.org/10.1016/j.oret.2019.09.017.

Kashani, A. H., Lebkowski, J. S., Rahhal, F. M., Avery, R. L., Salehi-Had, H., Chen, S., et al. (2021). One-year follow-up in a phase 1/2a clinical trial of an allogeneic RPE cell bioengineered implant for advanced dry age-related macular degeneration. *Translational Vision Science & Technology*, *10*, 13. https://doi.org/10.1167/tvst.10.10.13.

Kashani, A. H., Lebkowski, J. S., Hinton, D. R., Zhu, D., Faynus, M. A., Chen, S., et al. (2022). Survival of an HLA-mismatched, bioengineered RPE implant in dry age-related macular degeneration. *Stem Cell Reports*, *17*, 448–458. https://doi.org/10.1016/j.stemcr.2022.01.001.

Klimanskaya, I., Hipp, J., Rezai, K. A., West, M., Atala, A., & Lanza, R. (2004). Derivation and comparative assessment of retinal pigment epithelium from human embryonic stem cells using transcriptomics. *Cloning and Stem Cells*, *6*, 217–245. https://doi.org/10.1089/clo.2004.6.217.

Konagaya, S., Ando, T., Yamauchi, T., Suemori, H., & Iwata, H. (2015). Long-term maintenance of human induced pluripotent stem cells by automated cell culture system. *Scientific Reports*, *5*, 16647. https://doi.org/10.1038/srep16647.

Kruczek, K., Gonzalez-Cordero, A., Goh, D., Naeem, A., Jonikas, M., Blackford, S. J. I., et al. (2017). Differentiation and transplantation of embryonic stem cell-derived cone photoreceptors into a mouse model of end-stage retinal degeneration. *Stem Cell Reports*, *8*, 1659–1674. https://doi.org/10.1016/j.stemcr.2017.04.030.

Lee, S., Huh, J. Y., Turner, D. M., Lee, S., Robinson, J., Stein, J. E., et al. (2018). Repurposing the cord blood bank for haplobanking of HLA-homozygous iPSCs and their usefulness to multiple populations. *Stem Cells*, *36*, 1552–1566. https://doi.org/10.1002/stem.2865.

Li, S., Liu, Y., Wang, L., Wang, F., Zhao, T., Li, Q., et al. (2021). A phase I clinical trial of human embryonic stem cell-derived retinal pigment epithelial cells for early-stage Stargardt macular degeneration: 5-years' follow-up. *Cell Proliferation*, *54*(9), e13100. https://doi.org/10.1111/cpr.13100.

Liu, Y., Chen, S. J., Li, S. Y., Qu, L. H., Meng, X. H., Wang, Y., et al. (2017). Long-term safety of human retinal progenitor cell transplantation in retinitis pigmentosa patients. *Stem Cell Research & Therapy*, *8*, 209. https://doi.org/10.1186/s13287-017-0661-8.

Liu, Z., Ilmarinen, T., Tan, G. S. W., Hongisto, H., Wong, E. Y. M., Tsai, A. S. H., et al. (2021). Submacular integration of hESC-RPE monolayer xenografts in a surgical non-human primate model. *Stem Cell Research & Therapy*, *12*, 423. https://doi.org/10.1186/s13287-021-02395-6.

Lopez, R., Gouras, P., Kjeldbye, H., Sullivan, B., Reppucci, V., Brittis, M., et al. (1989). Transplanted retinal pigment epithelium modifies the retinal degeneration in the RCS rat. *Investigative Ophthalmology & Visual Science*, *30*, 586–588.

Lu, B., Malcuit, C., Wang, S., Girman, S., Francis, P., Lemieux, L., et al. (2009). Long-term safety and function of RPE from human embryonic stem cells in preclinical models of macular degeneration. *Stem Cells*, *27*, 2126–2135. https://doi.org/10.1002/stem.149.

Ludwig, A. L., & Gamm, D. M. (2021). Outer retinal cell replacement: Putting the pieces together. *Translational Vision Science & Technology, 10*, 15. https://doi.org/10.1167/tvst.10.10.15.

Lund, R. D., Wang, S., Klimanskaya, I., Holmes, T., Ramos-Kelsey, R., Lu, B., et al. (2006). Human embryonic stem cell-derived cells rescue visual function in dystrophic RCS rats. *Cloning and Stem Cells, 8*, 189–199. https://doi.org/10.1089/clo.2006.8.189.

Maeda, T., Sugita, S., Kurimoto, Y., & Takahashi, M. (2021). Trends of stem cell therapies in age-related macular degeneration. *Journal of Clinical Medicine, 10*, 1785. https://doi.org/10.3390/jcm10081785.

Maguire, A. M., Russell, S., Wellman, J. A., Chung, D. C., Yu, Z.-F., Tillman, A., et al. (2019). Efficacy, safety, and durability of voretigene neparvovec-rzyl in RPE65 mutation-associated inherited retinal dystrophy: Results of phase 1 and 3 trials. *Ophthalmology, 126*, 1273–1285. https://doi.org/10.1016/j.ophtha.2019.06.017.

Maguire, A. M., Russell, S., Chung, D. C., Yu, Z.-F., Tillman, A., Drack, A. V., et al. (2021). Durability of voretigene neparvovec for biallelic RPE65-mediated inherited retinal disease: Phase 3 results at 3 and 4 years. *Ophthalmology, 128*, 1460–1468. https://doi.org/10.1016/j.ophtha.2021.03.031.

Mandai, M., Watanabe, A., Kurimoto, Y., Hirami, Y., Morinaga, C., Daimon, T., et al. (2017). Autologous induced stem-cell-derived retinal cells for macular degeneration. *The New England Journal of Medicine, 376*, 1038–1046. https://doi.org/10.1056/NEJMoa1608368.

Mandai, M., Fujii, M., Hashiguchi, T., Sunagawa, G. A., Ito, S.-I., Sun, J., et al. (2017). iPSC-derived retina transplants improve vision in rd1 End-stage retinal-degeneration mice. *Stem Cell Reports, 8*, 1112–1113. https://doi.org/10.1016/j.stemcr.2017.03.024.

McGill, T. J., Cottam, B., Lu, B., Wang, S., Girman, S., Tian, C., et al. (2012). Transplantation of human central nervous system stem cells—Neuroprotection in retinal degeneration. *The European Journal of Neuroscience, 35*, 468–477. https://doi.org/10.1111/j.1460-9568.2011.07970.x.

Mehat, M. S., Sundaram, V., Ripamonti, C., Robson, A. G., Smith, A. J., Borooah, S., et al. (2018). Transplantation of human embryonic stem cell-derived retinal pigment epithelial cells in macular degeneration. *Ophthalmology, 125*, 1765–1775. https://doi.org/10.1016/j.ophtha.2018.04.037.

Mitchell, P., Liew, G., Gopinath, B., & Wong, T. Y. (2018). Age-related macular degeneration. *Lancet, 392*, 1147–1159. https://doi.org/10.1016/S0140-6736(18)31550-2.

Mitchell, P., Wang, J. J., Smith, W., & Leeder, S. R. (2002). Smoking and the 5-year incidence of age-related maculopathy: The Blue Mountains Eye Study. *Archives of Ophthalmology, 120*, 1357–1363. https://doi.org/10.1001/archopht.120.10.1357.

Morizur, L., Herardot, E., Monville, C., & Ben M'Barek, K. (2020). Human pluripotent stem cells: A toolbox to understand and treat retinal degeneration. *Molecular and Cellular Neuroscience, 107*, 103523. https://doi.org/10.1016/j.mcn.2020.103523.

Mullen, R. J., & LaVail, M. M. (1976). Inherited retinal dystrophy: Primary defect in pigment epithelium determined with experimental rat chimeras. *Science, 192*, 799–801. https://doi.org/10.1126/science.1265483.

Nakajima, F., Tokunaga, K., & Nakatsuji, N. (2007). Human leukocyte antigen matching estimations in a hypothetical bank of human embryonic stem cell lines in the Japanese population for use in cell transplantation therapy. *Stem Cells, 25*, 983–985. https://doi.org/10.1634/stemcells.2006-0566.

Nandrot, E. F., & Dufour, E. M. (2010). Mertk in daily retinal phagocytosis: A history in the making. *Advances in Experimental Medicine and Biology, 664*, 133–140. https://doi.org/10.1007/978-1-4419-1399-9_16.

Okumura, N., Kagami, T., Watanabe, K., Kadoya, S., Sato, M., & Koizumi, N. (2019). Feasibility of a cryopreservation of cultured human corneal endothelial cells. *PLoS One, 14*, e0218431. https://doi.org/10.1371/journal.pone.0218431.

Ortin-Martinez, A., Tsai, E. L. S., Nickerson, P. E., Bergeret, M., Lu, Y., Smiley, S., et al. (2017). A reinterpretation of cell transplantation: GFP transfer from donor to host photoreceptors. *Stem Cells*, *35*, 932–939. https://doi.org/10.1002/stem.2552.

Ortin-Martinez, A., Yan, N. E., Tsai, E. L. S., Comanita, L., Gurdita, A., Tachibana, N., et al. (2021). Photoreceptor nanotubes mediate the in vivo exchange of intracellular material. *The EMBO Journal*, *40*(22), e107264. https://doi.org/10.15252/embj.2020107264.

Padi, S., Manescu, P., Schaub, N., Hotaling, N., Simon, C., Bharti, K., et al. (2020). Comparison of Artificial Intelligence based approaches to cell function prediction. *Informatics in Medicine Unlocked*, *18*, 100270. https://doi.org/10.1016/j.imu.2019.100270.

Pappas, D. J., Gourraud, P.-A., Le Gall, C., Laurent, J., Trounson, A., DeWitt, N., et al. (2015). Proceedings: Human leukocyte antigen haplo-homozygous induced pluripotent stem cell haplobank modeled after the california population: Evaluating matching in a multiethnic and admixed population. *Stem Cells Translational Medicine*, *4*, 413–418. https://doi.org/10.5966/sctm.2015-0052.

Pasch, J., Schiefer, A., Heschel, I., & Rau, G. (1999). Cryopreservation of keratinocytes in a monolayer. *Cryobiology*, *39*, 158–168. https://doi.org/10.1006/cryo.1999.2197.

Paull, D., Sevilla, A., Zhou, H. Kim Hahn, A., Kim, H., Napolitano, C., et al. (2015). Automated, high-throughput derivation, characterization and differentiation of induced pluripotent stem cells. Nature Methods 12, 885–892. https://doi.org/10.1038/nmeth.3507.

Pearson, R. A., Barber, A. C., Rizzi, M., Hippert, C., Xue, T., West, E. L., et al. (2012). Restoration of vision after transplantation of photoreceptors. *Nature*, *485*, 99–103. https://doi.org/10.1038/nature10997.

Pearson, R. A., Gonzalez-Cordero, A., West, E. L., Ribeiro, J. R., Aghaizu, N., Goh, D., et al. (2016). Donor and host photoreceptors engage in material transfer following transplantation of post-mitotic photoreceptor precursors. *Nature Communications*, 7, 13029. https://doi.org/10.1038/ncomms13029.

Pennington, B. O., Bailey, J. K., Faynus, M. A., Hinman, C., Hee, M. N., Ritts, R., et al. (2021). Xeno-free cryopreservation of adherent retinal pigmented epithelium yields viable and functional cells in vitro and in vivo. *Scientific Reports*, *11*, 6286. https://doi.org/10.1038/s41598-021-85631-6.

Petrash, C. C., Palestine, A. G., & Canto-Soler, M. V. (2021). Immunologic rejection of transplanted retinal pigmented epithelium: Mechanisms and strategies for prevention. *Frontiers in Immunology*, *12*, 621007. https://doi.org/10.3389/fimmu.2021.621007.

Petrus-Reurer, S., Kumar, P., Padrell Sánchez, S., Aronsson, M., André, H., Bartuma, H., et al. (2020). Preclinical safety studies of human embryonic stem cell-derived retinal pigment epithelial cells for the treatment of age-related macular degeneration. *Stem Cells Translational Medicine*, *9*, 936–953. https://doi.org/10.1002/sctm.19-0396.

Petrus-Reurer, S., Winblad, N., Kumar, P., Gorchs, L., Chrobok, M., Wagner, A. K., et al. (2020). Generation of retinal pigment epithelial cells derived from human embryonic stem cells lacking human leukocyte antigen class I and II. *Stem Cell Reports*, *14*, 648–662. https://doi.org/10.1016/j.stemcr.2020.02.006.

Petrus-Reurer, S., Romano, M., Howlett, S., Jones, J. L., Lombardi, G., & Saeb-Parsy, K. (2021). Immunological considerations and challenges for regenerative cellular therapies. *Communications Biology*, *4*, 798. https://doi.org/10.1038/s42003-021-02237-4.

Pfeiffer, R. L., Marc, R. E., & Jones, B. W. (2020). Persistent remodeling and neurodegeneration in late-stage retinal degeneration. *Progress in Retinal and Eye Research*, *74*, 100771. https://doi.org/10.1016/j.preteyeres.2019.07.004.

Regent, F., Morizur, L., Lesueur, L., Habeler, W., Plancheron, A., Ben M'Barek, K., et al. (2019). Automation of human pluripotent stem cell differentiation toward retinal pigment epithelial cells for large-scale productions. *Scientific Reports*, *9*, 10646. https://doi.org/10.1038/s41598-019-47123-6.

Reichman, S., Slembrouck, A., Gagliardi, G., Chaffiol, A., Terray, A., Nanteau, C., et al. (2017). Generation of Storable retinal organoids and retinal pigmented epithelium from adherent human iPS cells in xeno-free and feeder-free conditions. *Stem Cells*, *35*, 1176–1188. https://doi.org/10.1002/stem.2586.

Rezazade Bazaz, M., & Dehghani, H. (2022). From DNA break repair pathways to CRISPR/Cas-mediated gene knock-in methods. *Life Sciences*, *295*, 120409. https://doi.org/10.1016/j.lfs.2022.120409.

Ribeiro, J., Procyk, C. A., West, E. L., O'Hara-Wright, M., Martins, M. F., Khorasani, M. M., et al. (2021). Restoration of visual function in advanced disease after transplantation of purified human pluripotent stem cell-derived cone photoreceptors. *Cell Reports*, *35*, 109022. https://doi.org/10.1016/j.celrep.2021.109022.

Riera, M., Fontrodona, L., Albert, S., Ramirez, D. M., Seriola, A., Salas, A., et al. (2016). Comparative study of human embryonic stem cells (hESC) and human induced pluripotent stem cells (hi PSC) as a treatment for retinal dystrophies. *Molecular Therapy. Methods & Clinical Development*, *3*, 16010. https://doi.org/10.1038/mtm.2016.10.

Riolobos, L., Hirata, R. K., Turtle, C. J., Wang, P.-R., Gornalusse, G. G., Zavajlevski, M., et al. (2013). HLA engineering of human pluripotent stem cells. *Molecular Therapy*, *21*, 1232–1241. https://doi.org/10.1038/mt.2013.59.

Sahel, J.-A., Boulanger-Scemama, E., Pagot, C., Arleo, A., Galluppi, F., Martel, J. N., et al. (2021). Partial recovery of visual function in a blind patient after optogenetic therapy. *Nature Medicine*, *27*, 1223–1229. https://doi.org/10.1038/s41591-021-01351-4.

Santos-Ferreira, T., Postel, K., Stutzki, H., Kurth, T., Zeck, G., & Ader, M. (2015). Daylight vision repair by cell transplantation. *Stem Cells*, *33*, 79–90. https://doi.org/10.1002/stem.1824.

Santos-Ferreira, T., Llonch, S., Borsch, O., Postel, K., Haas, J., & Ader, M. (2016). Retinal transplantation of photoreceptors results in donor-host cytoplasmic exchange. *Nature Communications*, 7, 13028. https://doi.org/10.1038/ncomms13028.

Santos-Ferreira, T., Völkner, M., Borsch, O., Haas, J., Cimalla, P., Vasudevan, P., et al. (2016). Stem cell-derived photoreceptor transplants differentially integrate into mouse models of cone-rod dystrophy. *Investigative Ophthalmology & Visual Science*, *57*, 3509–3520. https://doi.org/10.1167/iovs.16-19087.

Sauvé, Y., Girman, S. V., Wang, S., Lawrence, J. M., & Lund, R. D. (2001). Progressive visual sensitivity loss in the Royal College of Surgeons rat: Perimetric study in the superior colliculus. *Neuroscience*, *103*, 51–63. https://doi.org/10.1016/s0306-4522(00)00557-1.

Schaub, N. J., Hotaling, N. A., Manescu, P., Padi, S., Wan, Q., Sharma, R., et al. (2020). Deep learning predicts function of live retinal pigment epithelium from quantitative microscopy. *Journal of Clinical Investigation*, *130*, 1010–1023. https://doi.org/10.1172/JCI131187.

Scholl, H. P. N., Strauss, R. W., Singh, M. S., Dalkara, D., Roska, B., Picaud, S., et al. (2016). Emerging therapies for inherited retinal degeneration. *Science Translational Medicine*, *8*, 368rv6. https://doi.org/10.1126/scitranslmed.aaf2838.

Schwartz, S. D., Hubschman, J.-P., Heilwell, G., Franco-Cardenas, V., Pan, C. K., Ostrick, R. M., et al. (2012). Embryonic stem cell trials for macular degeneration: A preliminary report. *Lancet*, *379*, 713–720. https://doi.org/10.1016/S0140-6736(12)60028-2.

Schwartz, S. D., Regillo, C. D., Lam, B. L., Eliott, D., Rosenfeld, P. J., Gregori, N. Z., et al. (2015). Human embryonic stem cell-derived retinal pigment epithelium in patients with age-related macular degeneration and Stargardt's macular dystrophy: Follow-up of two open-label phase 1/2 studies. *Lancet*, *385*, 509–516. https://doi.org/10.1016/S0140-6736(14)61376-3.

Schwartz, S. D., Tan, G., Hosseini, H., & Nagiel, A. (2016). Subretinal transplantation of embryonic stem cell-derived retinal pigment epithelium for the treatment of macular

degeneration: An assessment at 4 years. *Investigative Ophthalmology & Visual Science, 57*, ORSFc1-9. https://doi.org/10.1167/iovs.15-18681.
Sharma, R., Bose, D., Maminishkis, A., & Bharti, K. (2020). Retinal pigment epithelium replacement therapy for age-related macular degeneration: Are we there yet? *Annual Review of Pharmacology and Toxicology, 60*, 553–572. https://doi.org/10.1146/annurev-pharmtox-010919-023245.
Sharma, R., Khristov, V., Rising, A., Jha, B. S., Dejene, R., Hotaling, N., et al. (2019). Clinical-grade stem cell-derived retinal pigment epithelium patch rescues retinal degeneration in rodents and pigs. *Science Translational Medicine, 11*, eaat5580. https://doi.org/10.1126/scitranslmed.aat5580.
Shirai, H., Mandai, M., Matsushita, K., Kuwahara, A., Yonemura, S., Nakano, T., et al. (2016). Transplantation of human embryonic stem cell-derived retinal tissue in two primate models of retinal degeneration. *Proceedings of the National Academy of Sciences of the United States of America, 113*, E81–E90. https://doi.org/10.1073/pnas.1512590113.
Singh, M. S., Charbel Issa, P., Butler, R., Martin, C., Lipinski, D. M., Sekaran, S., et al. (2013). Reversal of end-stage retinal degeneration and restoration of visual function by photoreceptor transplantation. *Proceedings of the National Academy of Sciences of the United States of America, 110*, 1101–1106. https://doi.org/10.1073/pnas.1119416110.
Singh, M. S., Balmer, J., Barnard, A. R., Aslam, S. A., Moralli, D., Green, C. M., et al. (2016). Transplanted photoreceptor precursors transfer proteins to host photoreceptors by a mechanism of cytoplasmic fusion. *Nature Communications*, 7, 13537. https://doi.org/10.1038/ncomms13537.
Singh, D., Chen, X., Xia, T., Ghiassi-Nejad, M., Tainsh, L., Adelman, R. A., et al. (2020). Partially differentiated neuroretinal cells promote maturation of the retinal pigment epithelium. *Investigative Ophthalmology & Visual Science, 61*, 9. https://doi.org/10.1167/iovs.61.13.9.
Singh, M. S., Park, S. S., Albini, T. A., Canto-Soler, M. V., Klassen, H., MacLaren, R. E., et al. (2020). Retinal stem cell transplantation: Balancing safety and potential. *Progress in Retinal and Eye Research, 75*, 100779. https://doi.org/10.1016/j.preteyeres.2019.100779.
Song, W. K., Park, K.-M., Kim, H.-J., Lee, J. H., Choi, J., Chong, S. Y., et al. (2015). Treatment of macular degeneration using embryonic stem cell-derived retinal pigment epithelium: Preliminary results in Asian patients. *Stem Cell Reports, 4*, 860–872. https://doi.org/10.1016/j.stemcr.2015.04.005.
Sparrow, J. R., Hicks, D., & Hamel, C. P. (2010). The retinal pigment epithelium in health and disease. *Current Molecular Medicine, 10*, 802–823. https://doi.org/10.2174/156652410793937813.
Stanzel, B. V., Liu, Z., Somboonthanakij, S., Wongsawad, W., Brinken, R., Eter, N., et al. (2014). Human RPE stem cells grown into polarized RPE monolayers on a polyester matrix are maintained after grafting into rabbit subretinal space. *Stem Cell Reports, 2*, 64–77. https://doi.org/10.1016/j.stemcr.2013.11.005.
Strauß, O. (2016). Pharmacology of the retinal pigment epithelium, the interface between retina and body system. *European Journal of Pharmacology, 787*, 84–93. https://doi.org/10.1016/j.ejphar.2016.03.066.
Strauss, O. (1995). The retinal pigment epithelium. In H. Kolb, E. Fernandez, & R. Nelson (Eds.), *Webvision: The organization of the retina and visual system*. Salt Lake City (UT): University of Utah Health Sciences Center.
Strauss, O. (2005). The retinal pigment epithelium in visual function. *Physiological Reviews, 85*, 845–881. https://doi.org/10.1152/physrev.00021.2004.
Sugita, S., Mandai, M., Kamao, H., & Takahashi, M. (2021). Immunological aspects of RPE cell transplantation. *Progress in Retinal and Eye Research, 84*, 100950. https://doi.org/10.1016/j.preteyeres.2021.100950.

Sung, Y., Lee, M. J., Choi, J., Jung, S. Y., Chong, S. Y., Sung, J. H., et al. (2021). Long-term safety and tolerability of subretinal transplantation of embryonic stem cell-derived retinal pigment epithelium in Asian Stargardt disease patients. *The British Journal of Ophthalmology*, *105*, 829–837. https://doi.org/10.1136/bjophthalmol-2020-316225.

Takagi, S., Mandai, M., Gocho, K., Hirami, Y., Yamamoto, M., Fujihara, M., et al. (2019). Evaluation of transplanted autologous induced pluripotent stem cell-derived retinal pigment epithelium in exudative age-related macular degeneration. *Ophthalmology. Retina.*, *3*, 850–859. https://doi.org/10.1016/j.oret.2019.04.021.

Takahashi, K., Tanabe, K., Ohnuki, M., Narita, M., Ichisaka, T., Tomoda, K., et al. (2007). Induction of pluripotent stem cells from adult human fibroblasts by defined factors. *Cell*, *131*, 861–872. https://doi.org/10.1016/j.cell.2007.11.019.

Taylor, C. J., Bolton, E. M., & Bradley, J. A. (2011). Immunological considerations for embryonic and induced pluripotent stem cell banking. *Philosophical Transactions of the Royal Society of London. Series B, Biological Sciences*, *366*, 2312–2322. https://doi.org/10.1098/rstb.2011.0030.

Taylor, C. J., Peacock, S., Chaudhry, A. N., Bradley, J. A., & Bolton, E. M. (2012). Generating an iPSC bank for HLA-matched tissue transplantation based on known donor and recipient HLA types. *Cell Stem Cell*, *11*, 147–152. https://doi.org/10.1016/j.stem.2012.07.014.

Terstegge, S., Laufenberg, I., Pochert, J., Schenk, S., Itskovitz-Eldor, J., Endl, E., et al. (2007). Automated maintenance of embryonic stem cell cultures. *Biotechnology and Bioengineering*, *96*(1), 195–201. https://doi.org/10.1002/bit.21061.

Tezel, T., & Ruff, A. (2021). Retinal cell transplantation in retinitis pigmentosa. *Taiwan Journal of Ophthalmology*, *11*, 336. https://doi.org/10.4103/tjo.tjo_48_21.

Thomas, R. J., Anderson, D., Chandra, A., Smith, N. M., Young, L. E., Williams, D., & Denning, C. (2009). Automated, scalable culture of human embryonic stem cells in feeder-free conditions. *Biotechnology and Bioengineering*, *102*(6), 1636–1644. https://doi.org/10.1002/bit.22187.

Thomas, B. B., Zhu, D., Lin, T.-C., Kim, Y. C., Seiler, M. J., Martinez-Camarillo, J. C., et al. (2018). A new immunodeficient retinal dystrophic rat model for transplantation studies using human-derived cells. *Graefes Archive for Clinical and Experimental Ophthalmology*, *256*, 2113–2125. https://doi.org/10.1007/s00417-018-4134-2.

Thompson, J. R., Worthington, K. S., Green, B. J., Mullin, N. K., Jiao, C., Kaalberg, E. E., et al. (2019). Two-photon polymerized poly (caprolactone) retinal cell delivery scaffolds and their systemic and retinal biocompatibility. *Acta Biomaterialia*, *94*, 204–218. https://doi.org/10.1016/j.actbio.2019.04.057.

Thomson, J. A., Itskovitz-Eldor, J., Shapiro, S. S., Waknitz, M. A., Swiergiel, J. J., Marshall, V. S., et al. (1998). Embryonic stem cell lines derived from human blastocysts. *Science*, *282*, 1145–1147. https://doi.org/10.1126/science.282.5391.1145.

Tu, H.-Y., Watanabe, T., Shirai, H., Yamasaki, S., Kinoshita, M., Matsushita, K., et al. (2019). Medium- to long-term survival and functional examination of human iPSC-derived retinas in rat and primate models of retinal degeneration. *eBioMedicine*, *39*, 562–574. https://doi.org/10.1016/j.ebiom.2018.11.028.

Tucker, B. A., Park, I.-H., Qi, S. D., Klassen, H. J., Jiang, C., Yao, J., et al. (2011). Transplantation of adult mouse iPS cell-derived photoreceptor precursors restores retinal structure and function in degenerative mice. *PLoS One*, *6*, e18992. https://doi.org/10.1371/journal.pone.0018992.

Turner, M., Leslie, S., Martin, N. G., Peschanski, M., Rao, M., Taylor, C. J., et al. (2013). Toward the development of a global induced pluripotent stem cell library. *Cell Stem Cell*, *13*, 382–384. https://doi.org/10.1016/j.stem.2013.08.003.

Verbakel, S. K., van Huet, R. A. C., Boon, C. J. F., den Hollander, A. I., Collin, R. W. J., Klaver, C. C. W., et al. (2018). Non-syndromic retinitis pigmentosa. *Progress in Retinal and Eye Research*, *66*, 157–186. https://doi.org/10.1016/j.preteyeres.2018.03.005.

Waldron, P. V., Di Marco, F., Kruczek, K., Ribeiro, J., Graca, A. B., Hippert, C., et al. (2018). Transplanted donor- or stem cell-derived cone photoreceptors can both integrate and undergo material transfer in an environment-dependent manner. *Stem Cell Reports, 10*, 406–421. https://doi.org/10.1016/j.stemcr.2017.12.008.

Wallace, V. A. (2011). Concise review: Making a retina—From the building blocks to clinical applications. *Stem Cells, 29*, 412–417. https://doi.org/10.1002/stem.602.

Weisz, J. M., Humayun, M. S., De Juan, E., Del Cerro, M., Sunness, J. S., Dagnelie, G., et al. (1999). Allogenic fetal retinal pigment epithelial cell transplant in a patient with geographic atrophy. *Retina, 19*, 540–545. https://doi.org/10.1097/00006982-199911000-00011.

West, E. L., Pearson, R. A., Barker, S. E., Luhmann, U. F. O., Maclaren, R. E., Barber, A. C., et al. (2010). Long-term survival of photoreceptors transplanted into the adult murine neural retina requires immune modulation. *Stem Cells, 28*, 1997–2007. https://doi.org/10.1002/stem.520.

Wong, W. L., Su, X., Li, X., Cheung, C. M. G., Klein, R., Cheng, C.-Y., et al. (2014). Global prevalence of age-related macular degeneration and disease burden projection for 2020 and 2040: A systematic review and meta-analysis. *The Lancet Global Health, 2*, e106–e116. https://doi.org/10.1016/S2214-109X(13)70145-1.

Worthington, K. S., Wiley, L. A., Kaalberg, E. E., Collins, M. M., Mullins, R. F., Stone, E. M., et al. (2017). Two-photon polymerization for production of human iPSC-derived retinal cell grafts. *Acta Biomaterialia, 55*, 385–395. https://doi.org/10.1016/j.actbio.2017.03.039.

Worthington, K. S., Do, A.-V., Smith, R., Tucker, B. A., & Salem, A. K. (2019). Two-photon polymerization as a tool for studying 3d printed topography-induced stem cell fate. *Macromolecular Bioscience, 19*, e1800370. https://doi.org/10.1002/mabi.201800370.

Yang, S., Zhou, J., & Li, D. (2021). Functions and diseases of the retinal pigment epithelium. *Frontiers in Pharmacology, 12*, 727870. https://doi.org/10.3389/fphar.2021.727870.

Zhou, R., & Caspi, R. R. (2010). Ocular immune privilege. *F1000 Biology Reports, 2*, 3. https://doi.org/10.3410/B2-3.

Zhu, J., Cifuentes, H., Reynolds, J., & Lamba, D. A. (2017). Immunosuppression via loss of IL2rγ enhances long-term functional integration of hESC-derived photoreceptors in the mouse retina. *Cell Stem Cell, 20*, 374–384.e5. https://doi.org/10.1016/j.stem.2016.11.019.

CHAPTER THREE

Cell transplantation to repair the injured spinal cord

Adam Hall[a,b], Tara Fortino[a,b], Victoria Spruance[a,b,c], Alessia Niceforo[a,b], James S. Harrop[d], Patricia E. Phelps[e], Catherine A. Priest[f], Lyandysha V. Zholudeva[a,b,g], and Michael A. Lane[a,b,*]

[a]Drexel University, Philadelphia, PA, United States
[b]Marion Murray Spinal Cord Research Center, Drexel University, Philadelphia, PA, United States
[c]Division of Kidney, Urologic, & Hematologic Diseases, National Institutes of Health, Bethesda, MD, United States
[d]Department of Neurological and Orthopedic Surgery, Thomas Jefferson University, Philadelphia, PA, United States
[e]Department of Integrative Biology & Physiology, UCLA, Los Angeles, CA, United States
[f]Neurona Therapeutics, South San Francisco, CA, United States
[g]Gladstone Institutes, San Francisco, CA, United States
[*]Corresponding author: e-mail address: mlane.neuro@gmail.com

Contents

International Review of Neurobiology, Volume 166
ISSN 0074-7742
https://doi.org/10.1016/bs.irn.2022.09.008

Abbreviations

ADL	activities of daily living
AIS	ASIA Impairment Scale
ANS	autonomic nervous system
ASIA	American Spinal Injury Association
BAEP	brain stem auditory evoked potential
BI	Bathal Index
BR-SCI-PT	Bryce/Ragnarsson SCI pain taxonomy
C#	cervical level of the spinal cord
CNS	central nervous system
COVS	Clinical Outcome Variable Scale
CSPGs	chondroitin sulphate proteoglycans
DTI	diffusion tensor imaging
E#	embryonic day #
ECM	extra cellular matrix
EMG	electromyogram
E-NCAM	embryonic neural cell adhesion molecule
ESCs	embryonic stem cells
FDA	Food and Drug Administration
FGF	fibroblast growth factor
FIM	Functional Independence Measure
FSC	fetal spinal cord tissue
GRP	glial restricted precursor
HLA	human leukocyte antigen
hNSPCs	human neural stem/precursor cells
HuCNS-SC	human central nervous system stem cells
IANR-SCIFRS	International Association of Neurorestoratology SCI Functional Rating Scale
iPSCs	induced pluripotent stem cells
ISAFSCI	International Standards to Document Remaining Autonomic Function After Spinal Cord Injury
ISNCSCI	International Standards for Neurological Classification of Spinal Cord Injury
KLS	Kunming Locomotion Scale
L#	lumbar level of the spinal cord
LEMS	lower extremities strength
MAS	Modified Ashworth Scale
MEP	Motor Evoked Potential
MPQ	McGill Pain Questionnaire
MRI	magnetic resonance imaging
MSC	mesenchymal stem cells
NBD	neurogenic bowel dysfunction
NBSS	Neurogenic Bladder Symptom Score
NEPs	neuroepithelial cells
NPCs	neural progenitor cells
NRP	neuronal restricted precursor
NSCs	neural stem cells

NSPCs	neural stem/precursor cells
OB-OECs	olfactory bulb-derived OECs
ODI	Oswestry disability questionnaire
OECs	olfactory ensheathing cells
OM-OECs	mucosal OECs
OPCs	oligodendrocyte progenitor cells
PNG	peripheral nerve graft
SCATS	Spinal Cord Assessment Tool for Spastic Reflexes
SCI	spinal cord injury
SCIM	Spinal Cord Independence Measure
SpIN	spinal interneuron
SpMN	spinal motor neuron
SSEP	somatosensory evoked potentials
T#	thoracic level of the spinal cord
TMS	transcranial magnetic stimulation
UEMS	upper extremities strength
VAS	Visual Analog Scale
WISCI	Walking Index of Spinal Cord Injury

1. Introduction

Spinal cord injuries (SCIs) are life-changing events that often irrevocably alter the lives of injured individuals and their family members. There are more than 300,000 people living with SCI in the U.S. with approximately 18,000 new cases per year (NSCISC, 2021). More than 60% of SCIs occur at the cervical spinal cord levels resulting in the most severe functional deficits, financial cost (millions of dollars), and a reduction in life expectancy. Non-invasive treatment options such as rehabilitation (Goldshmit, Lythgo, Galea, & Turnley, 2008; Lovett-Barr et al., 2012), or more invasive approaches such as neuromodulation (Young, 2015), can promote plasticity and some degree of functional improvement. Unfortunately, the extent of functional recovery is limited, as these interventions rely on spared neural substrates (not directly compromised by injury) and promote limited repair. In contrast, reparative strategies such as cell therapies, support significant growth and can even provide the building blocks to form new neural circuitry and are viewed with great promise. This chapter will emphasize the distinct characteristics, goals, and outcomes of cell therapies under development in pre-clinical research, with some consideration for how they are being translated to clinical trials to treat SCI.

2. Transplantation for spinal cord repair

The adult mammalian central nervous system (CNS) is incapable of significant spontaneous regeneration or repair after injury (Saunders, Fry, Lane, & Dziegielewska, 2002). However, even with these limitations, there is evidence for some plasticity, with anatomical reorganization following an injury (Reier et al., 2017; Zholudeva et al., 2018; Zholudeva et al., 2021), where the injured spinal cord represents a "new anatomy" comprising novel neural networks that contribute to functional recovery (Bareyre et al., 2004; Courtine et al., 2008, 2009). Recovery is inherently limited as it relies on neural components spared by the traumatic injury. Accordingly, there has been a great deal of effort to develop strategies to promote neural repair. Cell transplantation represents one leading strategy in both pre-clinical and clinical settings. Transplanting cells provides the components needed to support, enhance, and even guide neural repair (Fischer, Dulin, & Lane, 2020). The transplanted cells can provide a permissive bridge that fills the injury cavity (Figs. 1A and 2), provide donor neurons that act as relays integrating rostral and caudal host tissue (Figs. 1A and 3), restore neurochemical input (e.g., dopamine, serotonin) to denervated networks (Fig. 1B), deliver cytokine synthesis that can limit tissue damage and/or promote repair, and provide structural reorganization of the neurovasculature for angiogenic support.

The type of cells, tissues and biomaterials that have been used to treat SCI vary significantly based on the desired therapeutic effect. For the purpose of discussion throughout this chapter, we have provided definitions of commonly used cells (and their abbreviations) in Box 1. The remainder of this chapter will focus on three approaches for spinal cord repair: (i) *Bridging Strategies*: transplanting cells, tissues and/or biomaterials that provide a growth permissive/promoting bridge for repair (Figs. 1A and 2 and Table 1), (ii) *Neuronal Relays*: the transplantation of neural stem or progenitor cells that provide new neurons capable of forming novel interneuronal networks with damaged host pathways (Figs. 1B and 3; Table 2), and (iii) *Neuroprotection:* cells and materials to limit tissue damage after injury (Table 3).

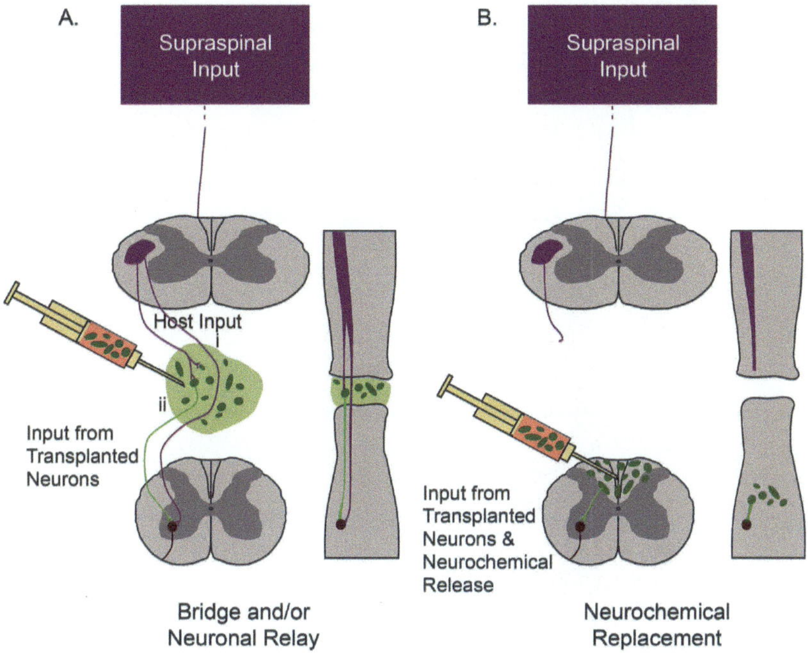

Fig. 1 Schematic diagram showing descending pathways from the brain and brainstem, that are compromised by traumatic spinal cord injury. Transplanted cells can be delivered to the site of injury, providing a growth supportive/promoting bridge that allows growth of host axons through the donor "bridge" (Ai), or delivering a donor population of new neurons that can receive host input and "relay" that connectivity onto host neurons below the injury (Aii). Alternatively, donor neurons or other engineered cells that secrete neurochemicals, can be transplanted distal to the injury site into spared spinal networks to restore this neurochemical input/balance (B; e.g., serotonergic neurons releasing 5-hydroxytryptophan).

3. Pro-regenerative transplants: Bridging for spinal cord repair

After a traumatic SCI, the internal milieu of the injured tissue undergoes a cascade of cellular and molecular changes (e.g., fibroglial interface) that limit neural repair. The resulting cystic cavity and the inhibitory fibroglial interface pose obstacles to neurite extension through the lesion site post-SCI. A primary objective of many cell, tissue or biomaterial transplants is to integrate with the injured host tissue to restore confluency, minimize reactive fibrogliosis, and provide a permissive substrate for host cell

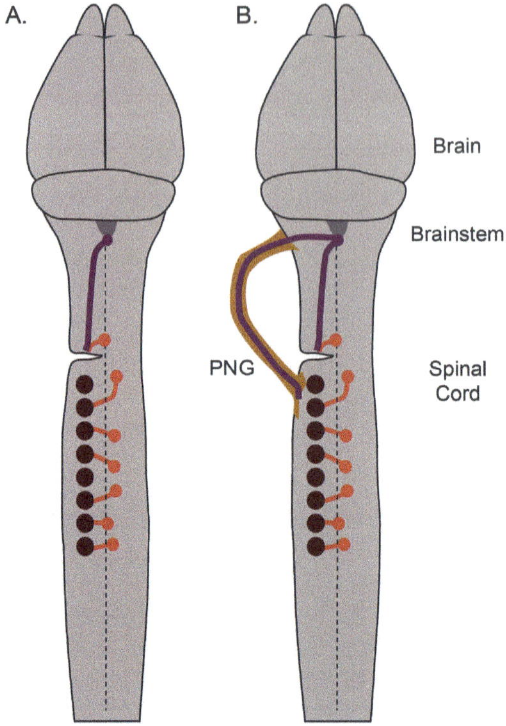

Fig. 2 Schematic diagram of the rodent brain, brainstem and spinal cord, following a lateral upper cervical spinal cord hemisection. Innervation of bulbospinal projections (purple, from cells in the brainstem) and spinal interneuronal input (red cells) to spinal motor neurons (brown) is compromised by injury, and the damaged axons cannot regenerate through the injury site. Pre-clinical studies have shown that implanting one end of a peripheral nerve into the brain or brainstem, and the other into the spinal cord caudal to the injury site, can provide a "bridge" along which host axons can grow. Peripheral nerve grafts (PNG) contain cellular and molecular components that support axon regeneration and have been tested pre-clinically extensively for over a century.

migration and neurite growth *into* and *out of* the donor region (Fawcett, 2008). For a transplanted bridge to be successful, the latter characteristic must occur. Transplanted bridges have been likened to a highway with an "on ramp" and "off ramp" (personal communication with Dr Mary Bunge). Many examples of transplanted bridges have successfully supported host growth into the donor material (taking the on ramp onto the highway), but not all examples successfully achieve host growth back out of the supportive material into host (taking the off ramp to the destination).

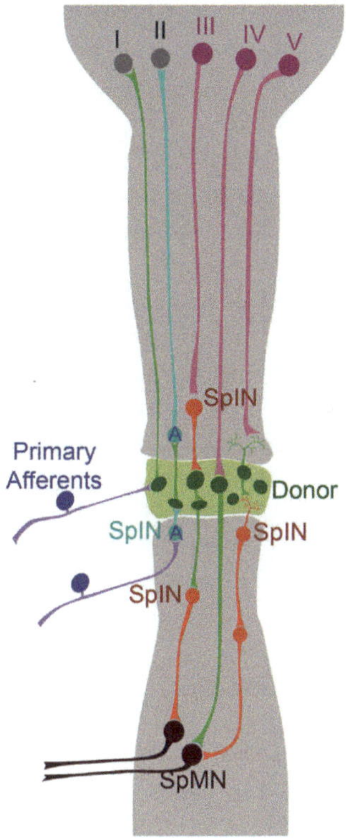

Fig. 3 Schematic diagram of the spinal cord (grey) following a traumatic injury that results in tissue loss and disruption of neural pathways. Transplant tissue and cells (green; donor) can produce neurons that are capable of receiving input from the host and extending axons to denervated host networks. This can contribute to neuronal relay formation and spinal repair in several ways. I: donor neurons can integrate with sensory networks, receiving primary afferent input and via ascending projections ("A") can relay the information to higher order structures. II: donor neurons can also integrate with sensory networks via an ascending propriospinal network (SpINs and ascending projection neurons). III: similarly, donor neurons can integrate with descending propriospinal networks that can relay information to denervated networks caudal to injury (e.g., spinal motor neurons; SpMN). IV: it is also possible that descending, supraspinal axons directly innervate donor neurons which in turn innervate SpMNs caudal to injury. V: another, perhaps underappreciated, alternative is that host neurons caudal to injury can extend neurites into the transplant, and donor cells, similarly, can extend dendrites into the host tissue. Such dendritic growth/plasticity bypasses the need for direct axonal innervation of either donor tissue or the host (i.e., dendritic plasticity). A: ascending; SpIN: spinal interneuron; SpMN: spinal motor neurons. *Modified from Nornes, H., Bjorklund, A., & Stenevi, U. (1984). Transplantation strategies in spinal cord regeneration, In: Sladek, J. R., Jr., Gash, D. M. (Eds.),* Neural transplants, development and function. *(pp. 407–421) New York, USA: Plenum Press; Reier, P. J., Houle, J. D., Jakeman, L., Winialski, D., & Tessler, A. (1988). Transplantation of fetal spinal cord tissue into acute and chronic hemisection and contusion lesions of the adult rat spinal cord.* Progress in Brain Research, 78, *173–179.*

BOX 1 Cell and tissues definition box.

Embryonic Stem Cells (ESCs): A *pluripotent* population of cells found in the inner cell mass of a blastocyst (Thomson et al., 1998) in early developing embryos. ESCs are amenable to differentiation into all three cellular germline lineages (mesoderm, ectoderm, endoderm) when exposed to growth factors (Schuldiner, Yanuka, Itskovitz-Eldor, Melton, & Benvenisty, 2000) and morphogens (Camacho-Aguilar & Warmflash, 2020).

Fetal spinal cord tissue (FSC): A primary source of developing tissue dissected from the spinal cord during embryonic stages of development, usually at embryonic day 13–14 (E13–14) in the rat (Lu et al., 2012). These tissues comprise approximately 80–90% neuronal and glial restricted progenitor cells (NRPs/GRPs) and 10% neuroepithelial cells, post-mitotic neuronal cells, and non-neuronal cells (Cai et al., 2002; Zholudeva et al., 2018). When prepared for transplantation, these tissues are either mechanically dissociated only (Reier, Stokes, Thompson, & Anderson, 1992; Spruance et al., 2018) or mechanically and chemically dissociated (Dulin et al., 2018; Lu et al., 2012) prior to transplantation. How these two preparations differentially affect the donor cell populations is not well-understood.

Induced Pluripotent Stem Cells (iPSCs): A population of *pluripotent* stem cells that can be engineered from somatic cells (e.g., skin fibroblasts) via genetic reprogramming strategies. These cells have the capacity to differentiate into all germline lineages, including various neuronal (Bell et al., 2019; Butts et al., 2019; Chipman, Toma, & Rafuse, 2012; Gupta et al., 2018) and glial (Ehrlich et al., 2017; Soubannier et al., 2020; Speicher, Wiendl, Meuth, & Pawlowski, 2019; Windrem et al., 2017) populations using specific cell culture methods.

Mesenchymal stem cells (MSCs): *Multipotent progenitor* cells initially derived from precursors associated with the embryonic mesoderm (Vodyanik et al., 2010). MSCs can be isolated from a range of sources including bone marrow (bone marrow stromal cells; BMSCs), umbilical cord tissue/blood, adipose, dental pulp, and amniotic fluid (Sarugaser, Hanoun, Keating, Stanford, & Davies, 2009). MSCs demonstrate significant renewal potential along with the capacity to differentiate into cells derived from each germ line lineage (endo-, meso-, and ectoderm) (Ullah, Subbarao, & Rho, 2015).

Neuroepithelial cells (NEPs): A population of neural stem cells found in the developing neural tube (e.g., FSCs) that are multipotent, highly proliferative, and morphologically homogenous (Cai et al., 2002; Reynolds et al., 2008).

Neural progenitor cells (NPCs): A population comprising both neural stem cells (NSCs) and precursor cells. Given this composition they have also been referred to as "neural stem/progenitor cells" (NSPCs). Given their NSC composition, these cells have a limited capacity for *multipotent* self-renewal and

BOX 1 Cell and tissues definition box.—cont'd

proliferation. The more lineage restricted composition has the capacity for *oligopotent (GRPs can differentiate into astrocytes or oligodendrocytes)* and *unipotent (NRPs primarily differentiate into neurons)* differentiation. NPCs can be derived from primary cell sources (e.g., FSC, developing brain) or from *pluripotent* stem cells. These cells can be cultured *in vitro* to engineer their differentiation towards their more lineage restricted progeny - *neural precursor cells* (Fischer et al., 2020; Lane, Lepore, & Fischer, 2017; Zholudeva, Jin, Qiang, Lane, & Fischer, 2021). A notable and important distinction in NPCs derived from different sources, or with different methods, is that the proportion of stem vs more mature precursor cells will vary, as will the relative cell maturity among the precursor population.

Neural precursor cells: A population of cells distinctively comprised of cultured/isolated neuronal and glial lineage-restricted precursor cells (NRPs and GRPs, respectively). They can be considered "oligopotent" cells with a limited capacity for self-renewal and differentiation. For example, glial restricted progenitors (GRPs) can self-renew, and/or differentiate into astrocytes (Davies et al., 2006) or oligodendrocytes (Herrera et al., 2001).

Neural stem cells (NSCs): A population *multipotent progenitors* cells with capacity to divide symmetrically to self-renew (Morshead et al., 1994), and differentiate to any neural cell in the mammalian central nervous system (neuronal or glial). Adult NSCs are found in the sub-ventricular and sub-granular zones in the adult brain, and surrounding the central canal of the spinal cord. NSCs have the capacity to differentiate into neurons (Vieira et al., 2018), astrocytes (Pous et al., 2020), oligodendrocytes (Li et al., 2019).

Olfactory ensheathing cells (OECs): Also known as olfactory ensheathing glia, these cells are unique Schwann-cell-like glia but also have characteristics of CNS glia. OECs are found in both the CNS (outer nerve and glomerular layers of the olfactory bulbs) and the peripheral mucosal lamina propria (nasal cavity; Doucette, 1990). The OECs obtained from these two locations have different properties and there is debate over which type would be most appropriate for transplantation studies (Doucette, 1990; Lu & Ashwell, 2002).

Oligodendrocyte "progenitor" cells (OPCs): Upon the expression of the transcription factor *Olig2*, GRPs begin to differentiate towards the oligodendrocyte lineage - the mature myelinating cells of the central nervous system. While commonly referred to as "progenitors" in the literature, they are perhaps more accurately considered oligo- or unipotent precursor cells.

Schwann cells: Myelin producing glial cells in the peripheral nervous system.

Table 1 Tissues, cells, and biomaterials used to provide a ***pro-regenerative*** bridge for spinal cord repair.

Transplant material		References	
		Transplants made alone	Transplanted in combination with another treatment (e.g., another donor cell/tissue or material)
Tissue	Peripheral nerve	Aguayo, David, and Bray (1981), David and Aguayo (1981, 1985), Decherchi and Gauthier (2000, 2002), Gauthier, Rega, Lammari-Barreault, and Polentes (2002), Houle (1991), Oudega, Varon, and Hagg (1994), Ramon y Cajal (1991), and Richardson, McGuinness, and Aguayo (1980)	Cote, Amin, Tom, and Houle (2011), Cote et al. (2010), Huang et al. (2006), Tom and Houle (2008), and Tom, Kadakia, Santi, and Houle (2009)
Cells	Schwann cells	Anderson et al. (2017), Bastidas et al. (2017), Duncan, Aguayo, Bunge, and Wood (1981) Guest, Santamaria, and Benavides (2013), and Hill, Moon, Wood, and Bunge (2006)	Bunge (2001), Calancie et al. (2009), Fouad et al. (2005), Guest, Rao, Olson, Bunge, and Bunge (1997), Lee, Wu, Arinzeh, and Bunge (2017), Liu et al. (2017), Moon, Leasure, Gage, and Bunge (2006), Oraee-Yazdani et al. (2021), Oraee-Yazdani et al. (2016), Oudega et al. (2001), Pearse et al. (2004), Tuszynski et al. (1998), Williams, Henao, Pearse, and Bunge (2015), Xu, Chen, Guenard, Kleitman, and Bunge (1997), and Xu, Guenard, Kleitman, Aebischer, and Bunge (1995)
	Neural stem cells (NSCs) or progenitor cells (NPCs)*	Amemori et al. (2013), Amemori et al. (2015), and Kumagai et al. (2009)	Hejcl et al. (2013) and Ruzicka et al. (2013)
	Glial-restricted precursors (GRPs)	Fischer (1985), Goulao et al. (2018), Haas and Fischer (2013), Han, Liu, Tyler-Polsz, Rao, and Fischer (2004), Hayakawa et al. (2015), Hill et al. (2004), Jin et al. (2011), and Yuan et al. (2016)	Cao et al. (2005), Dugan and Shumsky (2015), Mitsui, Neuhuber, and Fischer (2011), and Nout et al. (2011)
	Oligodendroglial precursor cell (OPC)	??	
	Astrocytes	Davies et al. (2006, 2008, 2011), and Pencalet et al. (2006)	Wu et al. (2013)
	Microglia	Rabchevsky and Streit (1997)	

Olfactory ensheathing cells (OECs)	Andrews and Stelzner (2007), Barbour, Plant, Harvey, and Plant (2013), Chuah et al. (2004), Collazos-Castro, Muneton-Gomez, and Nieto-Sampedro (2005), Feron et al. (2005), Gomez et al. (2003), Huang et al. (2006), Huang, Xi, Chen, Zhang, and Liu (2012), Kristensson (1976), Lakatos, Barnett, and Franklin (2003), Li, Decherchi, and Raisman (2003); Li, Field, and Raisman (1997), Lu, Feron, Ho, Mackay-Sim, and Waite (2001); Lu, Feron, Mackay-Sim, and Waite (2002), Polentes, Stamegna, Nieto-Sampedro, and Gauthier (2004), Ramer et al. (2004), Ramon-Cueto, Cordero, Santos-Benito, and Avila (2000), Richter, Fletcher, Liu, Tetzlaff, and Roskams (2005), Sasaki, Lankford, Zemedkun, and Kocsis (2004), Stamegna et al. (2011), Steward et al. (2006), Takeoka et al. (2011, 2010); Takeoka, Kubasak, Zhong, Roy, and Phelps (2009),and Ziegler et al. (2011)	Amemori, Jendelova, Ruzickova, Arboleda, and Sykova (2010), Deumens et al. (2006), Fouad et al. (2005), Gomes et al. (2018), Guo, Wang, Wang, and Qu (2015), Kristensson (1976), Kubasak et al. (2008), Moon et al. (2006), Pearse, Pereira, et al. (2004), Pearse et al. (2007), Plant, Christensen, Oudega, and Bunge (2003), Ramon-Cueto, Plant, Avila, and Bunge (1998), Ruitenberg et al. (2005), Ruitenberg et al. (2002), Ruitenberg et al. (2003), Silva, Gimble, Sousa, Reis, and Salgado (2013), Sun, Ye, Zhang, Wu, and Huang (2013), and Thompson et al. (2018)
Fibroblasts		Jin, Tessler, Fischer, and Houle (2000), Liu, Himes, Murray, Tessler, and Fischer (2002), Liu et al. (1998), Liu et al. (1999), Mitsui, Fischer, Shumsky, and Murray (2005), Tobias et al. (2001), Tobias et al. (2005), Tobias et al. (2003), Tuszynski et al. (1994)
Mesenchymal stem cells (MSCs): including bone marrow stromal cells and umbilical cord stem cells.	Cizkova, Rosocha, Vanicky, Jergova, and Cizek (2006), Hasegawa et al. (2005), Hofstetter et al. (2002), Lu, Jones, and Tuszynski (2007), Neuhuber, Timothy Himes, Shumsky, Gallo, and Fischer (2005), Novikova, Brohlin, Kingham, Novikov, and Wiberg (2011), Torres-Espin et al. (2013), Yang et al. (2008), Zurita and Vaquero (2006)	Atkinson et al. (1984), Chen et al. (2020), Gransee, Zhan, Sieck, and Mantilla (2015), Kumagai et al. (2013), Oraee-Yazdani et al. (2021, 2016)

Continued

Table 1 Tissues, cells, and biomaterials used to provide a ***pro-regenerative*** bridge for spinal cord repair.—cont'd

		References	
Transplant material		**Transplants made alone**	**Transplanted in combination with another treatment (e.g., another donor cell/tissue or material)**
Other materials	Biomaterials and scaffolds	Cai, Peng, Nelson, Eberhart, and Smith (2005), Dumont et al. (2019), Gros, Sakamoto, Blesch, Havton, and Tuszynski (2010), Hejcl et al. (2008), Novikova, Novikov, and Kellerth (2003), Sakiyama-Elbert, Johnson, Hodgetts, Plant, and Harvey (2012), Slotkin et al. (2017), Stokols and Tuszynski (2004, 2006), Sullivan et al. (2021), Yeh et al. (2020)	Hejcl et al. (2013), Hosseini, Sharafkhah, Koohi-Hosseinabadi, and Semsar-Kazerooni (2016), Houle and Ziegler (1994), Kim, Kim, Kim, and Ha (2016), Koffler et al. (2019), Lee et al. (2017), Li et al. (2016), Oudega et al. (2001), Rauch et al. (2009), Shin et al. (2018), Sykova, Jendelova, Urdzikova, Lesny, and Hejcl (2006), Teng et al. (2002), Thompson and Sakiyama-Elbert (2018), Thompson et al. (2018), Wang et al. (2017), Wei et al. (2010), Willerth, Rader, and Sakiyama-Elbert (2008), Willerth and Sakiyama-Elbert (2008), Zweckberger, Ahuja, Liu, Wang, and Fehlings (2016), Zweckberger, Liu, Wang, Forgione, and Fehlings (2015)

Table 2 Transplantation of tissues and cells to provide ***novel neuronal connections & an anatomical relay*** for spinal cord repair.

Transplant material		References	
		Transplants alone	Transplanted in combination with another treatment (e.g., another donor cell/tissue or material)
Tissue	Embryonic brain and brainstem tissue	Bjorklund, Katzman, Stenevi, and West (1971), Das (1983, 1983b, 1987), Nornes, Bjorklund, and Stenevi (1983), Nornes, Buchanan, and Bjorklund (1988) and Nygren, Olson, and Seiger (1977)	
	FSC: "Fetal" spinal cord; obtained from embryonic spinal cord tissue—transplanted either as strips (1 mm diameter, 2–3 mm length) or blocks of tissue (1 mm diameter), or *mechanically dissociated* into cell suspension.	Anderson, Howland, and Reier (1995), Anderson, Reier, Wirth, Theele, and Brown (1991), Bregman et al. (1993), Bregman and Reier (1986), Giovanini, Reier, Eskin, and Anderson (1997), Giovanini, Reier, Eskin, Wirth, and Anderson (1997), Horner, Reier, and Stokes (1996), Houle and Reier (1988, 1989), Jakeman and Reier (1991), Jakeman et al. (1989), Lee et al. (2014), Reier (1985), Reier, Anderson, Thompson, and Stokes (1992), Reier, Bregman, and Wujek (1985, 1986), Reier, Bregman, and Wujek (1986), Reier, Perlow, and Guth (1983), Reier, Anderson, et al. (1992), Spruance et al. (2018), Stokes and Reier (1991, 1992), Tessler, Himes, Houle, and Reier (1988), Tessler, Himes, Rogahn, Houle, and Reier (1988), Theele, Schrimsher, and Reier (1996), Thompson, Reier, Parmer, and Lucas (1993), Thompson et al. (2001), White et al. (2010), Wictorin and Bjorklund (1992), and Wirth et al. (2001); Wirth 3rd et al. (1992)	Beaumont, Houle, Peterson, and Gardiner (2004), Bregman et al. (2002), Houle and Johnson (1989), Houle, Morris, Skinner, Garcia-Rill, and Peterson (1999), and Peterson, Murphy, Dupont-Versteegden, and Houle (2000)
Cells	FSC-derived neural progenitor cells (NPCs), composed of NRPs, GRPs and NEPs—FSC is *chemically dissociated* to single cell suspension, but *without cell culture*	Adler, Lee-Kubli, Kumamaru, Kadoya, and Tuszynski (2017), Brock, Graham, Staufenberg, Im, and Tuszynski (2018), Dulin et al. (2018), Kadoya et al. (2016), Kumamaru, Lu, Rosenzweig, Kadoya, and Tuszynski (2019), Lu et al. (2019), Lu, Graham, Wang, Wu, and Tuszynski (2014), Lu et al. (2012), and Poplawski et al. (2020)	Koffler et al. (2019)
	FSC-derived Neural progenitor cells (NPCs); composed of NRPs, GRPs and NEPs—FSC is *chemically dissociated* and cultured for a minimum of 2 days	Bonner, Blesch, Neuhuber, and Fischer (2010), Bonner et al. (2011), Haas and Fischer (2014), Han, Kang, Mujtaba, Rao, and Fischer (2002), Jin, Bouyer, Shumsky, Haas, and Fischer (2016), Jin, Sura, and Fischer (2012), Kumamaru et al. (2019), Lepore, Bakshi, Swanger, Rao, and Fischer (2005), Lepore and Fischer (2005), Lepore et al. (2004), Lepore et al. (2006), Medalha, Jin, Yamagami, Haas, and Fischer (2014), Merianda et al. (2017), Mitsui, Shumsky, Lepore, Murray, and Fischer (2005), Mujtaba, Han, Fischer, Sandgren, and Rao (2002), Nakamura et al. (2005), Ogawa et al. (2002), Rosenzweig et al. (2018), Wu, Mujtaba, Han, Fischer, and Rao (2002), and Zholudeva, Iyer, et al. (2018)	Mitsui et al. (2011)

Continued

Table 2 Transplantation of tissues and cells to provide ***novel neuronal connections & an anatomical relay*** for spinal cord repair.—cont'd

	References	
Transplant material	**Transplants alone**	**Transplanted in combination with another treatment (e.g., another donor cell/tissue or material)**
Brain-derived Neural Progenitor cells (NPCs)—*mechanically and/or chemically dissociated* and cultured for a minimum of 2 days	Watanabe et al. (2004)	
ES-derived Neural Stem and Progenitor Cells (NSCs, NPCs)	Kadoya et al. (2016) and Lu et al. (2012)	
ES-derived Neural *Precursor* Cells (lineage restricted and selected)	Thompson et al. (2018) and Zholudeva, Iyer, et al. (2018)	
iPSC-derived neural stem and progenitor cells (NSCs, NPCs)	Kadoya et al. (2016), Kitagawa et al. (2022), Lu et al. (2014), Nori et al. (2011), Tsuji et al. (2010), and Yokota et al. (2015)	Rodriguez-Jimenez et al. (2021)
iPSC-derived Neural *Precursor* Cells (lineage restricted and selected)	Butts et al. (2017)	

Table 3 Transplantation of tissues and cells to provide *neuroprotection* of spinal cord tissue.

Transplant material		References	
		Transplants alone	Transplanted in combination with another treatment (e.g., other cells, biomaterial)
Tissue	Embryonic spinal cord tissues (FSC)	Bregman (1986) and Bregman and Reier (1986)	
Cells	Embryonic tissue (FSC)-derived Neural (neuronal and/or glial) progenitor cells (NPCs, containing NSCs)	Hasegawa et al. (2005)	Hosseini et al. (2016)
	Neonatal (early postnatal) Tissue derived NPCs—dissected, mechanically and chemically dissociated and cultured		Giraldo et al. (2021), Martinez-Rojas et al. (2022)
	Adult tissue-derived NPCs	Wilcox, Satkunendrarajah, Zuccato, Nassiri, and Fehlings (2014)	
	Embryonic stem cells and ES-derived NPCs		Araujo et al. (2017)
	iPSCs and iPSC-derived NPCs	Ruzicka et al. (2013)	Bonilla et al. (2021)
	Olfactory Ensheathing Cells	Khankan et al. (2016), Takeoka et al. (2011)	
	Mesenchymal Stem Cells (MSCs; e.g., bone marrow stromal cells, adipose stem cells, umbilical cord stem cells)	Arboleda et al. (2011), Brock et al. (2016), Hodgetts, Simmons, and Plant (2013), Kim, Kim, Kweon, and Kang (2020), Ruzicka et al. (2017), Ryu et al. (2012), Torres-Espin et al. (2013) and Urdzikova et al. (2014)	Bonilla et al. (2021), Chen et al. (2020), Hosseini et al. (2018) and Vanecek et al. (2012)

In general terms, these transplants can be considered "pro-regenerative," even though they may have other effects as well (e.g., neuroprotection). The primary goal is often promoting growth of host neurites across the injury site, which can be either axon regeneration (growth from the proximal stump of an injured axon) or neuroplastic axon sprouting (collateral sprouting from a spared, or a preserved portion of an injured axon), and/or increased dendritic sprouting into donor regions. Any combination of these could support repair of compromised pathways, new network formation, and functional improvement.

A vast range of cells, tissues, and biomaterials have been used as a bridging strategy to treat the injured spinal cord (Table 1). This includes the implantation of peripheral nerves (David & Aguayo, 1981; Richardson et al., 1980), Schwann cells (Bunge, 2001; Bunge, Monje, Khan, & Wood, 2017), neural stem and progenitor cells (NSCs, NPCs) and lineage-restricted precursor cells that derive from them (Fischer et al., 2020), astrocytes (Davies et al., 2006, 2008), olfactory ensheathing cells (OECs; a.k.a. olfactory ensheathing glia; Ekberg & St John, 2014; Gilmour, Reshamwala, Wright, Ekberg, & St John, 2020; Yao et al., 2018), mesenchymal cells (e.g., mesenchymal stem cells (MSCs), bone marrow stromal cells, umbilical cord stem cells), fibroblasts (Liu, Murray, Tessler, & Fischer, 2000; Murray et al., 2002), and biomaterials and scaffolds (McCreedy & Sakiyama-Elbert, 2012; Pakulska, Ballios, & Shoichet, 2012; Sakiyama-Elbert et al., 2012; Shoichet, Tate, Baumann, & LaPlaca, 2008; Tam, Fuehrmann, Mitrousis, & Shoichet, 2014).

3.1 Peripheral nerve bridge

Early experiments of Tello and Cajal (Ramon y Cajal, 1991), and others in the late 1800s, found that peripheral nerve could support growth of axons even following traumatic injury. The landmark studies of Sam David and Albert Aguayo (David & Aguayo, 1981) then built on this pioneering work almost a century later, used neural tracing to confirm that peripheral nerve can be used as a viable bridge to support long distance axon growth. In these experiments, one end of a peripheral nerve was implanted into the brainstem of an adult rat, while the other end was implanted into the cervical spinal cord (Fig. 2). Axons from brainstem neurons extended into the nerve bridge, growing the full length of the nerve, and then re-entered the spinal cord at the other end. This strategy has been repeated across multiple research groups, all with some success (see Table 1). However, it is not without limitations. A significant issue that most regenerating axons face is exiting the implanted bridge to innervate host spinal cord tissue. This may in part be due to growth properties that are intrinsic to the neuron extending its axon into the bridge, or a lack of guidance cues within the spinal cord (as opposed to the favorable terrain of the bridge). Another identified limitation is gliosis at the site of nerve insertion into the CNS. Sugar and Gerard reported a "tremendous accumulation of glia and macrophages which results from the injury and the presence of a foreign body" (Sugar & Gerard, 1940). Associated with this has been the increased, injury-induced, production of chondroitin sulphate proteoglycans (CSPGs) within the spinal cord tissue that limits axon outgrowth from the nerve bridge. To address this, there has been some effort to degrade CSPGs at the host-graft interface with the enzyme chondroitinase (Houle et al., 2006, 2009; Tom et al., 2009; Tom & Houle, 2008; Tom, Kadakia, et al., 2009).

3.2 Schwann cells

Like peripheral nerve bridges, Schwann cells can support axon growth into the graft site, and are easily obtained from a patient directly for autologous transplantation. While they also face some challenges supporting growth out of the graft site and back into the host tissue, this may be again addressed with a combinatorial strategy that includes chondroitinase treatment (Fouad et al., 2005). First studied for their myelinating properties, and potential for remyelinating demyelinated axons (Duncan et al., 1981), they soon became widely recognized as a key growth-supporting component of peripheral nerve bridges. Pioneered by Richard and Mary Bunge's teams at the Miami Project to Cure Paralysis, transplantation of Schwann cells into SCI epicenters has been used to provide a bridge for repair for several decades. It was soon found that donor Schwann cells best survived and supported axon growth when seeded into an injectable biomaterial or implantable scaffold (Xu, Guenard, Kleitman, & Bunge, 1995). Pre-clinical studies, led in a large part by the Miami Project teams, found that Schwann cells represented an effective means for providing a bridge to spinal cord repair, that did not have adverse side effects (e.g., tumorigenicity). In October 2014, the Food and Drug Administration (FDA) approved the use of Schwann cells for treating human SCI in a Phase 1 clinical trial led by investigators at the Miami Project Bunge et al., 2017). While clinical testing is underway (Anderson, Guest, et al., 2017; Monje, Deng, & Xu, 2021), parallel pre-clinical research will also help to guide the future use and potential of Schwann cells for spinal cord repair.

3.3 Olfactory ensheathing cells

Olfactory ensheathing cells (OECs) have important growth promoting properties as they surround the axons of the olfactory sensory neurons in the olfactory epithelium which undergo continual adult neurogenesis (Graziadei & Monti Graziadei, 1985). OECs are developmentally related to Schwann cells as both are derived from the neural crest (Forni, Taylor-Burds, Melvin, Williams, & Wray, 2011), but OECs also share phenotypic characteristics of astrocytes, oligodendrocytes and microglia. OECs are found in both the peripheral and central components of the olfactory system. Peripherally, they are located in the mucosal lamina propria where they ensheath large number of axons. These mucosal OECs (OM-OECs) can be easily obtained from the patient for autologous transplantation, but currently can't be differentiated from Schwann cells in culture. Olfactory bulb-derived OECs (OB-OECs), however, are often chosen in pre-clinical studies as they do not contain Schwann cells and therefore can be immunopurified. OB-OECs integrate well with host tissues despite the fibroglial scar, and facilitate the astrocyte border formation, and axonal outgrowth into the injury site (Barbour et al., 2013; Lakatos et al., 2003).

OECs produce a range of benefits after transplantation, including secretion of growth and neurotrophic factors (Roet & Verhaagen, 2014) which enhance axonal sprouting and regeneration (Richter et al., 2005), and angiogenesis (Lopez-Vales, Garcia-Alias, Fores, Navarro, & Verdu, 2004; Ramer et al., 2004). Compared to controls, many studies report substantial remodeling of the injury site following OEC transplantation, including a reduction in the injury site size, reduced astrogliosis, and less inhibitory chondroitin sulphate proteoglycan expression (Mayeur et al., 2013; Zhai et al., 2020). OECs mediate regeneration through cell-cell contact (Chung et al., 2004; Windus et al., 2011, 2010) enabling a variety of other beneficial functions, including bridge formation (Li, Field, & Raisman, 1998), modulation of the immune response to injury including phagocytosis of cellular debris (Nazari et al., 2021) and microbes (Panni et al., 2013), and remyelination of damaged axons (Imaizumi, Lankford, Waxman, Greer, & Kocsis, 1998; Lakatos et al., 2003; Radtke et al., 2004). Pre-clinical data also support the idea that OEC transplants, often in combination with exercise or treadmill step training, can lead to motor recovery (Liu et al., 2014; Nakhjavan-Shahraki et al., 2018; Watzlawick et al., 2016). Significant variability in study design and cellular purification techniques make comparisons between studies difficult.

The two "subtypes" of OECs have many features in common, but for clinical application, the OM-OECs would appear to be the best choice for an allogeneic transplantation as obtaining these is much less invasive than the methods required to obtain OB-OECs. The OB-OECs, however, appear to have greater and more consistent therapeutic efficacy, with fewer side-effects than OM-OECs. This is likely due to differences in gene expression (Guerout et al., 2011), as well as the methods of isolation and purification (Franssen, de Bree, & Verhaagen, 2007). Combining OECs with other cell types, such as olfactory fibroblasts (Miah, Ferretti, & Choi, 2021), Schwann cells (Fouad et al., 2005; Pearse et al., 2007; Sun et al., 2013; Zhang et al., 2017) or neural stem cells (Kang, Hu, Wang, & Wang, 2015; Luo et al., 2013; Wang et al., 2010) show benefits of cell survival of non-OEC transplants but still lack functional recovery (Table 1).

The earliest clinical studies transplanted OECs from human embryonic olfactory bulbs or mucosa biopsies into the spinal cord of humans and caused substantial controversy with no evidence of improvement. The first phase I trial was carried out by Feron et al. (2005) to test the safety and feasibility of autologous OM-OEC transplantation into the spinal cord. Three years later, the 3 treated paraplegics had no evidence of post-traumatic syringomyelia, tumors or other adverse findings, but also did not show evidence of functional recovery (Mackay-Sim et al., 2008). A second human study also transplanted OM-OECs into the spinal cord of 3 paraplegic patients and showed the autologous OM-OEC transplants were safe and feasible (Tabakow et al., 2013). One of the transplanted patients that received a transplant had an asymptomatic syringomyelia while another had improvement in sensory and motor function below the injury level (Tabakow et al., 2013). Interestingly, a person with an almost complete transection of his thoracic spinal cord received both autologous nerve and OB-OEC transplants due to his extensive nasal pathology (Tabakow et al., 2013). Assessments one year later showed increased trunk stability, partial voluntary movement in the legs and improved sensation, with some controversy (Guest & Dietrich, 2015). Ongoing research into the use of OECs for spinal cord repair is focusing more on the purity of cell selection, preparation of the cells under appropriate conditions for optimal survival, and a more rigorous characterization of the genetic and molecular phenotypes of the donor cells.

3.4 Central glia: The astrocytes, oligodendrocytes, and their precursors

Astrocytes, oligodendrocytes, and microglia have all been explored as donor cells for spinal cord repair. Their contributions to repair are quite diverse, not surprisingly, given the diversity in their roles regulating/supporting cytoarchitecture (including the neurovascular unit and extracellular matrix), neural activity, and immune functions within the CNS. In general terms, however, glial cells and their progenitors can support and accordingly provide a supportive bridge of axonal outgrowth and host cell migration. The focus of this section will be on glial precursors that give rise to astrocytes and oligodendrocytes, both of neural lineage. Very little attention has been given to the reparative role transplanted microglia might play (Akhmetzyanova et al., 2018; Kobashi et al., 2020; Kou et al., 2018; Li et al., 2020; Rabchevsky & Streit, 1997), but some evidence has suggested that they too may promote axonal growth post-SCI (Rabchevsky & Streit, 1997).

Transplants of stem, progenitor, or precursor cells that produce a glial lineage have been successful in providing a growth-permissive substrate, with evidence for reduced cavitation and post-injury gliosis, and increased host neurite extension. It is important to note that depending on the source these glial cells are derived, the end-product can contain a mixture of stem and precursor cells. Accordingly, the proportion of each and the maturity of the glial cell population can differ substantially, even when cells are "selected" for greater purity (e.g., immunopanning as described below). While their neuroprotective properties are highlighted below, they have been shown to support the growth of surrounding tissue (Davies et al., 2006; Goulao et al., 2018; Wu et al., 2013). This has been attributed to their growth factor release, extracellular matrix deposition, and their ability to align astrocytic processes to support axon outgrowth (Davies et al., 2006, 2008, 2011; Wu et al., 2013). Early research transplanting adult (Pencalet et al., 2006) and neo-natal (Joosten, Veldhuis, & Hamers, 2004; Wang, Chuah, Yew, Leung, & Tsang, 1995) astrocytes was met with varying degrees of success. The field has since recognized that developing glial restricted precursors (GRPs) isolated from developing spinal cord (obtained on embryonic day 13–14; Davies et al., 2006, 2008; Haas, Neuhuber, Yamagami, Rao, & Fischer, 2012; Jin et al., 2011; Mujtaba et al., 1999), developing brain tissue ("Q-cells"; Sandrock et al., 2010), neural stem cells (Hasegawa et al., 2005; Lien, Tuszynski, & Lu, 2019; Xu et al., 2011), ESCs (Mujtaba et al., 1999;), or iPSCs (Hayashi et al., 2011; Li et al., 2015) could generate a neurogenic and permissive microenvironment/bridge.

Pre-clinical studies evaluating the transplantation of GRPs have most often used cells derived from culturing dissected and chemically dissociated embryonic spinal cord-derived NPCs (comprising neuroepithelial cells and lineage-restricted precursors) (Kalyani, Hobson, & Rao, 1997; Kalyani & Rao, 1998; Mayer-Proschel, Kalyani, Mujtaba, & Rao, 1997; Rao & Mayer-Proschel, 1997; Rao, Noble, & Mayer-Proschel, 1998). These studies were pioneered by Mahendra Rao's and Itzhak Fischer's teams, thoroughly characterized the lineage-restricted precursor populations (neuronal and glial; NRP and GRP) derived from developing embryonic spinal cord tissues and associated cultured NPCs. In addition to removing neurotrophin-3 from the culture media, each freeze-thaw cycle, and each passage of cells, reduces the presence of stem cells (NEPs) and neuronal precursors (NRPs), selecting for GRPs (Zholudeva, Jin, et al., 2021). These GRP-enriched NPC cultures can be further selected by immunopanning for glial cells as well.

Given that GRPs cells give rise to oligodendrocyte or astrocytic lineages (Mayer-Proschel et al., 1997; Rao et al., 1998; Rao & Mayer-Proschel, 1997), some investigators began exploring the therapeutic potential of transplanted oligodendroglial precursor cells (OPCs). Early use of neural differentiated ESCs for treating SCI revealed that these cells can survive and differentiate to produce neurons, astrocytes, and oligodendrocytes, the latter of which was speculated at the time to contribute to remyelination of the injured tissue (McDonald et al., 1999). OPCs have since been derived from fetal spinal cord (Wang et al., 2015; Yao et al., 2017), ESCs (All et al., 2012; Erceg et al., 2010; Keirstead et al., 2005; Priest, Manley, Denham, Wirth 3rd, & Lebkowski, 2015), and iPSCs (Kawabata et al., 2016; Kim et al., 2015; Nazari et al., 2021; Yang et al., 2018). While they have been found to support and promote axon outgrowth *in vitro*, the focus of *in vivo* studies transplanting OPCs has been on either their myelinating properties (e.g., replacing myelin lost as a result of primary demyelination, or providing new myelin for any new axonal growth; Faulkner & Keirstead, 2005; Franklin, Bayley, & Blakemore, 1996; Jeffery, Crang, O'Leary, Hodge, & Blakemore, 1999; Keirstead & Blakemore, 1999; Keirstead, Levine, & Blakemore, 1998; Keirstead et al., 2005; Manley, Priest, Denham, Wirth 3rd, & Lebkowski, 2017; Nistor, Totoiu, Haque, Carpenter, & Keirstead, 2005; Priest et al., 2015; Sharp, Frame, Siegenthaler, Nistor, & Keirstead, 2010; Totoiu, Nistor, Lane, & Keirstead, 2004), or their neuroprotective capacity (promoting tissue sparing; Priest et al., 2015; Sharp et al., 2010), rather than providing a growth supportive bridge for repair. There have been several reports of improved functional outcome after OPC transplantation, (Keirstead et al., 2005; Kim et al., 2015; Priest et al., 2015; Sharp et al., 2010), and based on early pre-clinical data the transplantation of ESC-derived OPCs became the first human "stem cell" trial in history in 2006, led by the biopharmaceutical company Geron

Corporation. The clinical testing of OPC transplantation for spinal cord repair continues today with Lineage Cell Therapeutics, Inc. (previously Asterias and BioTime; see Table 4).

Given the success with tissue-derived and ES-derived glial cell therapy, there has been more recent effort to assess whether iPSC-derived glial cells can achieve similar efficacy. iPSCs have been differentiated towards astrocytic (Hayashi et al., 2011; Li et al., 2015), and OPC identities (All et al., 2015; Fuhrmann et al., 2016; Kawabata et al., 2016; Patil et al., 2021), but to date these cells have been studied mostly for their neuroprotective effects (see below).

3.5 Other pro-regenerative "bridging" transplants

There are many other cell types that have been used to support bridging of the injured spinal cord, alone or in combination with biomaterials or scaffolds (see Table 1), such as fibroblasts and mesenchymal stem cells (MSCs). MSCs, which can be derived from a vast range of tissue types (e.g., umbilical cord, placental blood, bone marrow, adipose tissue), have been used pre-clinically and clinically to treat SCI. Given the diversity in sources, the donor cells used may vary in phenotype, and accordingly the therapeutic goal has varied significantly. While most often used to modulate immune responses post-injury (see Neuroprotection section below), pre-clinical investigation of MSC transplantation has revealed they can promote significant growth of axons (Hofstetter et al., 2002), enabling them to grow through conventionally inhibitory chondroitin sulfate proteoglycans (CSPGs) (Lu, Jones, & Tuszynski, 2005), and myelin (Wright et al., 2007). This is achieved in part through the secretion of adhesion molecules and integrins that contribute to fiber bundles that can act as a permissive bridge for axons and astrocytes to grow into (Hofstetter et al., 2002). MSCs have been shown to relieve the intensity (Watanabe et al., 2015) and progression (Liu, Li, Wang, Zhao, & Jiang, 2020) of neuropathic pain, making the treatment particularly appealing.

Fibroblasts have also been transplanted to provide a growth permissive bridge, but alone these cells do not promote significant repair. This should not be surprising given that fibroblasts infiltrate the lesion site in many cases post-SCI, contributing to the fibroglial "scar" around the injury (Krikorian, Guth, & Donati, 1981). Given their distribution at the lesion site already, scientists began to assess whether they could be used as a cellular means of drug delivery. Several studies have now shown that transplanting

Table 4 Clinical trials transplanting tissue and cells for spinal cord repair.

Company/ institution	Intervention	Inclusion criteria	Post-injury time	Phase of study	Outcomes	Duration	Patients treated	Trial identifier and clinical references
University of Florida (USA)	Developing spinal cord tissue (6–9 weeks)	>18 y.o. Post-traumatic syringomyelia ASI A-D	Chronic: ~11–30 years	Pilot Safety Study	Safety, graft survival, motor and sensory testing	1997 – 2000	8	Thompson et al. (2001) and Wirth et al. (2001)
Griffith University, Australia	Olfactory ensheathing cells (OECs, autologous) from the nasal mucosa	18–55 y.o. >6 months post-injury AIS A, Thoracic SCI (T4–T10)	Chronic: 6 months to 3 years	Phase I/IIa	Safety, MRI; COVS, Improvement in motor and sensory functions evaluated by AIS score, SSEP, TMS, modified Ashworth scale, FIM, ADL; psychosocial assessments	2002–2007	6	
Cairo University (Egypt)	Autologous bone marrow mesenchymal stem cells	10–36 y.o. No progress on physiotherapy for >6 months	Chronic: 10 months to 3 years	Phase I and II	Safety (adverse events), MRI; Improvement in motor, sensory, sphincteric functions evaluated by AIS score	2005–2008	80	NCT00816803 El-Kheir et al. (2014)
Yonsei University (Korea)	Human neural stem/progenitor cells—telencephalon derived (13 weeks gestation) "hNSPCs"	16–60 years AIS A-B Cervical	Acute to Chronic: <1 week to >6 months	Phase I/IIa Non-randomized, Open label	Graft survival, sensory and motor evaluation by ISNCSCI, SSEP, MEP	2005 – 2008	19	NCT0000879 Shin et al. (2015)

Continued

Table 4 Clinical trials transplanting tissue and cells for spinal cord repair.—cont'd

Company/ institution	Intervention	Inclusion criteria	Post-injury time	Phase of study	Outcomes	Duration	Patients treated	Trial identifier and clinical references
Translational Research Center for Medical Innovation, Kobe, Hyogo (Japan)	Autologous bone marrow stromal cells	15–60 y.o.	Acute (methylprednisolone administered within 8 h post-injury)	Phase I and II	Motor function, sensation	2008–2010	5	NCT00695149 Ohta et al. (2004) and Saito et al. (2012)
International Stemcell Services Limited (India)	Autologous bone marrow-derived stem cells	20–55 y.o. SCI at C4-T12	Acute (<2 weeks); Subacute (2–8 weeks) Chronic (>6 months)	Phase I and II	Safety, MRI; sensory and motor function evaluated by AIS, MEP	2008–2010	12	NCT01186679
Pharmicell Co. Ltd. (Korea)	Autologous bone marrow-derived mesenchymal stem cell transplantation	16–65 y.o. Traumatic SCI at cervical level; AIS B	>12 months	Phase II/III	Safety, MRI, DTI; sensory and motor function evaluated by AIS, MEP; SSEP	2008–2021	20	NCT01676441 Park et al. (2012)
Wroclaw Medical University (Poland)	Olfactory mucosa ensheathing cell grafting	16–65 y.o. Single SCI C5-L5; myelopathy confirmed by MRI, not exceeding 2 spinal cord segments; AIS-A	Subacute or chronic	Phase I	None listed	2008 –none listed	10 (recruiting)	NCT01231893
R-Bio (Korea)	Autologous adipose tissue-derived MSC transplantation	19–60 y.o. AIS A, B or C	>2 months	Phase I	Safety through physical examination, vital sign, laboratory test	2009–2010	8	NCT01274975 Ra et al. (2011)

Geron (USA)	Oligodendrocyte precursors "GRNOPC1"	18–65 y.o. AIS A-B Thoracic, T3-T11	Acute: 7–14 days	Phase I Non-randomized, Open label	Safety	2010 – 2011	5	NCT01217008 Bretzner, Gilbert, Baylis, and Brownstone (2011), Chapman and Scala (2012), Strauss (2010), and Wirth 3rd, Lebkowski, and Lebacqz (2011)
Hospital Sao Rafael (Brazil)	Autologous bone marrow mesenchymal stem cell transplantation	18–50 y.o. Traumatic SCI at thoracic or lumbar level; Frankel A	All	Phase I	Feasibility, safety, functional improvement in muscle strength (Frankel scale); sphincter control (anal and bladder)	2010–2012	14	NCT01325103 Mendonca et al. (2014)
General Hospital of Chinese Armed Police Forces (China)	Umbilical cord-derived mesenchymal stem cells	20–50 y.o.	Any	Phase II	BAEP and EMG	2011–2012	40	NCT01393977 Cheng et al. (2014)
China Spinal Cord Injury Network (China)	Umbilical cord blood mononuclear cell	18–60 y.o. AIS A; SCI C5-T11	Chronic >12 months	Phase I and II	Sensory and motor function evaluated with AIS score; SCIM, WISCI, Kunming Walking Score, MAS; Visual analog scale of pain	2011–2013	20	NCT01354483

Continued

Table 4 Clinical trials transplanting tissue and cells for spinal cord repair.—cont'd

Company/ institution	Intervention	Inclusion criteria	Post-injury time	Phase of study	Outcomes	Duration	Patients treated	Trial identifier and clinical references
China Spinal Cord Injury Network (China)	Umbilical cord blood cells	18–65 y.o. Non-penetrating injury; AIS A; C5-T11	<4 weeks	Phase I and II	Sensory and motor function evaluated with AIS score; WISCI; SCIM; Kunming locomotor scales; Modified Ashworth Scale for spasticity	2011–2014	16	NCT01471613
StemCells Inc., "Pathway" (USA)	Human brain-derived neural stem cells "HuCNS-SC"	18–60 y.o. AIS A-C Thoracic (T2-T12)	>12 weeks	Phase I and II Single blind Randomized	Safety, Graft survival, ISNCSCI	2011–2015	12	NCT01321333 Follow-up studies: NCT01725880 NCT03069404 Anderson, Guest, et al. (2017) and Levi et al. (2018)
Memorial Hermann Health System (USA)	Autologous bone marrow progenitor cells	1–15 y.o. AIS A-D	6 months to 4 years	Phase I	Sensory and motor function measured with AIS score; Standard Neuropathic Pain Rating Scale	2011–2016	10	NCT01328860
Bukwang Pharmaceutical (Korea)	Autologous adipose tissue-derived mesenchymal stem cells	19–70 y.o.	>4 weeks	Phase I	Anatomical evaluation with MRI; functional changes measured by electrophysiology	2012–2014	15	NCT01624779
The Miami Project to Cure Paralysis (USA)	Autologous human Schwann cells	18–60 y.o. Traumatic SCI at T3–11; ISNCSCI grade A	<30 days	Phase I	ISNCSCI, MRI, SCIM III, FIM; Measures of neuropathic pain	2012–2016	9	NCT01739023 Anderson, Guest, et al. (2017)

Clínica Las Condes. LIT INNOVA CORFO (Chile)	Autologous expanded mesenchymal stem cells	2–65 y.o. AIS A, B, C; Cervical C4, thoracic or lumbar	All	Phase II	Safety; functional improvement in muscle strength; improvement in sphincters control; spasticity control	2012–2014	30	NCT01694927 Moviglia et al. (2006), Pal et al. (2009), and Ra et al. (2011)
General Hospital of Chinese Armed Police Forces (China)	Umbilical stem mesenchymal stem cells	20–65 y.o. SCI confirmed by MRI, electromyogram, electrophysiology	Chronic: >1year	Phase III	Motor and sensory improvement assessed with ISNCSCI; AIS score; MPQ; BI; SSEP, MEP	2012–2015	300	NCT01873547
National Institute of Blood and Marrow Transplant (NIBMT), Pakistan	Autologous mesenchymal stem cells	18–50 y.o. >2 weeks post-injury	Sub-acute and chronic; traumatic SCI at thoracic level; ASIA A	Phase I	Safety, sensory and motor improvement with AIS score; FIM; muscle strength measured with Frankel scale	2013–2016	9	NCT02482194 Satti et al. (2016)
R-Bio (Korea)	Autologous adipose tissue-derived MSCs transplantation	19–70 y.o. AIS A, B or C	>3 months	Phase I and II	Safety; Sensory and motor function measured with ASIA; MRI; MEP, SSEP, ADL; ODI	2013–2016	15	NCT01769872
Sklifosovsky Institute of Emergency Care (Russia)	Allogeneic umbilical cord blood mononuclear cells	18–65 y.o. Contusion SCI	Acute <3 days	Phase I and II	Safety; Sensory and motor assessment with ASIA Scale, FIM, UEMS, LEMS; Ashworth scale for spasticity; urinary inconsistence; EMG	2013–2018	20	NCT04331405

Continued

Table 4 Clinical trials transplanting tissue and cells for spinal cord repair.—cont'd

Company/ institution	Intervention	Inclusion criteria	Post-injury time	Phase of study	Outcomes	Duration	Patients treated	Trial identifier and clinical references
Chaitanya Hospital (India)	Autologous bone marrow-derived cells	18–55 y.o. Traumatic SCI, confirmed by MRI, below C4	None listed	Phase I and II	Safety; improvement in overall sensory and motor control with AIS and Frankel scores	2014–2016	50	NCT01833975
Novagenesis Foundation (Russia)	Autologous neural stem cells with 3D matrix	18–50 y.o. AIS A Cervical, thoracic, lumbar	Acute, subacute, chronic	Phase I	Safety	2014–2018	30	NCT02326662
NeuralStem [now Seneca Biopharma] (USA)	Human fetal spinal cord-derived neural stem cells: NSI-566RSC "NeuroCells"	18–65 y.o. AIS A Thoracic (T2-T12) and Cervical (C5-C7)	Chronic: 1–2 years	Phase I Open label	Safety, Graft Survival, ISNCSCI exam, Bowel function, Bladder function	2014–Active	Est. 8	NCT01772810
Asterias Biotherapeutics (Then BioTime, Now Lineage Cell Therapeutics)	Oligodendrocyte precursors "AST-OPC1"	18–69 y.o. AIS A-B Cervical, C4-C7	Acute/subacute: 21–42 days	Phase I and II, Non-randomized, Open label	Safety	2015–2018	25	NCT02302157
The Miami Project to Cure Paralysis (USA)	Autologous human Schwann cells	18–65 y.o. Traumatic SCI; T2–12; AIS-A, B, or C C5-T1, AIS-A, B or C	>12 months	Phase I	Motor and sensory function measured with ISNCSCI; MRI, Neuropathic Pain Symptoms Inventory, ISCI Basic Pain dataset, Pain Diagram, Quantitative Sensory Testing; SCIM	2015–2019	8	NCT02354625 Maher, Anderson, Gant, and Cowan (2021)

Chinese Academy of Sciences (China)	NeuroRegen scaffold with bone marrow mononuclear cells or mesenchymal stem cells	18–65 y.o. SCI at C5-T12; AIS A	Chronic	Phase I	Safety; sensory and motor function evaluated with SSEP and MEP; AIS score; FIM; VAS	2015–2021	30	NCT02352077
Da Nang Hospital (Japan)	Autologous bone marrow-derived mononuclear cells	20–60 y.o. SCI AIS A-B	3 weeks to 1 year	Phase II	Safety, motor and sensory function evaluated with AIS score	2016–2019	30	NCT02923817
University of Jordan (Jordan)	Autologous bone marrow-derived mesenchymal stem cells Adipose tissue-derived mesenchymal stem cells	18–70 y.o. Complete SCI AIS A or B, incomplete C	>2 weeks	Phase I and II	Safety; sensory and motor improvement evaluated with AIS score; MRI	2016–2019	14	NCT02981576
NeuroRegen Scaffold (China)	Scaffold™ combined with mesenchymal stem cells or neural stem cells	18–65 y.o. AIS A cervical and thoracic (C5-T12)	Chronic	Phase I and II	Improvements evaluated with AIS, SSEP, MEP	2016–2020	30	NCT02688049
Chinese Academy of Sciences (China)	NeuroRegen Scaffold with bone marrow mononuclear cells	18–60 y.o. Complete SCI at thoracic level, AIS A	All	Phase I	Improvement in sensory and motor function evaluated with AIS score; SSEP; MEP	2016–2021	22	NCT02688062

Continued

Table 4 Clinical trials transplanting tissue and cells for spinal cord repair.—cont'd

Company/ institution	Intervention	Inclusion criteria	Post-injury time	Phase of study	Outcomes	Duration	Patients treated	Trial identifier and clinical references
Stem Cells Arabia (Jordan)	Bone marrow and leukapheresis stem cells	5–50 y.o. Traumatic with complete or partial damage confirmed by MRI	Chronic: 6–60 months post-injury	Phase II	Improvement in sensory and motor function evaluated with AIS score; stool and incontinence improvement; sexual erection status; independence and productivity improvement (all based on questionnaires)	2016–2021	50	NCT02687672
Nicholls Spinal Injury Foundation; Wroclaw Medical University	Combination: Autologous olfactory ensheathing cells (olfactory bulb-derived) with intralesional peripheral (sural) nerve graft	16–65 y.o. AIS A C5–T10	Chronic	Phase I and II, Non-randomized, Open label, Single Group assignment	Improvement in sensory and motor function evaluated with AIS score; Berg Balance Score; Deep sensation, Vibratory test, Joint Sense, Ashworth scale, SCATS spasticity test; WISC test, SCIM III, EPT Test, urodynamic testing, MRI	2016 – 2023	2	NCT03933072

University of Jordan (Jordan)	Autologous bone marrow-derived mesenchymal stem cells Wharton Jelly-derived mesenchymal stem cells	18–70 y.o. AIS A, B or incomplete C; SCI at C5-T11 Complete transection	>12 weeks post-injury	Phase I	Sensory and motor function measured by AIS, ISNCSCI, SCIM III; SSEP	2017–2020	20	NCT04288934
Hospital Sao Rafael (Brazil)	Autologous bone marrow mesenchymal stem cells	18–65 y.o. Cervical level (C5-7) penetrating SCI; AIS A	>12 months	Phase I	Safety, absence of adverse events assessed by MRI; functional improvement assessed by AIS score	2017–2020	10	NCT02574572
Mayo Clinic (USA)	Autologous adipose-derived mesenchymal stem cells	18 y.o. and older Traumatic SCI AIS A or B	2 weeks to 1 year	Phase I	Safety; Sensory and motor function assessed with AIS; MEP; SSEP; MRI	2017–2023	10	NCT03308565 Bydon et al. (2020)
MD Stem Cells (USA)	Autologous bone marrow-derived stem cells (in combination with exoskeleton, virtual reality)	18 y.o. and older	All	N/A	Sensory and motor evaluations with AIS, ANS self-reporting	2017–2024	40	NCT03225625
The First Affiliated Hospital of Dalian Medical University (China)	Umbilical cord blood-derived mesenchymal stem cells	20–65 y.o. No change in AIS sensory or motor score within 6 months	Chronic: >1year	N/A	Sensory and motor function evaluated with AIS, WISCI, SCIM, KLS, MAS, VAS	2018–2021	42	NCT04213131

Continued

Table 4 Clinical trials transplanting tissue and cells for spinal cord repair.—cont'd

Company/ institution	Intervention	Inclusion criteria	Post-injury time	Phase of study	Outcomes	Duration	Patients treated	Trial identifier and clinical references
Third Affiliated Hospital, Sun Yat-Sen University (China)	Umbilical cord mesenchymal stem cells	18–65 y.o. Complete or incomplete traumatic SCI; AIS A-D	>2 months	Phase I and II	Sensory and motor function measured by AIS score; IANR-SCIFRS; ISAFSCI; Penn Scale for spasms; Ashworth scale for muscle spasticity; Geffner Scale for bladder function; NBD; residual urine volume assessed by ultrasound	2018–2020	102	NCT02481440
Third Affiliated Hospital, Sun Yat-Sen University (China)	Allogeneic umbilical cord mesenchymal stem cells	18–65 y.o. Traumatic SCI; AIS A-D	Sub-acute (2 weeks to 2 months); Early stage chronic (2–12 months) Late stage chronic (>12 months)	Phase II	Improvement in sensory and motor function measured with AIS score; changes in EMG, urine measured by ultrasound	2018–2021	66	NCT03521323 Yang et al. (2020)
Third Affiliated Hospital, Sun Yat-Sen University (China)	Allogeneic umbilical cord-derived mesenchymal stem cells	18–65 y.o. Traumatic SCI, AIS A-D; Injury >12 months	Sub-acute (2 weeks to 2 months) Early stage chronic (2–12 months) Late stage chronic (>12 months)	Phase II	Improvement in motor and sensory function assessed with, IANR-SCIRFS scales, EMG, residual urine measured by ultrasound	2019–2021	43	NCT03505034

Third Affiliated Hospital, Sun Yat-Sen University (China)	Umbilical cord mesenchymal stem cells	18–65 y.o. Traumatic SCI, AIS A-D	2 weeks to 2 months	Phase II	Improvement in motor and sensory function evaluated with AIS; IANR-SCIRFS; EMG; residual urine by ultrasound	2019–2021	84	NCT03521336
Hospital Sao Rafael (Brazil)	Autologous mesenchymal cells	18–65 y.o. Blunt SCI at thoracolumbar level T1-L2; AIS A	Chronic >12 months	Phase II	Safety, measured with MRI; motor and sensory function evaluated with AIS score	2019–2022	40	NCT02574585
Shanghai Changzheng Hospital (China)	Autologous bone marrow-derived mononuclear cells	18–60 y.o. Traumatic SCI, AIS A-D	Acute, <2 weeks	Phase II	Safety; Sensory and motor function measured with AIS; MEP; SSEP; urine measured by ultrasound	2020–2023	45	NCT04528550
Mayo Clinic (USA)		18 y.o. and older traumatic AIS-A or B	>1 year	Phase II	Sensory and motor function measured with AIS; SSEP; NBSS; NBD	2020–2024	40	NCT04520373
S.Biomedics Co., Ltd. (Korea)	Neural precursor cells derived from human embryonic stem cell line	18–65 y.o. SCI at C4-C7 level; AIS A	7–60 days post-injury	Phase I and II	Safety; sensory and motor function evaluated with AIS, ISNCSCI, and SCIM scores	2021–2028	5	NCT04812431
Neuroplast (Spain)		18–65 y.o. Traumatic SCI AIS A, B or C; SCI level C4-T12	6–10 weeks	Phase II and III	Safety; sensory and motor evaluated with ISNCSCI, AIS scores	2022–2023	70	NCT03935724

Continued

Table 4 Clinical trials transplanting tissue and cells for spinal cord repair.—cont'd

Company/ institution	Intervention	Inclusion criteria	Post-injury time	Phase of study	Outcomes	Duration	Patients treated	Trial identifier and clinical references
StemCyte, Inc. (USA, Taiwan)	Umbilical cord blood mononuclear stem cells	18–60 y.o. Traumatic SCI C5-T11 confirmed by MRI	Chronic (>12 months)	Phase II	Sensory and motor recovery evaluated with AIS, WISCI II, SCIM III scores	2022–2024	18	NCT03979742
The Foundation for Orthopaedics and Regenerative Medicine (Antigua and Barbuda)	Cultured allogeneic adult umbilical cord-derived mesenchymal stem cells	Child, adult, older adult	All	Phase I	Safety; sensory and motor function evaluated with AIS score	2022–2026	20	NCT05152290

Table summarizing key clinical trials that transplanted tissue and cells with a goal of repair and/or neuroprotection after human spinal cord injury, with references to the ClinicalTrial registry and relevant publications. Please note that there is not presently a standard approved primary outcome measure for clinical trials in spinal cord injury, although there are several peer-reviewed guidelines (Anderson et al., 2005; Chhabra et al., 2019; Kwon et al., 2015; Steeves & Blight, 2012). Search terms used in the data base are as follows: "spinal cord injury" as disease/condition, "transplantation," "stem cell." Trials terminated with 0 participants were not included.

fibroblasts that are genetically modified to release growth factors can stimulate neurite outgrowth into the lesion site (Grill, Blesch, & Tuszynski, 1997; Liu et al., 2000; McTigue, Horner, Stokes, & Gage, 1998; Nakahara, Gage, & Tuszynski, 1996; Shumsky et al., 2003; Tobias et al., 2001; Tuszynski et al., 1994). Given their limited reparative potential otherwise, and their demonstrated safety, transplantation of fibroblasts has actually become a commonly used "negative" control to compare against the transplantation of other cell types.

4. Building novel neuronal pathways: Neuronal relays for repair

The transplantation of NPCs or neural tissues cannot only serve as a growth permissive/promoting bridge, but if they contain neuronal precursors, they may also provide an anatomical relay. Such transplants can synaptically integrate with growing host neurites (and vice versa), establishing a functional interneuronal connection across the injury site (Bonner & Steward, 2015). Evidence for synaptic integration between donor and host neurons has been observed in sensory (Bonner et al., 2011; Dulin et al., 2018), motor (Adler et al., 2017; Ceto, Sekiguchi, Takashima, Nimmerjahn, & Tuszynski, 2020; Lee et al., 2014; Spruance et al., 2018; White et al., 2010; Zholudeva, Iyer, et al., 2018), and autonomic networks (Fortino et al., 2022; Hou et al., 2020). Evidence for connectivity between host and donor neurons, or donor neurons and networks of host neurons, has also been demonstrated using transneuronal tracing and electrophysiological recordings.

4.1 Neural tissue transplantation

Experiments using transplantation of developing neural tissues to repair the injured adult spinal cord began in the 1930–1940s (Sugar & Gerard, 1940), though it was not until the 1980s that preclinical studies began to suggest significant promise for this treatment as a viable strategy (Reier et al., 1983). In fact, many of the early studies concluded that fetal spinal cord (FSC) tissue did not survive in the intact or injured adult CNS (Das, 1983; Sugar & Gerard, 1940). However, the advent of new research tools and improved technologies eventually showed that cells not only survive but remain capable of growing and making connections within the injured spinal cord under the right conditions (Jakeman & Reier, 1991; Reier et al., 1983, 1985; Reier, Bregman, & Wujek, 1986). From these studies, two main concepts emerged as being critical for the growth and survival of FSC tissue in an injured cavity, including the age of the donor embryo from which tissue is obtained and approximation of the transplant to the host vascular supply. Studies concluded that the ideal age of the donor embryo for FSC tissue was between E13 and E15 in rat (Reier et al., 1983) (E12–13 in mouse), coinciding with the mitotic potential of developing neurons, and that the use of older tissue (E16-E22) resulted in poor graft survival and no functional improvements (Das, 1983). While neonatal or older spinal tissues are sub-optimal as a donor cell source, they are logistically easier to obtain and do contain populations of neuroepithelial cells that can be isolated and expanded in cell culture under the right conditions (Giraldo et al., 2021; Moreno-Manzano et al., 2009) (discussed below in stem cell derived NPCs).

Pre-clinical investigations of the survival and differentiation of developing embryonic spinal cord tissue grafted into ectopic sites (Henschen, Hoffer, & Olson, 1985; Reier et al., 1983) or into spinal lesions of neonatal and adult rats (Bernstein & Goldberg, 1987; Bernstein, Hoovler, & Turtil, 1985; Reier, Bregman, & Wujek, 1986) demonstrated restoration of tissue confluency and donor-host integration. Building on this success (Jakeman & Reier, 1991; Reier et al., 1983, 1985; Reier, Bregman, & Wujek, 1986), attention was next directed to grafting into chronic spinal injuries (Houle & Reier, 1988, 1989) and spinal contusion injuries, though this model was not as extensively used as it is today (Reier, Anderson, et al., 1992; Reier, Houle, Jakeman, Winialski, & Tessler, 1988; Stokes & Reier, 1992; Thompson et al., 1993). This approach was subsequently extended to a compression (static-load) SCI model in cat (Anderson et al., 1991; Anderson et al., 1995; Reier, Stokes, et al., 1992), which included magnetic resonance imaging for assessing graft survival (Wirth 3rd et al., 1992; Wirth 3rd, Theele, Mareci, Anderson, & Reier, 1995), well before consideration about pre-clinical tests of invasive cell-based interventions for SCI being performed in large animal models (Kwon et al., 2015).

Several decades of pre-clinical and clinical testing have shown that FSC grafts are capable of surviving in the injured spinal cord, but few experiments have examined their specific neuronal composition from early graft development to maturity and long-term survival. It is known that FSC tissue undergoes substantial cell death within the first 4 days following intraspinal transplantation (Lepore et al., 2005;), with small islands of remaining cells resembling germinal neuroepthelium (Theele et al., 1996). Though the identity of these cells was not determined, other lines of evidence suggest they were neuronal and glial precursors (Kalyani, Piper, Mujtaba, Lucero, & Rao, 1998; Lepore et al., 2006; Mayer-Proschel et al., 1997). Subsequent graft development entails proliferation and differentiation of these cells to fill the injury site, and mature grafts show confluency with host tissues. Matured FSC grafts are also capable of achieving a significant degree of organotypic differentiation as indicated by the formation of morphologically-distinct regions with cytological and neurotransmitter characteristics that resemble the normal adult spinal cord (e.g., substantia gelatinosa-like regions; Jakeman et al., 1989; White et al., 2010). More recent studies show FSC grafts mature into a variety of interneuronal subtypes with evidence of anatomical and functional connectivity between donor and host cells, though the extent of connectivity is variable between recipients (e.g., Spruance et al., 2018). It is important to note that grafts of rat E13.5–14 FSC tissue are typically devoid of motor neurons (White et al., 2010), as the process of obtaining donor tissue essentially results in a mix of glial and interneuronal precursors. This by no means is a disadvantage given the important role of spinal interneurons are known to play in neuroplasticity and recovery post-SCI (Bareyre et al., 2004; Courtine et al., 2008; Flynn, Graham, Galea, & Callister, 2011; Zholudeva, Abraira, et al., 2021; Zholudeva & Lane, 2022); indeed, a donor population rich in interneuronal precursors may represent a therapeutic ideal. In addition, the combined transplantation of both neuronal and glial precursors, as opposed to isolated precursor populations, has been reported to promote survival and differentiation of glial- and neuronally-restricted donor cells following transplantation (Cao, Howard, Dennison, & Whittemore, 2002; Lepore & Fischer, 2005).

Success of FSC transplantation in different experimental SCI settings, including studies of human FSC xenografts (Akesson et al., 2001; Akesson, Kjaeldgaard, & Seiger, 1998; Gilerovich, Fedorova, & Otellin, 1991), led to a formal clinical translation in a small cohort of SCI subjects (reviewed in Reier, 2004). The primary purpose of that clinical experience was to test procedural safety and feasibility (Wirth et al., 2001), though some early measure of efficacy also was obtained (Thompson et al., 2001). Demonstration of safety and feasibility confirmed findings from two earlier, albeit more limited, clinical studies involving the same type of SCI recipients (Blagodatskii et al., 1994; Falci et al., 1997). These early translational experiences laid the foundation for forthcoming clinical trials testing safety of various neural stem or stem-like cells that have been generated given the ethical and logistical concerns regarding clinical utility of human embryonic CNS tissue (Table 4).

The future of cellular transplantation as a treatment for CNS injury lies in identifying a bankable, refined NPC population that can be prepared readily and universally. Unfortunately, the phenotype of neuronal and glial components of NPCs that are optimal for repair and functional restoration, and the ideal proportions of them, remain unknown. The early success with FSC transplantation has meant that it remains an important experimental tool for elucidating the requirements for a successful, clinically viable cell-based therapy. For example, comparing FSC tissue grafts with more refined NPCs can offer insights into the contributions of neural and non-neural constituents to graft efficacy and how refinement protocols can affect differentiation or selection of donor phenotypes. FSC tissue contains not only neuronal and glial progenitors but also some neural epithelial cells, extra cellular matrix (ECM), and non-neural cells such as endothelial cells. All these are essential components of spinal cord tissue. Culturing FSC tissue to isolate NPCs yields a refined population of donor cells that can be cryopreserved and banked for later use, but less is known about the impact of other components being omitted. Standard culturing methods also may affect the phenotype of donor neuronal and glial progenitor cells available (Zholudeva, Iyer, et al., 2018). With advances in neuroanatomical, genetic, and molecular methods, preclinical studies are now beginning to identify some of the differences between FSC and cultured NPCs. Identifying beneficial and detrimental components of donor tissues or cells can then be used to guide future transplantation studies, so that clinically viable cell sources contain the appropriate cellular composition for repair.

4.2 Neural progenitor cells (NPCs)

The transplantation of NPCs has been identified as a promising therapeutic strategy for SCI. Like FSC transplants, this therapeutic approach offers the potential for not only protecting and providing trophic support for host cell regeneration, but also for replacing lost or damaged cells. However, as they can be derived from many sources (FSC, ES, and iPS cells) this transplantation strategy moves away from acutely dissected grafts of FSC tissue and towards a more clinically relevant model of easily obtainable source of cells that can be expanded, cryopreserved, and utilized upon demand. An important consideration for the use of NPCs in the following sections is that the proportion of stem vs more mature precursor cells in the NPCs derived from different sources can differ a great deal depending on the methods used to isolate and prepare the cells for transplantation. Advances in single-cell analysis in recent years will begin to shed light on what these components are in future research.

Pioneering studies beginning in the 1940s paved the way for identification of crucial events in the development of the rat spinal cord. For example, the closure of the embryonic rat neural tube was seen at E10 and the earliest evidence for differentiation of these cells was evident at E11 (Altman & Bayer, 1984; Hamburger, 1948; Nornes & Das, 1974). More detailed characterization of the cellular phenotypes at E10.5 revealed that the early-formed neural tube is comprised of proliferating and morphologically homogeneous neuroepithelial cells (NEPs; the NSCs of the neural tube) (Kalyani et al., 1998). Not only can these cells be cultured under conditions that maintain their multipotency and self-renewal capacity, but they also can be driven to differentiate into neurons, glia, and astrocytes (Mayer-Proschel et al., 1997). These multipotent NEPs generate all the differentiated cells of the CNS via intermediate precursors that have been characterized by certain surface markers and corroborated with their differential spatial location within the developing spinal cord. Identification of two specific intermediate restricted precursors that have been shown to have a direct lineage relationship from the multipotent NEP cells, and that can be maintained in culture, are neuronal and glial restricted progenitors (NRPs and GRPs, respectively) (Mayer-Proschel et al., 1997; Rao & Mayer-Proschel, 1997). GRPs are cells restricted to the glial lineage that have the potential of differentiating into oligodendrocytes and type 1 astrocytes both *in vitro* and *in vivo* (reviewed in Rao, 1999). NRPs are cells morphologically distinct from NEPs and GRPs, expressing a specific cell surface marker (embryonic neural cell adhesion molecule; E-NCAM), and have been shown to differentiate into Beta III-expressing neurons *in vitro* (Mayer-Proschel et al., 1997). While lineage restricted, both NRPs and GRPs are capable of self-renewal over multiple passages in the presence of fibroblast growth factor (FGF), and can be differentiated into more mature neuronal and glial phenotypes using specific combinations of growth factors (Mayer-Proschel et al., 1997; Rao & Mayer-Proschel, 1997).

Characterization of fetal multipotent NEPs and their subsequent differentiation into mature neurons and glial cells via intermediate precursor cells has also been made in the uncultured embryonic spinal cord. Such studies demonstrated E-NCAM reactivity for NRPs (Mayer-Proschel et al., 1997) and A2B5 immunoreactivity for GRPs (Cai et al., 2002) to be spatially restricted at E13.5 in rat (E12–13 in mice) to regions of neuron and glial generation, respectively. These *in vivo* characterization studies provided confidence in corroborating *in vitro* results, especially considering the methods necessary for progenitor isolation.

There are two distinct approaches for preparing FSC-derived NPCs in pre-clinical transplantation studies. While these both chemically dissociate the tissues dissected from FSC, one approach then prepares cells for transplantation acutely (without any period of cell culture, but often along with growth factors in the delivery vehicle) (Lu et al., 2012; Robinson & Lu, 2017), and the other cultures dissociated cells for a minimum of 2 days (with growth factors and nutrient support) to further "refine" donor cell populations (Zholudeva, Jin, et al., 2021) (see Table 2). For this discussion, these two approaches are separated, as it is likely that the methods used to prepare the NPCs for transplantation, and the vehicle they are transplanted with, alter the donor cell population numbers and phenotypes present. While a detailed characterization of donor cells derived from each method is lacking, ongoing work using single cell sequencing is likely to offer insight in coming years. This will be especially crucial for the translational use of NPCs derived from stem cell banks (ES or iPS cells) described below.

4.3 FSC-derived NPCs prepared without cell culture

Work pioneered by Paul Lu and Mark Tuszynski used isolated and chemically dissociated donor cells from embryonic mouse (Adler et al., 2017; Dulin et al., 2018; Poplawski et al., 2020) and rat (Koffler et al., 2019; Kumamaru, Lu, Rosenzweig, & Tuszynski, 2018; Lu et al., 2012) spinal cord tissues (FSC) without further manipulation (e.g., culture) prior to transplantation. As was shown with FSC transplantation, donor NPCs survived, integrated with surrounding injured spinal cord, produced populations of neurons and glia, and then supported functional improvement. These transplants produced vast numbers of neurites that extended for many millimeters into the host tissue, through the spinal cord and even to the olfactory regions of the brain. Notably, however, survival of tissues prepared in this way was supplemented by delivery with a cocktail of growth factors to support donor cell survival and integration. One reason for this could be that chemical dissociation (unlike the mechanical dissociation of FSCs described above) results in significant cell damage or stress, and removal of supportive extracellular matrices, that then necessitates growth factor support. Since the original studies, this growth factor cocktail has been reduced to 4 essential factors: brain-derived neurotrophic factor, basic-fibroblastic growth factor, vascular endothelial growth factor, and MDL28170, a cell death inhibitor (Robinson & Lu, 2017). Aside from the use of growth factors and supportive materials (i.e., fibrinogen), these transplants very closely resembled the parallel work being conducted with cultured populations of FSC-derived NPCs. These donor cells have also been referred to as neural/stem progenitor cells, as the proportions of both populations at the time of transplantation, and the effect of growth factor delivery on donor cell survival, proliferation and differentiation is still being investigated.

The Lu and Tuszynski teams have since shown that these donor FSC-derived NPCs integrate appropriately with the injured host networks. Dulin et al. (2018) showed that not only did donor NPCs yield specific interneuronal subtypes found in the dorsal horn, but they also received input from relevant host sensory fibers. Kumamaru et al. (2018) then also showed that corticospinal axons innervated phenotypically appropriate donor neurons. It is important to note, that in several of these studies, the investigators also tested human embryonic spinal cord-derived NPCs that did undergo a period of cell culture toward NPC phenotypes for transplantation. These donor cells (NST-566RSC; NeuralStem Inc.) have also been tested in non-human primate models of SCI, in which it was shown that transplanted cells integrated with recipient networks and improved functional outcomes (Rosenzweig et al., 2018).

4.4 FSC-derived NPCs prepared using cell culture

After establishing a process of isolating and culturing of NEPs, NRPs and GRPs from FSC tissue to generate consistent and defined populations of progenitor cells (Cai et al., 2002), a number of studies investigated the phenotypic fate of these cells post-transplantation into the adult rat spinal cord. Although transplantation of pluripotent NEPs into the normal or lesioned adult spinal cord limits differentiation of NEPs to a glial lineage (Cao et al., 2001), transplanting neuronal-restricted precursors into the normal (Han et al., 2002) and contused adult spinal cord yields some mature neuronal phenotypes but also many undifferentiated cells, two to eight weeks post-transplantation (Cao et al., 2001). Subsequent studies of transplantation of mixed NRP and GRP cell cultures demonstrated the necessity of including GRPs to ensure neuronal survival and differentiation post-transplantation (Lepore & Fischer, 2005).

One of the most successful experiments demonstrating the potential of transplantation of a mixed population of NRPs and GRPs to treat the injured spinal cord was demonstrated in a dorsal column hemisection model (Bonner et al., 2011). In this study, NRPs and GRPs transplanted into the injury site formed a functional relay demonstrated by the survival and differentiation of progenitor cells into mature neurons. Additionally, synapse formation between lesioned host axons and grafted-derived neurons was confirmed anatomically and electrophysiologically. These promising results have been replicated by others, using similar partial lesion models of SCI (Nakamura et al., 2005).

The partial section injury models used in prior work is an intentionally precise but mild white matter injury, which provides an excellent "proof-of-principle" but may lack the severity of human injuries. In contrast, when a similar approach was employed in a severe injury (a complete transection), poor survival of grafted cells was observed despite utilizing an assortment of lesion methods, cell matrices and even with delays between injury and transplantation (Medalha et al., 2014) confirming existing challenges for using this NPC therapy. Section injuries also lack the neuropathology seen in other injury models that more closely resembles human injury (e.g., contusion or compression models). Transplantation of FSC-derived NPCs in such models have shown therapeutic efficacy (Watanabe et al., 2004; Zholudeva, Iyer, et al., 2018), although ongoing research in this area has shown that the extent of donor cell survival and overall therapeutic efficacy can still be enhanced. Several investigators are now exploring ways to enhance donor cell survival, integration and associated functional improvement (Zholudeva & Lane, 2019a, 2019b, 2022). This again raises the important question of whether there are components of FSC tissue lost in the culturing process that are crucial for survival and repair of the injured cord.

4.5 Neural stem and progenitor cells from other tissues

The focus of the discussion so far has been on NPCs derived from developing spinal tissues, as these source tissues and derived NPCs remain a "gold-standard" for spinal cord repair, comprising all the essential components to build functional spinal tissue. While spinally-derived donor cells have long been considered an optimal source for spinal repair, the transplantation of developing brain and brainstem tissue has showed some promising results, especially in paradigms that provided supplemental neurochemical support to denervated spinal networks (Bjorklund, Nornes, & Gage, 1986; Nornes et al., 1988; Privat, Mansour, Pavy, Geffard, & Sandillon, 1986). While the survival and integration of brain-derived tissues have been described as less optimal for spinal cord repair, they are still a valuable source of neural stem cells (Uchida et al., 2000). Neural stem cells have been isolated from adult and fetal sub-ventricular zone (Reynolds & Weiss, 1992), sub-granular zones (Obernier & Alvarez-Buylla, 2019), and spinal cord (Weiss et al., 1996). They can also be engineered from other available populations of stem cells; e.g., trans-differentiated from multipotent MSCs, or differentiated from pluripotent ESCs or iPSCs (Fig. 4). Regardless of the source, culturing NSCs towards differentiated cells produces NPCs: comprising sub-populations of both NSCs and more lineage-restricted neural precursor cells (as defined in Box 1). When described in the literature for use in treating SCI, they are often referred to as either NSCs, NPCs or "NSPCs"

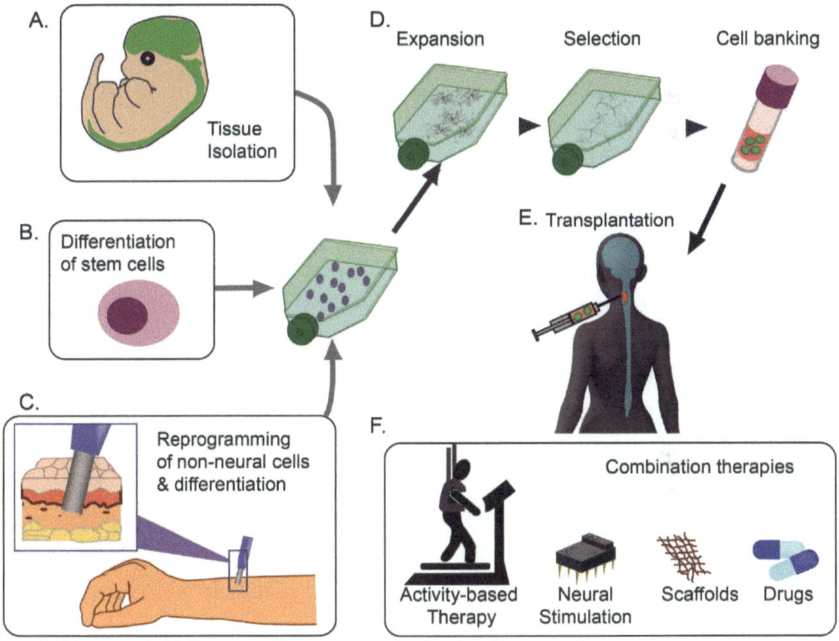

Fig. 4 Schematic diagram highlighting potential sources of neural stem and precursor cells: the developing neural tissues (A), pluripotent embryonic stem cells (B), and induced pluripotent stem cells (C; exemplified by skin fibroblast de-differentiation/reprogramming). These cells can be engineered to produce neural stem and progenitor cells that can then be expanded (D), specific subsets selected (e.g., NRPs and GRPs), and cryopreserved for "cell banking." These cell stores can be thawed and prepared for transplantation into the injured spinal cord (E) alone, or in combination with additional treatments (E), including rehabilitation and activity-based therapy, neural interfacing and neuromodulation, application of scaffolds/biomaterials, or additional pharmaceuticals to enhance efficacy. *Figure modified from Zholudeva, L. V., & Lane, M. A. (2022).* Spinal interneurons: Plasticity after spinal cord injury, *1st ed. Academic Press.*

(Neural stem/progenitor cells). However, for the purpose of the discussion in this chapter, we will collectively refer to them as NPCs (Box 1).

The following sections will highlight pre-clinical work that has used isolated and cultured bankable populations of NPCs derived from (i) developing brain tissues, (ii) ESCs, and (iii) iPSCs. While like the FSC-derived cells, they can all be considered NPCs, it is highly likely that the cell phenotypes produced from each originating source will differ somewhat, even if similar methods are used to engineer and differentiate them.

4.6 Brain-derived NPCs

With the discovery that stem cells could be isolated from developing neural tissues, research began to explore methods for isolating and selecting for these cells (Uchida et al., 2000), that could be expanded in cell culture and/or cryopreserved for future use. One well-studied example has been the line of cells isolated from human fetal brain tissue, subsequently characterized by the Cummings and Anderson teams, that led to the clinical testing by StemCells Inc. (Table 4). Expansion of these cells in culture yielded neurospheres of human NPCs that could be dissociated to single cells for transplantation into the injured spinal cord (Cummings et al., 2005; Salazar, Uchida, Hamers, Cummings, & Anderson, 2010). Much like FSC-derived NPC transplants, these studies often aimed to transplant a mixture of stem and NPCs to facilitate tissue repair. Consistent with this idea, transplanted NSCs provided donor neuronal populations that would synaptically integrate with the surrounding host cell populations (Cummings et al., 2005), glial tissue that provided a permissive bridge (Lien et al., 2019), and produced neurotrophic factors (Cheng et al., 2017).

Although there are now several examples of developing neural spinal cord or brain tissues being translated to clinical studies, it is likely that given logistical and ethical considerations, and the advances in cellular engineering, stem cell-derived NPCs will become the more translationally relevant alternative for future clinical translation. Accordingly, there has been significant effort in recent decades to engineer NPCs from pluripotent stem cells (ESCs and iPSCs).

4.7 NPCs derived from pluripotent stem cells

Early efforts to transplant embryonically derived-stem cells into the injured spinal cord demonstrated that they could yield both neurons and glia (McDonald et al., 1999). These studies showed that transplantation of partially trypsinized embryoid bodies from mouse ESCs into thoracic contusion injuries in adult rats resulted in survival of donor glia (astrocytes, oligodendrocytes), and to a lesser extent neurons. These undifferentiated donor cells, however, were reliant on the internal milieu of the injured spinal cord to regulate their survival and differentiation. Pioneering work by Mahendra Rao's team also showed, however, that neuronal and glial precursor cells could be obtained from ESCs under appropriate conditions (Mujtaba et al., 1999). With a growing understanding of the factors that mediating cell differentiation (e.g., cell-cell contract, growth factors release), greater effort was made to engineer ESCs towards relevant neural fates (Kirkeby et al., 2012; Nistor et al., 2005; Parsons et al., 2011; Peljto & Wichterle, 2011; Robertson, Gip, & Schaffer, 2008; Willerth, Arendas, Gottlieb, & Sakiyama-Elbert, 2006).

Building on the successes of NPC transplantation from tissue derived sources, recent efforts have compared the use of NPCs derived from different sources for spinal cord repair. Initial series of experiments from the Lu and Tuszynski teams compared NPCs derived from rodent embryonic spinal cord (FSC), human embryonic spinal cord NSCs (NST-566RSC; NeuralStem Inc), and human ESC lines (HUES7; Harvard University) (Lu et al., 2012). Subsequent work, and ongoing studies, focused more on the use of the WiCell H9 ESC line, confirming that derived NPCs were able to survive, integrate and support neuronal network repair in a similar manner to that seen with NPCs derived from embryonic mouse or rat tissue (Kumamaru et al., 2018; Lu et al., 2017). There ongoing assessment of donor cell sources also revealed migration of donor ESC-derived glial cells that appeared to support integration and repair (Lien et al., 2019).

4.8 iPSC-derived NPCs

iPSCs are rapidly becoming one of the most appealing sources for cellular therapies as they can be derived from readily accessible (e.g., skin fibroblasts or blood), autologous tissue, avoiding possible ethical concerns with using ESCs, and potentially reducing the chance of immune rejection (Nagoshi, Okano, & Nakamura, 2020). The growing interest in iPSCs as a therapeutic cell source has enabled generation of allogeneic human leukocyte antigen (HLA) matched cell banks, which also reduce risk of donor cell immune rejection (Tsuji et al., 2019). Due to the replicative potential of these pluripotent cells, they represent a renewable reservoir that can be available to a broad population of patients (Tsuji et al., 2019).

One of the earliest reports of iPSC-derived transplants used for treating SCI was engineered NPC-rich neurospheres introduced into injured mice

(Tsuji et al., 2010). Though donor cells differentiated into glia and neurons, with some remyelination and functional motor recovery, there was some evidence of teratoma formation (Tsuji et al., 2010). The Okano team built on these studies by optimizing cell preparation (Nori et al., 2011; Sugai et al., 2016), and there have been extensive efforts made to ensure safety and limit the risk of tumorgenicity (Iida et al., 2017; Kojima et al., 2019; Okubo et al., 2016). Transplanting more refined NPC donor cells into a marmoset model of SCI, they demonstrated motor recovery without tumorigenicity (Kobayashi et al., 2012). Similar efficacy has been reported by others, with extensive tissue confluency, neurogenesis, axonal extension, synaptic integration, and functional recovery following transplants into rats (Lu, Woodruff, et al., 2014; Poplawski et al., 2018; Rodriguez-Jimenez et al., 2021; Strnadel et al., 2018). It should be noted that a reported therapeutic characteristic of donor iPSC-NPCs has been associated with differentiation into oligodendrocytes and myelination (Salewski et al., 2015; Suzuki et al., 2017; Yasuda et al., 2011). Accordingly, there has also been some effort made to focus more selectively on iPSC-derived OPCs (Kawabata et al., 2016). Therapeutic efficacy from iPSC-NPC transplantation is not limited to myelination however, with evidence for synaptic integration between host and donor neurons also contributing to functional improvement (Kitagawa et al., 2022; Yokota et al., 2015). Also worth noting is that cells that are differentiated to be similar to cells of the developing spinal cord, or "spinalized donor cell phenotypes," have been found to be more efficacious for treating SCI than cells more closely associated with brain/brainstem regions (Kajikawa et al., 2020). Given the pre-clinical success to date, the first human iPSC-NPCs clinical trial has been initiated, with a goal of treating sub-acute thoracic SCIs (Sugai et al., 2021; Tsuji et al., 2019). Even with this demonstrated therapeutic efficacy, and translation to clinical trials, ongoing research aims to further enhance this efficacy and optimize outcomes (Ito et al., 2021; Kawai et al., 2021; Shibata et al., 2022).

5. Cell transplantation for neuroimmune modulation and neuroprotection

Nearly all cell types that have been transplanted to promote spinal cord repair have been reported to have protective properties, and many donor cell types express trophic factors or anti-inflammatory cytokines, or can mediate the inflammatory processes post-SCI via cell-cell contact. An overview of cell transplants that have been used for modulating inflammation and/or limiting tissue damage is provided in Table 3.

Aside from providing a bridge for repair, or a donor interneuronal relay for novel neuronal networks, transplanted NPCs release a vast range of cytokines that can have pro-reparative effects on the host tissue. Perhaps one of the most extensively characterized examples is the neuroprotective effects of donor cells at the injury site. Transplanted cells can release pro-reparative, neurotrophic, and anti-inflammatory cytokines that alter the internal milieu of the injury site, limiting secondary tissue degeneration, attenuating infiltration of inflammatory cells from the periphery, and potentially stimulating existing inflammatory cells towards pro-reparative phenotypes. The mechanisms through which these effects could be mediated are vast, but include (i) reduced N-methyl-D-aspartate (NMDA)-mediated excitotoxicity (Papazian, Kyrargyri, Evangelidou, Voulgari-Kokota, & Probert, 2018), altered immune cell function (Ren et al., 2018), (ii) reduced production of reactive oxygen species production (Riemann et al., 2018), and caspase inhibition (Gashmardi, Hosseini, Mehrabani, Edalatmanesh, & Khodabandeh, 2017; Liu, Ding, Zhang, & Wang, 2013), limiting apoptosis, and (iv) altered trophic factor expression (e.g., Brain Derived Neurotrophic Factor; Kermani & Hempstead, 2007), Vascular Endothelial Growth Factor (Yu et al., 2016), and Hypoxia-inducible factor-1α (Hashimoto & Shibasaki, 2015). GRP components alone have also been shown to bias inflammatory cells towards an anti-inflammatory phenotype (Goulao et al., 2019b; Hasegawa et al., 2005; Shih, Lacagnina, Leuer-Bisciotti, & Proschel, 2014; Wang et al., 2022), limit excitotoxicity, and attenuate gliosis (Hasegawa et al., 2005; Jin et al., 2011; Wang et al., 1995; Wu et al., 2013). As a result of these anti-inflammatory/neuroprotective functions, they have also been reported to limit the extent of secondary tissue damage post-SCI (Jin et al., 2011; Li et al., 2015). Building on reported neuroprotective properties of donor NPCs, more recent use of iPSC-derived NPC has seen similar protective results (Romanyuk et al., 2015). There has been some recent effort to use iPSC-derived glial cells to achieve neuroprotection. iPSC-derived astrocyte transplants have shown limited therapeutic efficacy alone (Hayashi et al., 2011), but once engineered to overexpress GLT1 (the glutamate transporter responsible for limiting excitotoxicity), they were able to limit lesion size and increase diaphragm output (Li et al., 2015) in animals with SCI. iPSC-OPCs have also shown considerable promise for treating SCI, showing increased myelination of spared axons (All et al., 2015; Kawabata et al., 2016), decreased lesion sizes (Patil et al., 2021), and improved functional recovery after transplantation (All et al., 2015; Kawabata et al., 2016). However, these treatments did not demonstrate functional improvements in a chronic injury model (Patil et al., 2021), and caused teratomas in one study (Fuhrmann et al., 2016).

One donor cell population frequently used for its neuroprotective properties has been MSCs. Due to their plentiful supply, diverse differentiation capacity, and minimal transplant-associated adverse side effects (Cofano et al., 2019), MSCs have been long considered a viable and translationally relevant candidate cell therapy (Caplan, 1990). Transplanted MSCs evoke a broad array of pro-reparative reactions, including neurotrophic factor secretion (Hofer & Tuan, 2016), and anti-inflammatory cytokine release (Planat-Benard, Varin, & Casteilla, 2021) that attenuate host inflammation, apoptosis and tissue degeneration associated with secondary injury (Boido et al., 2014). MSCs also display powerful angiogenic properties (De Luca et al., 2011), leading to structural reorganization of the surrounding tissueand vasculature repair, further limiting degeneration and promoting repair (Vawda & Fehlings, 2013). One therapeutic advantage that investigators have reported with MSC transplantation over other cell types is that they can achieve efficacy even when not delivered directly into the injured spinal tissue. Some have attributed this to their ability to migrate to the site of injury (Filippi et al., 2016; Zachar, Bacenkova, & Rosocha, 2016), but an alternative consideration is that they can achieve efficacy without being present in the nervous system (e.g., by targeting cells in the spleen or liver).

MSCs are among one of the most translated cell products for treating SCI, with a vast number of several clinical studies currently ongoing worldwide (Table 4). This may be due in part to the ease with which MSCs can be obtained, but also due to the relatively acute nature of their intended administration for treatment. Two recent meta-analyses of human transplant trials (Chen et al., 2021; Johnson, Pickard, & Johnson, 2021) have found that the MSCs have contributed to consistent recovery, with autologous transplants and intrathecal delivery methods being perhaps the most efficacious. However, given (i) the variable donor cell source, (ii) potential donor cell phenotype, (iii) extensive distribution through the body after delivery, (iv) the range of potential therapeutic actions, and (v) the documented evidence for some unsuccessful studies, caution should be used in broadly labeling MSC therapy as successful for SCI.

To address concerns with variable outcomes, cell source, cell preparation, and mechanisms of action, a joint collaboration with scientists at Athersys and Case Western Reserve University utilized a more refined population of MSC-like cells that could be used to treat wide range of neurological diseases and injuries (Busch et al., 2011; DePaul et al., 2015; Jones et al., 2019). The resulting allogeneic MultiStem product can be delivered peripherally and acutely following traumatic SCI. While some cells did accumulate in the spinal cord lesion site, the majority of cells accumulated in the spleen, and regulated ongoing systemic immune activity. While the MultiStem cells did not appear to survive for long periods of time, they had significant effects on attenuation of immune response to injury, that resulted in improved anatomical and functional outcomes. MultiStem is in clinical trial for non-SCI neurological disorders, but with ongoing preclinical development the product could be translated to SCI trials in coming years.

6. Considerations for contraindications and adverse effects

Safety remains one of the highest priorities for all therapeutic development. The development of cellular therapeutics from pluripotent cell sources has a history of association with some significant adverse effects, such as tumor formation (teratoma). Past experience has taught us, however, that cell therapies have become far more effectively prepared to limit risks of such adverse events. Pre-clinical assessments are being evaluated more critically, and it has led to more careful development of treatments with safety and efficacy being more rigorously assessed.

Clinical translation of cell therapies has also provided valuable insight into how cell therapies can be developed and administered. Clinical testing with the Geron and NeuralStem trials led to the development of neurosurgical approaches to maximally deliver large number of cells into the injured spinal cord while preserving neural tracts and assuring the upmost safety (Boulis & Federici, 2011; Riley et al., 2009). Clinical professionals and scientists will continue to learn from ongoing translation efforts, which will reverse-translate to revision and re-development of more effective treatments, delivered by the most effective means, under the right conditions.

7. Closing remarks

As highlighted in this chapter, there is a vast range of cell therapy strategies that have been developed to treat SCI over the past several decades. A key component of therapeutic development for SCI has been identifying the treatment goal. This requires a comprehensive understanding of the disorder being treated, the systems/networks being targeted, and what ongoing neuroplastic changes are taking place, to select the strategy for repair. Unlike some neurological diseases, SCI can occur to anybody, at any time, and occurs in a vast range of ways (e.g., contusion, compression, laceration), at different spinal levels and to varying degrees in every injured individual. Similarly, the anatomical disruption, and the potential for neuroplastic changes post-injury, will likely varying significantly depending on the age and health of the individual, and the access to rapid medical intervention and care. Accordingly, "spinal cord injury" is not a disorder with often predictable consequences. It is perhaps not surprising, therefore, that a range of cell therapies have been developed to treat multiple aspects of SCI.

As highlighted here, there are several cell therapy strategies that have been developed to treat traumatic SCI, with different approaches for promoting repair and recovery, and many of these have been translated for clinical testing. Crucial to the development of more effective cell therapies in future is a bidirectional translational effort (see Fig. 5), with scientists and

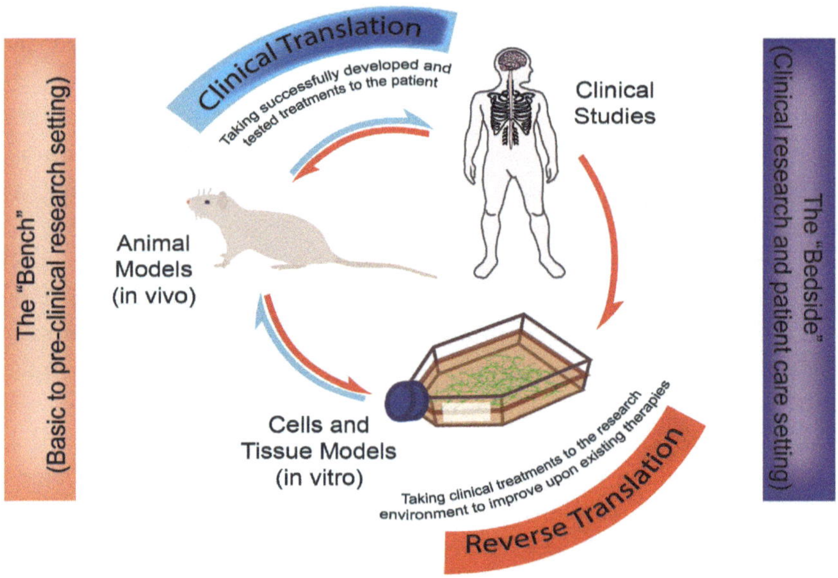

Fig. 5 This flow diagram indicates the translation path from *in vitro* cell culture studies, to *in vivo* testing of transplantable cells, to the translation of treatments to the clinic for testing in people living with SCI. This "forward" progression (in blue) toward the "Bedside" needs to be matched by reverse translation efforts (in red), taking information learned from each step back to the "Bench" for refinement, redevelopment, and optimization.

clinical professionals maintaining effective communication, and under consultation with those living with SCI. Translated cell therapies should still be taken back to the bench for optimization wherever possible. This is also true at all levels of translation. Once a cell product has been developed and tested in animals, in addition to finding a translational path towards clinical testing, the product can also be taken back to the bench for further engineering and refinement in cell culture. While each step the translational spectrum is needed to optimize translated treatments, there is a need for parallel development.

Though the chapter has outlined a range of cell therapies with evidence for therapeutic efficacy, it is also widely acknowledged that there is no "silver bullet" or complete cure. Accordingly, in parallel to efforts to optimize existing cell therapies, significant attention is being given to how cell therapies can be combined, or how non-cellular therapies (e.g., activity-based therapies, neuromodulation) can be paired with transplantation for synergistic effects.

Despite the challenges and limitations faced by all cell therapies, advances in neuroanatomy, developmental neurobiology, stem cell biology, single cell sequencing, proteomics and cellular engineering are rapidly improving our appreciation for spinal cord anatomy. These principles are guiding a new era of cell transplantation which harnesses the strengths of translationally-relevant PSCs, enabling engineering of directed, pro-reparative donor cell phenotypes. Continued technological advances are shaping the future of cell therapies for all neurological injury and disease. As the chapters in this book can attest to, there is an extremely promising future ahead.

References

Adler, A. F., Lee-Kubli, C., Kumamaru, H., Kadoya, K., & Tuszynski, M. H. (2017). Comprehensive monosynaptic rabies virus mapping of host connectivity with neural progenitor grafts after spinal cord injury. *Stem Cell Reports*, *8*, 1525–1533.

Aguayo, A. J., David, S., & Bray, G. M. (1981). Influences of the glial environment on the elongation of axons after injury: Transplantation studies in adult rodents. *The Journal of Experimental Biology*, *95*, 231–240.

Akesson, E., Holmberg, L., Jonhagen, M. E., Kjaeldgaard, A., Falci, S., Sundstrom, E., et al. (2001). Solid human embryonic spinal cord xenografts in acute and chronic spinal cord cavities: A morphological and functional study. *Experimental Neurology*, *170*, 305–316.

Akesson, E., Kjaeldgaard, A., & Seiger, A. (1998). Human embryonic spinal cord grafts in adult rat spinal cord cavities: Survival, growth, and interactions with the host. *Experimental Neurology*, *149*, 262–276.

Akhmetzyanova, E. R., Mukhamedshina, Y. O., Zhuravleva, M. N., Galieva, L. R., Kostennikov, A. A., Garanina, E. E., et al. (2018). Transplantation of microglia in the area of spinal cord injury in an acute period increases tissue sparing, but not functional recovery. *Frontiers in Cellular Neuroscience*, *12*, 507.

All, A. H., Bazley, F. A., Gupta, S., Pashai, N., Hu, C., Pourmorteza, A., et al. (2012). Human embryonic stem cell-derived oligodendrocyte progenitors aid in functional recovery of sensory pathways following contusive spinal cord injury. *PLoS One*, *7*, e47645.

All, A. H., Gharibani, P., Gupta, S., Bazley, F. A., Pashai, N., Chou, B. K., et al. (2015). Early intervention for spinal cord injury with human induced pluripotent stem cells oligodendrocyte progenitors. *PLoS One*, *10*, e0116933.

Altman, J., & Bayer, S. A. (1984). The development of the rat spinal cord. *Advances in Anatomy, Embryology, and Cell Biology*, *85*, 1–164.

Amemori, T., Jendelova, P., Ruzickova, K., Arboleda, D., & Sykova, E. (2010). Co-transplantation of olfactory ensheathing glia and mesenchymal stromal cells does not have synergistic effects after spinal cord injury in the rat. *Cytotherapy*, *12*, 212–225.

Amemori, T., Romanyuk, N., Jendelova, P., Herynek, V., Turnovcova, K., Prochazka, P., et al. (2013). Human conditionally immortalized neural stem cells improve locomotor function after spinal cord injury in the rat. *Stem Cell Research & Therapy*, *4*, 68.

Amemori, T., Ruzicka, J., Romanyuk, N., Jhanwar-Uniyal, M., Sykova, E., & Jendelova, P. (2015). Comparison of intraspinal and intrathecal implantation of induced pluripotent stem cell-derived neural precursors for the treatment of spinal cord injury in rats. *Stem Cell Research & Therapy*, *6*, 257.

Anderson, D. K., Beattie, M., Blesch, A., Bresnahan, J., Bunge, M., Dietrich, D., et al. (2005). Recommended guidelines for studies of human subjects with spinal cord injury. *Spinal Cord*, *43*, 453–458.

Anderson, K. D., Guest, J. D., Dietrich, W. D., Bartlett Bunge, M., Curiel, R., Dididze, M., et al. (2017). Safety of autologous human schwann cell transplantation in subacute thoracic spinal cord injury. *Journal of Neurotrauma*, *34*, 2950–2963.

Anderson, D. K., Howland, D. R., & Reier, P. J. (1995). Fetal neural grafts and repair of the injured spinal cord. *Brain Pathology*, *5*, 451–457.

Anderson, D. K., Reier, P. J., Wirth Iii, E. D., Theele, D. P., Mareci, T., & Brown, S. A. (1991). Delayed grafting of fetal CNS tissue into chronic compression lesions of the adult cat spinal cord. *Restorative Neurology and Neuroscience*, *2*, 309–325.

Anderson, D. K., Reier, P. J., Wirth, E. D. I., Theele, D. P., & Brown, S. A. (1991). Delayed grafting of fetal CNS tissue into chronic compression lesions of the adult spinal cord. 2: 309–325. *Restorative Neurology and Neuroscience*, *2*, 309–325.

Andrews, M. R., & Stelzner, D. J. (2007). Evaluation of olfactory ensheathing and schwann cells after implantation into a dorsal injury of adult rat spinal cord. *Journal of Neurotrauma*, *24*, 1773–1792.

Araujo, M. R., Kyrylenko, S., Spejo, A. B., Castro, M. V., Ferreira Junior, R. S., Barraviera, B., et al. (2017). Transgenic human embryonic stem cells overexpressing FGF2 stimulate neuroprotection following spinal cord ventral root avulsion. *Experimental Neurology*, *294*, 45–57.

Arboleda, D., Forostyak, S., Jendelova, P., Marekova, D., Amemori, T., Pivonkova, H., et al. (2011). Transplantation of predifferentiated adipose-derived stromal cells for the treatment of spinal cord injury. *Cellular and Molecular Neurobiology*, *31*, 1113–1122.

Atkinson, K., Biggs, J., Darveniza, P., Boland, J., Concannon, A., & Dodds, A. (1984). Cyclosporin-associated central nervous system toxicity after allogeneic bone marrow transplantation. *Transplantation*, *38*, 34–37.

Barbour, H. R., Plant, C. D., Harvey, A. R., & Plant, G. W. (2013). Tissue sparing, behavioral recovery, supraspinal axonal sparing/regeneration following sub-acute glial transplantation in a model of spinal cord contusion. *BMC Neuroscience*, *14*, 106.

Bareyre, F. M., Kerschensteiner, M., Raineteau, O., Mettenleiter, T. C., Weinmann, O., & Schwab, M. E. (2004). The injured spinal cord spontaneously forms a new intraspinal circuit in adult rats. *Nature Neuroscience*, *7*, 269 277.

Bastidas, J., Athauda, G., De La Cruz, G., Chan, W. M., Golshani, R., Berrocal, Y., et al. (2017). Human Schwann cells exhibit long-term cell survival, are not tumorigenic and promote repair when transplanted into the contused spinal cord. *Glia, 65*, 1278–1301.

Beaumont, E., Houle, J. D., Peterson, C. A., & Gardiner, P. F. (2004). Passive exercise and fetal spinal cord transplant both help to restore motoneuronal properties after spinal cord transection in rats. *Muscle & Nerve, 29*, 234–242.

Bell, S., Hettige, N. C., Silveira, H., Peng, H., Wu, H., Jefri, M., et al. (2019). Differentiation of human induced pluripotent stem cells (iPSCs) into an effective model of forebrain neural progenitor cells and mature neurons. *Bio-Protocol, 9*, e3188.

Bernstein, J. J., & Goldberg, W. J. (1987). Fetal spinal cord homografts ameliorate the severity of lesion-induced hind limb behavioral deficits. *Experimental Neurology, 98*, 633–644.

Bernstein, J. J., Hoovler, D. W., & Turtil, S. (1985). Initial growth of transplanted E11 fetal cortex and spinal cord in adult rat spinal cord. *Brain Research, 343*, 336–345.

Bjorklund, A., Katzman, R., Stenevi, U., & West, K. A. (1971). Development and growth of axonal sprouts from noradrenaline and 5-hydroxytryptamine neurones in the rat spinal cord. *Brain Research, 31*, 21–33.

Bjorklund, A., Nornes, H., & Gage, F. H. (1986). Cell suspension grafts of noradrenergic locus coeruleus neurons in rat hippocampus and spinal cord: Reinnervation and transmitter turnover. *Neuroscience, 18*, 685–698.

Blagodatskii, M. D., Sufianov, A. A., Larionov, S. N., Kibort, R. V., Seminskii, I., & Manokhin, P. A. (1994). The transplantation of embryonic nerve tissue in syringomyelia: Initial clinical experience. *Zhurnal Voprosy Neĭrokhirurgii Imeni N. N. Burdenko*, 27–29.

Boido, M., Piras, A., Valsecchi, V., Spigolon, G., Mareschi, K., Ferrero, I., et al. (2014). Human mesenchymal stromal cell transplantation modulates neuroinflammatory milieu in a mouse model of amyotrophic lateral sclerosis. *Cytotherapy, 16*, 1059–1072.

Bonilla, P., Hernandez, J., Giraldo, E., Gonzalez-Perez, M. A., Alastrue-Agudo, A., Elkhenany, H., et al. (2021). Human-induced neural and mesenchymal stem cell therapy combined with a curcumin nanoconjugate as a spinal cord injury treatment. *International Journal of Molecular Sciences, 22*.

Bonner, J. F., Blesch, A., Neuhuber, B., & Fischer, I. (2010). Promoting directional axon growth from neural progenitors grafted into the injured spinal cord. *Journal of Neuroscience Research, 88*, 1182–1192.

Bonner, J. F., Connors, T. M., Silverman, W. F., Kowalski, D. P., Lemay, M. A., & Fischer, I. (2011). Grafted neural progenitors integrate and restore synaptic connectivity across the injured spinal cord. *The Journal of Neuroscience: The Official Journal of the Society for Neuroscience, 31*, 4675–4686.

Bonner, J. F., & Steward, O. (2015). Repair of spinal cord injury with neuronal relays: From fetal grafts to neural stem cells. *Brain Research, 1619*, 115–123.

Boulis, N., & Federici, T. (2011). Surgical approach and safety of spinal cord stem cell transplantation. *Neurosurgery, 68*, E599–E600.

Bregman, B. S. (1986). Neural tissue transplants modify central neurons' responses to damage. In M. E. Goldberger, A. Gorio, & M. Murray (Eds.), *Development and plasticity of the mammalian spinal cord* (pp. 271–299). Padova: Liviana Press.

Bregman, B. S., Coumans, J. V., Dai, H. N., Kuhn, P. L., Lynskey, J., McAtee, M., et al. (2002). Transplants and neurotrophic factors increase regeneration and recovery of function after spinal cord injury. *Progress in Brain Research, 137*, 257–273.

Bregman, B. S., Kunkel-Bagden, E., Reier, P. J., Dai, H. N., McAtee, M., & Gao, D. (1993). Recovery of function after spinal cord injury: Mechanisms underlying transplant-mediated recovery of function differ after spinal cord injury in newborn and adult rats. *Experimental Neurology, 123*, 3–16.

Bregman, B. S., & Reier, P. J. (1986). Neural tissue transplants rescue axotomized rubrospinal cells from retrograde death. *The Journal of Comparative Neurology, 244*, 86–95.

Bretzner, F., Gilbert, F., Baylis, F., & Brownstone, R. M. (2011). Target populations for first-in-human embryonic stem cell research in spinal cord injury. *Cell Stem Cell*, *8*, 468–475.
Brock, J. H., Graham, L., Staufenberg, E., Collyer, E., Koffler, J., & Tuszynski, M. H. (2016). Bone marrow stromal cell intraspinal transplants fail to improve motor outcomes in a severe model of spinal cord injury. *Journal of Neurotrauma*, *33*, 1103–1114.
Brock, J. H., Graham, L., Staufenberg, E., Im, S., & Tuszynski, M. H. (2018). Rodent neural progenitor cells support functional recovery after cervical spinal cord contusion. *Journal of Neurotrauma*, *35*, 1069–1078.
Bunge, M. B. (2001). Bridging areas of injury in the spinal cord. *The Neuroscientist*, *7*, 325–339.
Bunge, M. B., Monje, P. V., Khan, A., & Wood, P. M. (2017). From transplanting Schwann cells in experimental rat spinal cord injury to their transplantation into human injured spinal cord in clinical trials. *Progress in Brain Research*, *231*, 107–133. https://doi.org/10.1016/bs.pbr.2016.12.012. Epub 2017 Mar 31. PMID: 285543945.
Busch, S. A., Hamilton, J. A., Horn, K. P., Cuascut, F. X., Cutrone, R., Lehman, N., et al. (2011). Multipotent adult progenitor cells prevent macrophage-mediated axonal dieback and promote regrowth after spinal cord injury. *The Journal of Neuroscience: The Official Journal of the Society for Neuroscience*, *31*, 944–953.
Butts, J. C., Iyer, N., White, N., Thompson, R., Sakiyama-Elbert, S., & McDevitt, T. C. (2019). V2a interneuron differentiation from mouse and human pluripotent stem cells. *Nature Protocols*, *14*, 3033–3058.
Butts, J. C., McCreedy, D. A., Martinez-Vargas, J. A., Mendoza-Camacho, F. N., Hookway, T. A., Gifford, C. A., et al. (2017). Differentiation of V2a interneurons from human pluripotent stem cells. *Proceedings of the National Academy of Sciences of the United States of America*, *114*, 4969–4974.
Bydon, M., Dietz, A. B., Goncalves, S., Moinuddin, F. M., Alvi, M. A., Goyal, A., et al. (2020). CELLTOP clinical trial: First report from a phase 1 trial of autologous adipose tissue-derived mesenchymal stem cells in the treatment of paralysis due to traumatic spinal cord injury. *Mayo Clinic Proceedings*, *95*, 406–414.
Cai, J., Peng, X., Nelson, K. D., Eberhart, R., & Smith, G. M. (2005). Permeable guidance channels containing microfilament scaffolds enhance axon growth and maturation. *Journal of Biomedical Materials Research. Part A*, *75*, 374–386.
Cai, J., Wu, Y., Mirua, T., Pierce, J. L., Lucero, M. T., Albertine, K. H., et al. (2002). Properties of a fetal multipotent neural stem cell (NEP cell). *Developmental Biology*, *251*, 221–240.
Calancie, B., Madsen, P. W., Wood, P., Marcillo, A. E., Levi, A. D., & Bunge, R. P. (2009). A guidance channel seeded with autologous Schwann cells for repair of cauda equina injury in a primate model. *The Journal of Spinal Cord Medicine*, *32*, 379–388.
Camacho-Aguilar, E., & Warmflash, A. (2020). Insights into mammalian morphogen dynamics from embryonic stem cell systems. *Current Topics in Developmental Biology*, *137*, 279–305.
Cao, Q.-L., Howard, R. M., Dennison, J. B., & Whittemore, S. R. (2002). Differentiation of engrafted neuronal-restricted precursor cells is inhibited in the traumatically injured spinal cord. *Experimental Neurology*, *177*, 349–359.
Cao, Q., Xu, X. M., Devries, W. H., Enzmann, G. U., Ping, P., Tsoulfas, P., et al. (2005). Functional recovery in traumatic spinal cord injury after transplantation of multineurotrophin-expressing glial-restricted precursor cells. *The Journal of Neuroscience: The Official Journal of the Society for Neuroscience*, *25*, 6947–6957.
Cao, Q. L., Zhang, Y. P., Howard, R. M., Walters, W. M., Tsoulfas, P., & Whittemore, S. R. (2001). Pluripotent stem cells engrafted into the normal or lesioned adult rat spinal cord are restricted to a glial lineage. *Experimental Neurology*, *167*, 48–58.
Caplan, A. I. (1990). Stem cell delivery vehicle. *Biomaterials*, *11*, 44–46.

Ceto, S., Sekiguchi, K. J., Takashima, Y., Nimmerjahn, A., & Tuszynski, M. H. (2020). Neural stem cell grafts form extensive synaptic networks that integrate with host circuits after spinal cord injury. *Cell Stem Cell, 27*, 430–440. e435.

Chapman, A. R., & Scala, C. C. (2012). Evaluating the first-in-human clinical trial of a human embryonic stem cell-based therapy. *Kennedy Institute of Ethics Journal, 22*, 243–261.

Chen, W. C., Liu, W. F., Bai, Y. Y., Zhou, Y. Y., Zhang, Y., Wang, C. M., et al. (2021). Transplantation of mesenchymal stem cells for spinal cord injury: A systematic review and network meta-analysis. *Journal of Translational Medicine, 19*, 178.

Chen, X., Wu, J., Sun, R., Zhao, Y., Li, Y., Pan, J., et al. (2020). Tubular scaffold with microchannels and an H-shaped lumen loaded with bone marrow stromal cells promotes neuroregeneration and inhibits apoptosis after spinal cord injury. *Journal of Tissue Engineering and Regenerative Medicine, 14*, 397–411.

Cheng, Z., Bosco, D. B., Sun, L., Chen, X., Xu, Y., Tai, W., et al. (2017). Neural stem cell-conditioned medium suppresses inflammation and promotes spinal cord injury recovery. *Cell Transplantation, 26*, 469–482.

Cheng, H., Liu, X., Hua, R., Dai, G., Wang, X., Gao, J., et al. (2014). Clinical observation of umbilical cord mesenchymal stem cell transplantation in treatment for sequelae of thoracolumbar spinal cord injury. *Journal of Translational Medicine, 12*, 253.

Chhabra, H. S., Sarda, K., Jotwani, G., Gourie-Devi, M., Kaptanoglu, E., Charlifue, S., et al. (2019). Stem cell/cellular interventions in human spinal cord injury: Is it time to move from guidelines to regulations and legislations? Literature review and Spinal Cord Society position statement. *European Spine Journal, 28*, 1837–1845.

Chipman, P. H., Toma, J. S., & Rafuse, V. F. (2012). Generation of motor neurons from pluripotent stem cells. *Progress in Brain Research, 201*, 313–331.

Chuah, M. I., Choi-Lundberg, D., Weston, S., Vincent, A. J., Chung, R. S., Vickers, J. C., et al. (2004). Olfactory ensheathing cells promote collateral axonal branching in the injured adult rat spinal cord. *Experimental Neurology, 185*, 15–25.

Chung, R. S., Woodhouse, A., Fung, S., Dickson, T. C., West, A. K., Vickers, J. C., et al. (2004). Olfactory ensheathing cells promote neurite sprouting of injured axons in vitro by direct cellular contact and secretion of soluble factors. *Cellular and Molecular Life Sciences, 61*, 1238–1245.

Cizkova, D., Rosocha, J., Vanicky, I., Jergova, S., & Cizek, M. (2006). Transplants of human mesenchymal stem cells improve functional recovery after spinal cord injury in the rat. *Cellular and Molecular Neurobiology, 26*, 1167–1180.

Cofano, F., Boido, M., Monticelli, M., Zenga, F., Ducati, A., Vercelli, A., et al. (2019). Mesenchymal stem cells for spinal cord injury: Current options, limitations, and future of cell therapy. *International Journal of Molecular Sciences, 20*.

Collazos-Castro, J. E., Muneton-Gomez, V. C., & Nieto-Sampedro, M. (2005). Olfactory glia transplantation into cervical spinal cord contusion injuries. *Journal of Neurosurgery. Spine, 3*, 308–317.

Cote, M. P., Amin, A. A., Tom, V. J., & Houle, J. D. (2011). Peripheral nerve grafts support regeneration after spinal cord injury. *Neurotherapeutics, 8*, 294–303.

Cote, M. P., Hanna, A., Lemay, M. A., Ollivier-Lanvin, K., Santi, L., Miller, K., et al. (2010). Peripheral nerve grafts after cervical spinal cord injury in adult cats. *Experimental Neurology, 225*, 173–182.

Courtine, G., Gerasimenko, Y., van den Brand, R., Yew, A., Musienko, P., Zhong, H., et al. (2009). Transformation of nonfunctional spinal circuits into functional states after the loss of brain input. *Nature Neuroscience, 12*, 1333–1342.

Courtine, G., Song, B., Roy, R. R., Zhong, H., Herrmann, J. E., Ao, Y., et al. (2008). Recovery of supraspinal control of stepping via indirect propriospinal relay connections after spinal cord injury. *Nature Medicine, 14*, 69–74.

Cummings, B. J., Uchida, N., Tamaki, S. J., Salazar, D. L., Hooshmand, M., Summers, R., et al. (2005). Human neural stem cells differentiate and promote locomotor recovery in spinal cord-injured mice. *Proceedings of the National Academy of Sciences of the United States of America, 102*, 14069–14074.

Das, G. D. (1983). Neural transplantation in the spinal cord of adult rats. Conditions, survival, cytology and connectivity of the transplants. *Journal of the Neurological Sciences, 62*, 191–210.

Das, G. D. (1983b). Neural transplantation in the spinal cord of the adult mammal. In C. C. Kao, R. P. Bunge, & P. J. Reier (Eds.), *Spinal cord reconstruction* (pp. 367–396). New York: Raven Press.

Das, G. D. (1987). Neural transplantation in normal and traumatized spinal cord. *Annals of the New York Academy of Sciences, 495*, 53–70.

David, S., & Aguayo, A. J. (1981). Axonal elongation into peripheral nervous system "bridges" after central nervous system injury in adult rats. *Science, 214*, 931–933.

David, S., & Aguayo, A. J. (1985). Axonal regeneration after crush injury of rat central nervous system fibres innervating peripheral nerve grafts. *Journal of Neurocytology, 14*, 1–12.

Davies, J. E., Huang, C., Proschel, C., Noble, M., Mayer-Proschel, M., & Davies, S. J. (2006). Astrocytes derived from glial-restricted precursors promote spinal cord repair. *Journal of Biology, 5*, 7.

Davies, J. E., Proschel, C., Zhang, N., Noble, M., Mayer-Proschel, M., & Davies, S. J. (2008). Transplanted astrocytes derived from BMP- or CNTF-treated glial-restricted precursors have opposite effects on recovery and allodynia after spinal cord injury. *Journal of Biology, 7*, 24.

Davies, S. J., Shih, C. H., Noble, M., Mayer-Proschel, M., Davies, J. E., & Proschel, C. (2011). Transplantation of specific human astrocytes promotes functional recovery after spinal cord injury. *PLoS One, 6*, e17328.

De Luca, A., Gallo, M., Aldinucci, D., Ribatti, D., Lamura, L., D'Alessio, A., et al. (2011). Role of the EGFR ligand/receptor system in the secretion of angiogenic factors in mesenchymal stem cells. *Journal of Cellular Physiology, 226*, 2131–2138.

Decherchi, P., & Gauthier, P. (2000). Regrowth of acute and chronic injured spinal pathways within supra-lesional post-traumatic nerve grafts. *Neuroscience, 101*, 197–210.

Decherchi, P., & Gauthier, P. (2002). Regeneration of acutely and chronically injured descending respiratory pathways within post-traumatic nerve grafts. *Neuroscience, 112*, 141–152.

DePaul, M. A., Palmer, M., Lang, B. T., Cutrone, R., Tran, A. P., Madalena, K. M., et al. (2015). Intravenous multipotent adult progenitor cell treatment decreases inflammation leading to functional recovery following spinal cord injury. *Scientific Reports, 5*, 16795.

Deumens, R., Koopmans, G. C., Honig, W. M., Hamers, F. P., Maquet, V., Jerome, R., et al. (2006). Olfactory ensheathing cells, olfactory nerve fibroblasts and biomatrices to promote long-distance axon regrowth and functional recovery in the dorsally hemisected adult rat spinal cord. *Experimental Neurology*.

Doucette, R. (1990). Glial influences on axonal growth in the primary olfactory system. *Glia, 3*, 433–449.

Dugan, E. A., & Shumsky, J. S. (2015). A combination therapy of neural and glial restricted precursor cells and chronic quipazine treatment paired with passive cycling promotes quipazine-induced stepping in adult spinalized rats. *The Journal of Spinal Cord Medicine, 38*, 792–804.

Dulin, J. N., Adler, A. F., Kumamaru, H., Poplawski, G. H. D., Lee-Kubli, C., Strobl, H., et al. (2018). Injured adult motor and sensory axons regenerate into appropriate organotypic domains of neural progenitor grafts. *Nature Communications, 9*, 84.

Dumont, C. M., Carlson, M. A., Munsell, M. K., Ciciriello, A. J., Strnadova, K., Park, J., et al. (2019). Aligned hydrogel tubes guide regeneration following spinal cord injury. *Acta Biomaterialia, 86*, 312–322.

Duncan, I. D., Aguayo, A. J., Bunge, R. P., & Wood, P. M. (1981). Transplantation of rat Schwann cells grown in tissue culture into the mouse spinal cord. *Journal of the Neurological Sciences*, *49*, 241–252.

Ehrlich, M., Mozafari, S., Glatza, M., Starost, L., Velychko, S., Hallmann, A. L., et al. (2017). Rapid and efficient generation of oligodendrocytes from human induced pluripotent stem cells using transcription factors. *Proceedings of the National Academy of Sciences of the United States of America*, *114*, E2243–E2252.

Ekberg, J. A., & St John, J. A. (2014). Crucial roles for olfactory ensheathing cells and olfactory mucosal cells in the repair of damaged neural tracts. *Anatomical Record (Hoboken)*, *297*, 121–128.

El-Kheir, W. A., Gabr, H., Awad, M. R., Ghannam, O., Barakat, Y., Farghali, H. A., et al. (2014). Autologous bone marrow-derived cell therapy combined with physical therapy induces functional improvement in chronic spinal cord injury patients. *Cell Transplantation*, *23*, 729–745.

Erceg, S., Ronaghi, M., Oria, M., Rosello, M. G., Arago, M. A., Lopez, M. G., et al. (2010). Transplanted oligodendrocytes and motoneuron progenitors generated from human embryonic stem cells promote locomotor recovery after spinal cord transection. *Stem Cells*, *28*, 1541–1549.

Falci, S., Holtz, A., Akesson, E., Azizi, M., Ertzgaard, P., Hultling, C., et al. (1997). Obliteration of a posttraumatic spinal cord cyst with solid human embryonic spinal cord grafts: First clinical attempt. *Journal of Neurotrauma*, *14*, 875–884.

Faulkner, J., & Keirstead, H. S. (2005). Human embryonic stem cell-derived oligodendrocyte progenitors for the treatment of spinal cord injury. *Transplant Immunology*, *15*, 131–142.

Fawcett, J. W. (2008). Bridging spinal cord injuries. *Journal of Biology*, *7*, 25.

Feron, F., Perry, C., Cochrane, J., Licina, P., Nowitzke, A., Urquhart, S., et al. (2005). Autologous olfactory ensheathing cell transplantation in human spinal cord injury. *Brain*, *128*, 2951–2960.

Filippi, M., Boido, M., Pasquino, C., Garello, F., Boffa, C., & Terreno, E. (2016). Successful in vivo MRI tracking of MSCs labeled with gadoteridol in a spinal cord injury experimental model. *Experimental Neurology*, *282*, 66–77.

Fischer, D. A. (1985). Long-term management of the ventilator patient in the home. *Cleveland Clinic Quarterly*, *52*, 303–306.

Fischer, I., Dulin, J. N., & Lane, M. A. (2020). Transplanting neural progenitor cells to restore connectivity after spinal cord injury. *Nature Reviews. Neuroscience*, *21*(7), 366–383. https://doi.org/10.1038/s41583-020-0314-2. Epub 2020 Jun 9.3.

Flynn, J. R., Graham, B. A., Galea, M. P., & Callister, R. J. (2011). The role of propriospinal interneurons in recovery from spinal cord injury. *Neuropharmacology*, *60*, 809–822.

Fortino, T. A., Randelman, M. L., Hall, A. A., Singh, J., Bloom, D. C., Engel, E., et al. (2022). Transneuronal tracing to map connectivity in injured and transplanted spinal networks. *Experimental Neurology*, *351*, 113990.

Forni, P. E., Taylor-Burds, C., Melvin, V. S., Williams, T., & Wray, S. (2011). Neural crest and ectodermal cells intermix in the nasal placode to give rise to GnRH-1 neurons, sensory neurons, and olfactory ensheathing cells. *The Journal of Neuroscience*, *31*(18), 6915–6927.

Fouad, K., Schnell, L., Bunge, M. B., Schwab, M. E., Liebscher, T., & Pearse, D. D. (2005). Combining Schwann cell bridges and olfactory-ensheathing glia grafts with chondroitinase promotes locomotor recovery after complete transection of the spinal cord. *The Journal of Neuroscience*, *25*, 1169–1178.

Franklin, R. J., Bayley, S. A., & Blakemore, W. F. (1996). Transplanted CG4 cells (an oligodendrocyte progenitor cell line) survive, migrate, and contribute to repair of areas of demyelination in X-irradiated and damaged spinal cord but not in normal spinal cord. *Experimental Neurology*, *137*, 263–276.

Franssen, E. H., de Bree, F. M., & Verhaagen, J. (2007). Olfactory ensheathing glia: Their contribution to primary olfactory nervous system regeneration and their regenerative potential following transplantation into the injured spinal cord. *Brain Research Reviews*, *56*(1), 236–258.

Fuhrmann, T., Tam, R. Y., Ballarin, B., Coles, B., Elliott Donaghue, I., van der Kooy, D., et al. (2016). Injectable hydrogel promotes early survival of induced pluripotent stem cell-derived oligodendrocytes and attenuates longterm teratoma formation in a spinal cord injury model. *Biomaterials*, *83*, 23–36.

Gashmardi, N., Hosseini, S. E., Mehrabani, D., Edalatmanesh, M. A., & Khodabandeh, Z. (2017). Impacts of bone marrow stem cells on caspase-3 levels after spinal cord injury in mice. *Iranian Journal of Medical Sciences*, *42*, 593–598.

Gauthier, P., Rega, P., Lammari-Barreault, N., & Polentes, J. (2002). Functional reconnections established by central respiratory neurons regenerating axons into a nerve graft bridging the respiratory centers to the cervical spinal cord. *Journal of Neuroscience Research*, *70*, 65–81.

Gilerovich, E. G., Fedorova, E. A., & Otellin, V. A. (1991). Transplantation of human embryonal nervous tissue into the spinal cord of adult rats. *Neuroscience and Behavioral Physiology*, *21*, 193–198.

Gilmour, A. D., Reshamwala, R., Wright, A. A., Ekberg, J. A. K., & St John, J. A. (2020). Optimizing olfactory ensheathing cell transplantation for spinal cord injury repair. *Journal of Neurotrauma*, *37*, 817–829.

Giovanini, M. A., Reier, P. J., Eskin, T. A., & Anderson, D. K. (1997). MAP2 expression in the developing human fetal spinal cord and following xenotransplantation. *Cell Transplantation*, *6*, 339–346.

Giovanini, M. A., Reier, P. J., Eskin, T. A., Wirth, E., & Anderson, D. K. (1997). Characteristics of human fetal spinal cord grafts in the adult rat spinal cord: Influences of lesion and grafting conditions. *Experimental Neurology*, *148*, 523–543.

Giraldo, E., Nebot, V. J., Dordevic, S., Requejo-Aguilar, R., Alastrue-Agudo, A., Zagorodko, O., et al. (2021). A rationally designed self-immolative linker enhances the synergism between a polymer-rock inhibitor conjugate and neural progenitor cells in the treatment of spinal cord injury. *Biomaterials*, *276*, 121052.

Goldshmit, Y., Lythgo, N., Galea, M. P., & Turnley, A. M. (2008). Treadmill training after spinal cord hemisection in mice promotes axonal sprouting and synapse formation and improves motor recovery. *Journal of Neurotrauma*, *25*, 449–465.

Gomes, E. D., Mendes, S. S., Assuncao-Silva, R. C., Teixeira, F. G., Pires, A. O., Anjo, S. I., et al. (2018). Co-transplantation of adipose tissue-derived stromal cells and olfactory ensheathing cells for spinal cord injury repair. *Stem Cells*, *36*, 696–708.

Gomez, V. M., Averill, S., King, V., Yang, Q., Perez, E. D., Chacon, S. C., et al. (2003). Transplantation of olfactory ensheathing cells fails to promote significant axonal regeneration from dorsal roots into the rat cervical cord. *Journal of Neurocytology*, *32*, 53–70.

Goulao, M., Ghosh, B., Urban, M. W., Sahu, M., Mercogliano, C., Charsar, B. A., et al. (2019a). Astrocyte progenitor transplantation promotes regeneration of bulbospinal respiratory axons, recovery of diaphragm function, and a reduced macrophage response following cervical spinal cord injury. *Glia*, *67*(3), 452–466. https://doi.org/10.1002/glia.23555. Epub 2018 Dec 11.

Goulao, M., Ghosh, B., Urban, M. W., Sahu, M., Mercogliano, C., Charsar, B. A., et al. (2019b). Astrocyte progenitor transplantation promotes regeneration of bulbospinal respiratory axons, recovery of diaphragm function, and a reduced macrophage response following cervical spinal cord injury. *Glia*, *67*, 452–466.

Gransee, H. M., Zhan, W. Z., Sieck, G. C., & Mantilla, C. B. (2015). Localized delivery of brain derived neurotrophic factor-expressing mesenchymal stem cells enhances functional recovery following cervical spinal cord injury. *Journal of Neurotrauma*, *32*, 185–193.

Graziadei, P. P., & Monti Graziadei, G. A. (1985). Neurogenesis and plasticity of the olfactory sensory neurons. *Annals of the New York Academy of Sciences*, *457*, 127–142.

Grill, R. J., Blesch, A., & Tuszynski, M. H. (1997). Robust growth of chronically injured spinal cord axons induced by grafts of genetically modified NGF-secreting cells. *Experimental Neurology*, *148*, 444–452.

Gros, T., Sakamoto, J. S., Blesch, A., Havton, L. A., & Tuszynski, M. H. (2010). Regeneration of long-tract axons through sites of spinal cord injury using templated agarose scaffolds. *Biomaterials*, *31*, 6719–6729.

Guerout, N., Duclos, C., Drouot, L., Abramovici, O., Bon-Mardion, N., Lacoume, Y., et al. (2011). Transplantation of olfactory ensheathing cells promotes axonal regeneration and functional recovery of peripheral nerve lesion in rats. *Muscle & Nerve*, *43*, 543–551.

Guest, J., & Dietrich, W. D. (2015). Commentary regarding the recent publication by Tabakow et al., "Functional regeneration of supraspinal connections in a patient with transected spinal cord following transplantation of bulbar olfactory ensheathing cells with peripheral nerve bridging". *Journal of Neurotrauma*, *32*, 1176–1178.

Guest, J. D., Rao, A., Olson, L., Bunge, M. B., & Bunge, R. P. (1997). The ability of human Schwann cell grafts to promote regeneration in the transected nude rat spinal cord. *Experimental Neurology*, *148*, 502–522.

Guest, J., Santamaria, A. J., & Benavides, F. D. (2013). Clinical translation of autologous Schwann cell transplantation for the treatment of spinal cord injury. *Current Opinion in Organ Transplantation*, *18*, 682–689.

Guo, S. G., Wang, C. J., Wang, Y. X., & Qu, C. Q. (2015). Transplantation of hyperthermic preconditioning olfactory ensheathing cells combined with neural stem cells in the treatment of central nerve injury. *Journal of Biological Regulators and Homeostatic Agents*, *29*, 677–682.

Gupta, S., Sivalingam, D., Hain, S., Makkar, C., Sosa, E., Clark, A., et al. (2018). Deriving dorsal spinal sensory interneurons from human pluripotent stem cells. *Stem Cell Reports*, *10*, 390–405.

Haas, C., & Fischer, I. (2013). Human astrocytes derived from glial restricted progenitors support regeneration of the injured spinal cord. *Journal of Neurotrauma*, *30*, 1035–1052.

Haas, C., & Fischer, I. (2014). Transplanting neural progenitors to build a neuronal relay across the injured spinal cord. *Neural Regeneration Research*, *9*, 1173–1176.

Haas, C., Neuhuber, B., Yamagami, T., Rao, M., & Fischer, I. (2012). Phenotypic analysis of astrocytes derived from glial restricted precursors and their impact on axon regeneration. *Experimental Neurology*, *233*, 717–732.

Hamburger, V. (1948). The mitotic patterns in the spinal cord of the chick embryo and their relation to histogenetic processes. *The Journal of Comparative Neurology*, *88*, 221–283.

Han, S. S., Kang, D. Y., Mujtaba, T., Rao, M. S., & Fischer, I. (2002). Grafted lineage-restricted precursors differentiate exclusively into neurons in the adult spinal cord. *Experimental Neurology*, *177*, 360–375.

Han, S. S., Liu, Y., Tyler-Polsz, C., Rao, M. S., & Fischer, I. (2004). Transplantation of glial-restricted precursor cells into the adult spinal cord: survival, glial-specific differentiation, and preferential migration in white matter. *Glia*, *45*, 1–16.

Hasegawa, K., Chang, Y. W., Li, H., Berlin, Y., Ikeda, O., Kane-Goldsmith, N., et al. (2005). Embryonic radial glia bridge spinal cord lesions and promote functional recovery following spinal cord injury. *Experimental Neurology*, *193*, 394–410.

Hashimoto, T., & Shibasaki, F. (2015). Hypoxia-inducible factor as an angiogenic master switch. *Frontiers in Pediatrics*, *3*, 33.

Hayakawa, K., Haas, C., Jin, Y., Bouyer, J., Otsuka, T., & Fischer, I. (2015). Glial restricted precursors maintain their permissive properties after long-term expansion but not following exposure to pro-inflammatory factors. *Brain Research*, *1629*, 113–125.

Hayashi, K., Hashimoto, M., Koda, M., Naito, A. T., Murata, A., Okawa, A., et al. (2011). Increase of sensitivity to mechanical stimulus after transplantation of murine induced pluripotent stem cell-derived astrocytes in a rat spinal cord injury model. *Journal of Neurosurgery. Spine*, *15*, 582–593.

Hejcl, A., Lesny, P., Pradny, M., Michalek, J., Jendelova, P., Stulik, J., et al. (2008). Biocompatible hydrogels in spinal cord injury repair. *Physiological Research*, *57*(Suppl 3), S121–S132.

Hejcl, A., Ruzicka, J., Kapcalova, M., Turnovcova, K., Krumbholcova, E., Pradny, M., et al. (2013). Adjusting the chemical and physical properties of hydrogels leads to improved stem cell survival and tissue ingrowth in spinal cord injury reconstruction: A comparative study of four methacrylate hydrogels. *Stem Cells and Development*, *22*, 2794–2805.

Henschen, A., Hoffer, B., & Olson, L. (1985). Spinal cord grafts in oculo: Survival, growth, histological organization and electrophysiological characteristics. *Experimental Brain Research*, *60*, 38–47.

Herrera, J., Yang, H., Zhang, S. C., Proschel, C., Tresco, P., Duncan, I. D., et al. (2001). Embryonic-derived glial-restricted precursor cells (GRP cells) can differentiate into astrocytes and oligodendrocytes in vivo. *Experimental Neurology*, *171*, 11–21.

Hill, C. E., Moon, L. D., Wood, P. M., & Bunge, M. B. (2006). Labeled Schwann cell transplantation: Cell loss, host Schwann cell replacement, and strategies to enhance survival. *Glia*, *53*, 338–343.

Hill, C. E., Proschel, C., Noble, M., Mayer-Proschel, M., Gensel, J. C., Beattie, M. S., et al. (2004). Acute transplantation of glial-restricted precursor cells into spinal cord contusion injuries: Survival, differentiation, and effects on lesion environment and axonal regeneration. *Experimental Neurology*, *190*, 289–310.

Hodgetts, S. I., Simmons, P. J., & Plant, G. W. (2013). Human mesenchymal precursor cells (Stro-1(+)) from spinal cord injury patients improve functional recovery and tissue sparing in an acute spinal cord injury rat model. *Cell Transplantation*, *22*, 393–412.

Hofer, H. R., & Tuan, R. S. (2016). Secreted trophic factors of mesenchymal stem cells support neurovascular and musculoskeletal therapies. *Stem Cell Research & Therapy*, *7*, 131.

Hofstetter, C. P., Schwarz, E. J., Hess, D., Widenfalk, J., El Manira, A., Prockop, D. J., et al. (2002). Marrow stromal cells form guiding strands in the injured spinal cord and promote recovery. *Proceedings of the National Academy of Sciences of the United States of America*, *99*, 2199–2204.

Horner, P. J., Reier, P. J., & Stokes, B. T. (1996). Quantitative analysis of vascularization and cytochrome oxidase following fetal transplantation in the contused rat spinal cord. *The Journal of Comparative Neurology*, *364*, 690–703.

Hosseini, S. M., Sani, M., Haider, K. H., Dorvash, M., Ziaee, S. M., Karimi, A., et al. (2018). Concomitant use of mesenchymal stem cells and neural stem cells for treatment of spinal cord injury: A combo cell therapy approach. *Neuroscience Letters*, *668*, 138–146.

Hosseini, S. M., Sharafkhah, A., Koohi-Hosseinabadi, O., & Semsar-Kazerooni, M. (2016). Transplantation of neural stem cells cultured in alginate scaffold for spinal cord injury in rats. *Asian Spine Journal*, *10*, 611–618.

Hou, S., Saltos, T. M., Mironets, E., Trueblood, C. T., Connors, T. M., & Tom, V. J. (2020). Grafting embryonic raphe neurons reestablishes serotonergic regulation of sympathetic activity to improve cardiovascular function after spinal cord injury. *The Journal of Neuroscience: The Official Journal of the Society for Neuroscience*, *40*, 1248–1264.

Houle, J. D. (1991). Demonstration of the potential for chronically injured neurons to regenerate axons into intraspinal peripheral nerve grafts. *Experimental Neurology*, *113*, 1–9.

Houle, J. D., Amin, A., Cote, M. P., Lemay, M., Miller, K., Sandrow, H., et al. (2009). Combining peripheral nerve grafting and matrix modulation to repair the injured rat spinal cord. *Journal of Visualized Experiments*, *33*, 1324. https://doi.org/10.3791/1324.

Houle, J. D., & Johnson, J. E. (1989). Nerve growth factor (NGF)-treated nitrocellulose enhances and directs the regeneration of adult rat dorsal root axons through intraspinal neural tissue transplants. *Neuroscience Letters*, *103*, 17–23.

Houle, J. D., Morris, K., Skinner, R. D., Garcia-Rill, E., & Peterson, C. A. (1999). Effects of fetal spinal cord tissue transplants and cycling exercise on the soleus muscle in spinalized rats. *Muscle & Nerve*, *22*, 846–856.

Houle, J. D., & Reier, P. J. (1988). Transplantation of fetal spinal cord tissue into the chronically injured adult rat spinal cord. *The Journal of Comparative Neurology*, *269*, 535–547.

Houle, J. D., & Reier, P. J. (1989). Regrowth of calcitonin gene-related peptide (CGRP) immunoreactive axons from the chronically injured rat spinal cord into fetal spinal cord tissue transplants. *Neuroscience Letters*, *103*, 253–258.

Houle, J. D., Tom, V. J., Mayes, D., Wagoner, G., Phillips, N., & Silver, J. (2006). Combining an autologous peripheral nervous system "bridge" and matrix modification by chondroitinase allows robust, functional regeneration beyond a hemisection lesion of the adult rat spinal cord. *The Journal of Neuroscience: The Official Journal of the Society for Neuroscience*, *26*, 7405–7415.

Houle, J. D., & Ziegler, M. K. (1994). Bridging a complete transection lesion of adult rat spinal cord with growth factor-treated nitrocellulose implants. *Journal of Neural Transplantation & Plasticity*, *5*, 115–124.

Huang, H., Chen, L., Wang, H., Xi, H., Gou, C., Zhang, J., et al. (2006). Safety of fetal olfactory ensheathing cell transplantation in patients with chronic spinal cord injury. A 38-month follow-up with MRI. *Zhongguo Xiu Fu Chong Jian Wai Ke Za Zhi*, *20*, 439–443.

Huang, H., Xi, H., Chen, L., Zhang, F., & Liu, Y. (2012). Long-term outcome of olfactory ensheathing cell therapy for patients with complete chronic spinal cord injury. *Cell Transplantation*, *21*(Suppl 1), S23–S31.

Iida, T., Iwanami, A., Sanosaka, T., Kohyama, J., Miyoshi, H., Nagoshi, N., et al. (2017). Whole-genome dna methylation analyses revealed epigenetic instability in tumorigenic human iPS cell-derived neural stem/progenitor cells. *Stem Cells*, *35*, 1316–1327.

Imaizumi, T., Lankford, K. L., Waxman, S. G., Greer, C. A., & Kocsis, J. D. (1998). Transplanted olfactory ensheathing cells remyelinate and enhance axonal conduction in the demyelinated dorsal columns of the rat spinal cord. *The Journal of Neuroscience*, *18*, 6176–6185.

Ito, S., Nagoshi, N., Kamata, Y., Kojima, K., Nori, S., Matsumoto, M., et al. (2021). LOTUS overexpression via ex vivo gene transduction further promotes recovery of motor function following human iPSC-NS/PC transplantation for contusive spinal cord injury. *Stem Cell Reports*, *16*, 2703–2717.

Jakeman, L. B., & Reier, P. J. (1991). Axonal projections between fetal spinal cord transplants and the adult rat spinal cord: A neuroanatomical tracing study of local interactions. *The Journal of Comparative Neurology*, *307*, 311–334.

Jakeman, L. B., Reier, P. J., Bregman, B. S., Wade, E. B., Dailey, M., Kastner, R. J., et al. (1989). Differentiation of substantia gelatinosa-like regions in intraspinal and intracerebral transplants of embryonic spinal cord tissue in the rat. *Experimental Neurology*, *103*, 17–33.

Jeffery, N. D., Crang, A. J., O'Leary, M. T., Hodge, S. J., & Blakemore, W. F. (1999). Behavioural consequences of oligodendrocyte progenitor cell transplantation into experimental demyelinating lesions in the rat spinal cord. *The European Journal of Neuroscience*, *11*, 1508–1514.

Jin, Y., Bouyer, J., Shumsky, J. S., Haas, C., & Fischer, I. (2016). Transplantation of neural progenitor cells in chronic spinal cord injury. *Neuroscience*, *320*, 69–82.

Jin, Y., Neuhuber, B., Singh, A., Bouyer, J., Lepore, A., Bonner, J., et al. (2011). Transplantation of human glial restricted progenitors and derived astrocytes into a contusion model of spinal cord injury. *Journal of Neurotrauma*, *28*, 579–594.

Jin, Y., Sura, K., & Fischer, I. (2012). Differential effects of distinct central nervous system regions on cell migration and axonal extension of neural precursor transplants. *Journal of Neuroscience Research*, *90*, 2065–2073.

Jin, Y., Tessler, A., Fischer, I., & Houle, J. D. (2000). Fibroblasts genetically modified to produce BDNF support regrowth of chronically injured serotonergic axons. *Neurorehabilitation and Neural Repair, 14*, 311–317.

Johnson, L. D. V., Pickard, M. R., & Johnson, W. E. B. (2021). The comparative effects of mesenchymal stem cell transplantation therapy for spinal cord injury in humans and animal models: A systematic review and meta-analysis. *Biology (Basel), 10*(3), 230. https://doi.org/10.3390/biology10030230.

Jones, J. M., DePaul, M. A., Gregory, C. R., Lang, B. T., Xie, H., Zhu, M., et al. (2019). Multipotent adult progenitor cells, but not tissue inhibitor of matrix metalloproteinase-3, increase tissue sparing and reduce urological complications following spinal cord injury. *Journal of Neurotrauma, 36*, 1416–1427.

Joosten, E. A., Veldhuis, W. B., & Hamers, F. P. (2004). Collagen containing neonatal astrocytes stimulates regrowth of injured fibers and promotes modest locomotor recovery after spinal cord injury. *Journal of Neuroscience Research, 77*, 127–142.

Kadoya, K., Lu, P., Nguyen, K., Lee-Kubli, C., Kumamaru, H., Yao, L., et al. (2016). Spinal cord reconstitution with homologous neural grafts enables robust corticospinal regeneration. *Nature Medicine, 22*, 479–487.

Kajikawa, K., Imaizumi, K., Shinozaki, M., Shibata, S., Shindo, T., Kitagawa, T., et al. (2020). Cell therapy for spinal cord injury by using human iPSC-derived region-specific neural progenitor cells. *Molecular Brain, 13*, 120.

Kalyani, A., Hobson, K., & Rao, M. S. (1997). Neuroepithelial stem cells from the embryonic spinal cord: Isolation, characterization, and clonal analysis. *Developmental Biology, 186*, 202–223.

Kalyani, A. J., Piper, D., Mujtaba, T., Lucero, M. T., & Rao, M. S. (1998). Spinal cord neuronal precursors generate multiple neuronal phenotypes in culture. *The Journal of Neuroscience: The Official Journal of the Society for Neuroscience, 18*, 7856–7868.

Kalyani, A. J., & Rao, M. S. (1998). Cell lineage in the developing neural tube. *Biochemistry and Cell Biology, 76*, 1051–1068.

Kang, X. W., Hu, J. L., Wang, S. K., & Wang, J. (2015). Effectiveness of muscle basal lamina carrying neural stem cells and olfactory ensheathing cells in spinal cord repair. *Genetics and Molecular Research, 14*, 13437–13455.

Kawabata, S., Takano, M., Numasawa-Kuroiwa, Y., Itakura, G., Kobayashi, Y., Nishiyama, Y., et al. (2016). Grafted human iPS cell-derived oligodendrocyte precursor cells contribute to robust remyelination of demyelinated axons after spinal cord injury. *Stem Cell Reports, 6*, 1–8.

Kawai, M., Imaizumi, K., Ishikawa, M., Shibata, S., Shinozaki, M., Shibata, T., et al. (2021). Long-term selective stimulation of transplanted neural stem/progenitor cells for spinal cord injury improves locomotor function. *Cell Reports, 37*, 110019.

Keirstead, H. S., & Blakemore, W. F. (1999). The role of oligodendrocytes and oligodendrocyte progenitors in CNS remyelination. *Advances in Experimental Medicine and Biology, 468*, 183–197.

Keirstead, H. S., Levine, J. M., & Blakemore, W. F. (1998). Response of the oligodendrocyte progenitor cell population (defined by NG2 labelling) to demyelination of the adult spinal cord. *Glia, 22*, 161–170.

Keirstead, H. S., Nistor, G., Bernal, G., Totoiu, M., Cloutier, F., Sharp, K., et al. (2005). Human embryonic stem cell-derived oligodendrocyte progenitor cell transplants remyelinate and restore locomotion after spinal cord injury. *The Journal of Neuroscience: The Official Journal of the Society for Neuroscience, 25*, 4694–4705.

Kermani, P., & Hempstead, B. (2007). Brain-derived neurotrophic factor: A newly described mediator of angiogenesis. *Trends in Cardiovascular Medicine, 17*, 140–143.

Khankan, R. R., Griffis, K. G., Haggerty-Skeans, J. R., Zhong, H., Roy, R. R., Edgerton, V. R., et al. (2016). Olfactory ensheathing cell transplantation after a complete spinal cord transection mediates neuroprotective and immunomodulatory mechanisms to facilitate regeneration. *The Journal of Neuroscience: The Official Journal of the Society for Neuroscience*, *36*, 6269–6286.

Kim, Y. C., Kim, Y. H., Kim, J. W., & Ha, K. Y. (2016). Transplantation of mesenchymal stem cells for acute spinal cord injury in rats: Comparative study between intralesional injection and scaffold based transplantation. *Journal of Korean Medical Science*, *31*, 1373–1382.

Kim, W. K., Kim, W. H., Kweon, O. K., & Kang, B. J. (2020). Intravenous administration of heat shock-treated MSCs can improve neuroprotection and neuroregeneration in canine spinal cord injury model. *Animals (Basel)*, *10*(11), 2164. https://doi.org/10.3390/ani10112164.

Kim, J. B., Lee, H., Arauzo-Bravo, M. J., Hwang, K., Nam, D., Park, M. R., et al. (2015). Oct4-induced oligodendrocyte progenitor cells enhance functional recovery in spinal cord injury model. *The EMBO Journal*, *34*, 2971–2983.

Kirkeby, A., Grealish, S., Wolf, D. A., Nelander, J., Wood, J., Lundblad, M., et al. (2012). Generation of regionally specified neural progenitors and functional neurons from human embryonic stem cells under defined conditions. *Cell Reports*, *1*, 703–714.

Kitagawa, T., Nagoshi, N., Kamata, Y., Kawai, M., Ago, K., Kajikawa, K., et al. (2022). Modulation by DREADD reveals the therapeutic effect of human iPSC-derived neuronal activity on functional recovery after spinal cord injury. *Stem Cell Reports*, *17*, 127–142.

Kobashi, S., Terashima, T., Katagi, M., Nakae, Y., Okano, J., Suzuki, Y., et al. (2020). Transplantation of M2-deviated microglia promotes recovery of motor function after spinal cord injury in mice. *Molecular Therapy*, *28*, 254–265.

Kobayashi, Y., Okada, Y., Itakura, G., Iwai, H., Nishimura, S., Yasuda, A., et al. (2012). Pre-evaluated safe human iPSC-derived neural stem cells promote functional recovery after spinal cord injury in common marmoset without tumorigenicity. *PLoS One*, *7*, e52787.

Koffler, J., Zhu, W., Qu, X., Platoshyn, O., Dulin, J. N., Brock, J., et al. (2019). Biomimetic 3D-printed scaffolds for spinal cord injury repair. *Nature Medicine*, *25*, 263–269.

Kojima, K., Miyoshi, H., Nagoshi, N., Kohyama, J., Itakura, G., Kawabata, S., et al. (2019). Selective ablation of tumorigenic cells following human induced pluripotent stem cell-derived neural stem/progenitor cell transplantation in spinal cord injury. *Stem Cells Translational Medicine*, *8*, 260–270.

Kou, D., Li, T., Liu, H., Liu, C., Yin, Y., Wu, X., et al. (2018). Transplantation of rat-derived microglial cells promotes functional recovery in a rat model of spinal cord injury. *Brazilian Journal of Medical and Biological Research*, *51*, e7076.

Krikorian, J. G., Guth, L., & Donati, E. J. (1981). Origin of the connective tissue scar in the transected rat spinal cord. *Experimental Neurology*, *72*, 698–707.

Kristensson, K. (1976). Experimental herpes simplex virus infection in the immature mouse brain. *Acta Neuropathologica*, *35*, 343–351.

Kubasak, M. D., Jindrich, D. L., Zhong, H., Takeoka, A., McFarland, K. C., Munoz-Quiles, C., et al. (2008). OEG implantation and step training enhance hindlimb-stepping ability in adult spinal transected rats. *Brain*, *131*, 264–276.

Kumagai, G., Okada, Y., Yamane, J., Nagoshi, N., Kitamura, K., Mukaino, M., et al. (2009). Roles of ES cell-derived gliogenic neural stem/progenitor cells in functional recovery after spinal cord injury. *PLoS One*, *4*, e7706.

Kumagai, G., Tsoulfas, P., Toh, S., McNiece, I., Bramlett, H. M., & Dietrich, W. D. (2013). Genetically modified mesenchymal stem cells (MSCs) promote axonal regeneration and prevent hypersensitivity after spinal cord injury. *Experimental Neurology*, *248*, 369–380.

Kumamaru, H., Lu, P., Rosenzweig, E. S., Kadoya, K., & Tuszynski, M. H. (2019). Regenerating corticospinal axons innervate phenotypically appropriate neurons within neural stem cell grafts. *Cell Reports*, *26*, 2329–2339. e2324.

Kumamaru, H., Lu, P., Rosenzweig, E. S., & Tuszynski, M. H. (2018). Activation of intrinsic growth state enhances host axonal regeneration into neural progenitor cell grafts. *Stem Cell Reports*, *11*, 861–868.

Kwon, B. K., Streijger, F., Hill, C. E., Anderson, A. J., Bacon, M., Beattie, M. S., et al. (2015). Large animal and primate models of spinal cord injury for the testing of novel therapies. *Experimental Neurology*, *269*, 154–168.

Lakatos, A., Barnett, S. C., & Franklin, R. J. (2003). Olfactory ensheathing cells induce less host astrocyte response and chondroitin sulphate proteoglycan expression than Schwann cells following transplantation into adult CNS white matter. *Experimental Neurology*, *184*, 237–246.

Lane, M. A., Lepore, A. C., & Fischer, I. (2017). Improving the therapeutic efficacy of neural progenitor cell transplantation following spinal cord injury. *Expert Review of Neurotherapeutics*, *17*, 433–440.

Lee, K. Z., Lane, M. A., Dougherty, B. J., Mercier, L. M., Sandhu, M. S., Sanchez, J. C., et al. (2014). Intraspinal transplantation and modulation of donor neuron electrophysiological activity. *Experimental Neurology*, *251*, 47–57.

Lee, Y. S., Wu, S., Arinzeh, T. L., & Bunge, M. B. (2017). Enhanced noradrenergic axon regeneration into schwann cell-filled PVDF-TrFE conduits after complete spinal cord transection. *Biotechnology and Bioengineering*, *114*, 444–456.

Lepore, A. C., Bakshi, A., Swanger, S. A., Rao, M. S., & Fischer, I. (2005). Neural precursor cells can be delivered into the injured cervical spinal cord by intrathecal injection at the lumbar cord. *Brain Research*, *1045*, 206–216.

Lepore, A. C., & Fischer, I. (2005). Lineage-restricted neural precursors survive, migrate, and differentiate following transplantation into the injured adult spinal cord. *Experimental Neurology*, *194*, 230–242.

Lepore, A. C., Han, S. S., Tyler-Polsz, C. J., Cai, J., Rao, M. S., & Fischer, I. (2004). Differential fate of multipotent and lineage-restricted neural precursors following transplantation into the adult CNS. *Neuron Glia Biology*, *1*, 113–126.

Lepore, A. C., Neuhuber, B., Connors, T. M., Han, S. S., Liu, Y., Daniels, M. P., et al. (2006). Long-term fate of neural precursor cells following transplantation into developing and adult CNS. *Neuroscience*, *139*, 513–530.

Levi, A. D., Okonkwo, D. O., Park, P., Jenkins, A. L., 3rd, Kurpad, S. N., Parr, A. M., et al. (2018). Emerging safety of intramedullary transplantation of human neural stem cells in chronic cervical and thoracic spinal cord injury. *Neurosurgery*, *82*, 562–575.

Li, Y., Decherchi, P., & Raisman, G. (2003). Transplantation of olfactory ensheathing cells into spinal cord lesions restores breathing and climbing. *The Journal of Neuroscience: The Official Journal of the Society for Neuroscience*, *23*, 727–731.

Li, Y., Field, P. M., & Raisman, G. (1997). Repair of adult rat corticospinal tract by transplants of olfactory ensheathing cells. *Science*, *277*, 2000–2002.

Li, Y., Field, P. M., & Raisman, G. (1998). Regeneration of adult rat corticospinal axons induced by transplanted olfactory ensheathing cells. *The Journal of Neuroscience*, *18*, 10514–10524.

Li, H., Ham, T. R., Neill, N., Farrag, M., Mohrman, A. E., Koenig, A. M., et al. (2016). A hydrogel bridge incorporating immobilized growth factors and neural stem/progenitor cells to treat spinal cord injury. *Advanced Healthcare Materials*, *5*, 802–812.

Li, Y., He, X., Kawaguchi, R., Zhang, Y., Wang, Q., Monavarfeshani, A., et al. (2020). Microglia-organized scar-free spinal cord repair in neonatal mice. *Nature*, *587*, 613–618.

Li, K., Javed, E., Scura, D., Hala, T. J., Seetharam, S., Falnikar, A., et al. (2015). Human iPS cell-derived astrocyte transplants preserve respiratory function after spinal cord injury. *Experimental Neurology*, *271*, 479–492.

Li, S., Zheng, J., Chai, L., Lin, M., Zeng, R., Lu, J., et al. (2019). Rapid and efficient differentiation of rodent neural stem cells into oligodendrocyte progenitor cells. *Developmental Neuroscience*, *41*, 79–93.

Lien, B. V., Tuszynski, M. H., & Lu, P. (2019). Astrocytes migrate from human neural stem cell grafts and functionally integrate into the injured rat spinal cord. *Experimental Neurology, 314*, 46–57.

Liu, J., Chen, P., Wang, Q., Chen, Y., Yu, H., Ma, J., et al. (2014). Meta analysis of olfactory ensheathing cell transplantation promoting functional recovery of motor nerves in rats with complete spinal cord transection. *Neural Regeneration Research, 9*, 1850–1858.

Liu, W., Ding, Y., Zhang, X., & Wang, L. (2013). Bone marrow stromal cells inhibit caspase-12 expression in rats with spinal cord injury. *Experimental and Therapeutic Medicine, 6*, 671–674.

Liu, Y., Himes, B. T., Murray, M., Tessler, A., & Fischer, I. (2002). Grafts of BDNF-producing fibroblasts rescue axotomized rubrospinal neurons and prevent their atrophy. *Experimental Neurology, 178*, 150–164.

Liu, Y., Himes, B. T., Tryon, B., Moul, J., Chow, S. Y., Jin, H., et al. (1998). Intraspinal grafting of fibroblasts genetically modified by recombinant adenoviruses. *Neuroreport, 9*, 1075–1079.

Liu, Y., Kim, D., Himes, B. T., Chow, S. Y., Schallert, T., Murray, M., et al. (1999). Transplants of fibroblasts genetically modified to express BDNF promote regeneration of adult rat rubrospinal axons and recovery of forelimb function. *The Journal of Neuroscience: The Official Journal of the Society for Neuroscience, 19*, 4370–4387.

Liu, M., Li, K., Wang, Y., Zhao, G., & Jiang, J. (2020). Stem cells in the treatment of neuropathic pain: Research progress of mechanism. *Stem Cells International, 2020*, 8861251.

Liu, Y., Murray, M., Tessler, A., & Fischer, I. (2000). Grafting of genetically modified fibroblasts into the injured spinal cord. *Progress in Brain Research, 128*, 309–319.

Liu, S., Sandner, B., Schackel, T., Nicholson, L., Chtarto, A., Tenenbaum, L., et al. (2017). Regulated viral BDNF delivery in combination with Schwann cells promotes axonal regeneration through capillary alginate hydrogels after spinal cord injury. *Acta Biomaterialia, 60*, 167–180.

Lopez-Vales, R., Garcia-Alias, G., Fores, J., Navarro, X., & Verdu, E. (2004). Increased expression of cyclo-oxygenase 2 and vascular endothelial growth factor in lesioned spinal cord by transplanted olfactory ensheathing cells. *Journal of Neurotrauma, 21*, 1031–1043.

Lovett-Barr, M. R., Satriotomo, I., Muir, G. D., Wilkerson, J. E., Hoffman, M. S., Vinit, S., et al. (2012). Repetitive intermittent hypoxia induces respiratory and somatic motor recovery after chronic cervical spinal injury. *The Journal of Neuroscience: The Official Journal of the Society for Neuroscience, 32*, 3591–3600.

Lu, J., & Ashwell, K. (2002). Olfactory ensheathing cells: Their potential use for repairing the injured spinal cord. *Spine (Phila Pa 1976), 27*, 887–892.

Lu, P., Ceto, S., Wang, Y., Graham, L., Wu, D., Kumamaru, H., et al. (2017). Prolonged human neural stem cell maturation supports recovery in injured rodent CNS. *The Journal of Clinical Investigation, 127*, 3287–3299.

Lu, J., Feron, F., Ho, S. M., Mackay-Sim, A., & Waite, P. M. (2001). Transplantation of nasal olfactory tissue promotes partial recovery in paraplegic adult rats. *Brain Research, 889*, 344–357.

Lu, J., Feron, F., Mackay-Sim, A., & Waite, P. M. (2002). Olfactory ensheathing cells promote locomotor recovery after delayed transplantation into transected spinal cord. *Brain, 125*, 14–21.

Lu, P., Gomes-Leal, W., Anil, S., Dobkins, G., Huie, J. R., Ferguson, A. R., et al. (2019). Origins of neural progenitor cell-derived axons projecting caudally after spinal cord injury. *Stem Cell Reports, 13*, 105–114.

Lu, P., Graham, L., Wang, Y., Wu, D., & Tuszynski, M. (2014). Promotion of survival and differentiation of neural stem cells with fibrin and growth factor cocktails after severe spinal cord injury. *Journal of Visualized Experiments, e50641*.

Lu, P., Jones, L. L., & Tuszynski, M. H. (2005). BDNF-expressing marrow stromal cells support extensive axonal growth at sites of spinal cord injury. *Experimental Neurology, 191*, 344–360.

Lu, P., Jones, L. L., & Tuszynski, M. H. (2007). Axon regeneration through scars and into sites of chronic spinal cord injury. *Experimental Neurology, 203*, 8–21.

Lu, P., Wang, Y., Graham, L., McHale, K., Gao, M., Wu, D., et al. (2012). Long-distance growth and connectivity of neural stem cells after severe spinal cord injury. *Cell, 150*, 1264–1273.

Lu, P., Woodruff, G., Wang, Y., Graham, L., Hunt, M., Wu, D., et al. (2014). Long-distance axonal growth from human induced pluripotent stem cells after spinal cord injury. *Neuron, 83*, 789–796.

Luo, Y., Zou, Y., Yang, L., Liu, J., Liu, S., Liu, J., et al. (2013). Transplantation of NSCs with OECs alleviates neuropathic pain associated with NGF downregulation in rats following spinal cord injury. *Neuroscience Letters, 549*, 103–108.

Mackay-Sim, A., Feron, F., Cochrane, J., Bassingthwaighte, L., Bayliss, C., Davies, W., et al. (2008). Autologous olfactory ensheathing cell transplantation in human paraplegia: A 3-year clinical trial. *Brain, 131*, 2376–2386.

Maher, J. L., Anderson, K. D., Gant, K. L., & Cowan, R. E. (2021). Development and deployment of an at-home strength and conditioning program to support a phase I trial in persons with chronic spinal cord injury. *Spinal Cord, 59*, 44–54.

Manley, N. C., Priest, C. A., Denham, J., Wirth, E. D., 3rd, & Lebkowski, J. S. (2017). Human embryonic stem cell-derived oligodendrocyte progenitor cells: Preclinical efficacy and safety in cervical spinal cord injury. *Stem Cells Translational Medicine, 6*, 1917–1929.

Martinez-Rojas, B., Giraldo, E., Grillo-Risco, R., Hidalgo, M. R., Lopez-Mocholi, E., Alastrue-Agudo, A., et al. (2022). NPC transplantation rescues sci-driven cAMP/EPAC2 alterations, leading to neuroprotection and microglial modulation. *Cellular and Molecular Life Sciences, 79*, 455.

Mayer-Proschel, M., Kalyani, A. J., Mujtaba, T., & Rao, M. S. (1997). Isolation of lineage-restricted neuronal precursors from multipotent neuroepithelial stem cells. *Neuron, 19*, 773–785.

Mayeur, A., Duclos, C., Honore, A., Gauberti, M., Drouot, L., do Rego, J. C., et al. (2013). Potential of olfactory ensheathing cells from different sources for spinal cord repair. *PLoS One, 8*, e62860.

McCreedy, D. A., & Sakiyama-Elbert, S. E. (2012). Combination therapies in the CNS: Engineering the environment. *Neuroscience Letters, 519*, 115–121.

McDonald, J. W., Liu, X. Z., Qu, Y., Liu, S., Mickey, S. K., Turetsky, D., et al. (1999). Transplanted embryonic stem cells survive, differentiate and promote recovery in injured rat spinal cord. *Nature Medicine, 5*, 1410–1412.

McTigue, D. M., Horner, P. J., Stokes, B. T., & Gage, F. H. (1998). Neurotrophin-3 and brain-derived neurotrophic factor induce oligodendrocyte proliferation and myelination of regenerating axons in the contused adult rat spinal cord. *The Journal of Neuroscience: The Official Journal of the Society for Neuroscience, 18*, 5354–5365.

Medalha, C. C., Jin, Y., Yamagami, T., Haas, C., & Fischer, I. (2014). Transplanting neural progenitors into a complete transection model of spinal cord injury. *Journal of Neuroscience Research, 92*, 607–618.

Mendonca, M. V., Larocca, T. F., de Freitas Souza, B. S., Villarreal, C. F., Silva, L. F., Matos, A. C., et al. (2014). Safety and neurological assessments after autologous transplantation of bone marrow mesenchymal stem cells in subjects with chronic spinal cord injury. *Stem Cell Research & Therapy, 5*, 126.

Merianda, T. T., Jin, Y., Kalinski, A. L., Sahoo, P. K., Fischer, I., & Twiss, J. L. (2017). Neural progenitor cells promote axonal growth and alter axonal mRNA localization

in adult neurons. *eNeuro*, *4*(1). https://doi.org/10.1523/ENEURO.0171-16.2017. ENEURO.0171-16.2017. eCollection 2017 Jan-Feb.
Miah, M., Ferretti, P., & Choi, D. (2021). Considering the cellular composition of olfactory ensheathing cell transplants for spinal cord injury repair: A review of the literature. *Frontiers in Cellular Neuroscience*, *15*, 781489.
Mitsui, T., Fischer, I., Shumsky, J. S., & Murray, M. (2005). Transplants of fibroblasts expressing BDNF and NT-3 promote recovery of bladder and hindlimb function following spinal contusion injury in rats. *Experimental Neurology*, *194*, 410–431.
Mitsui, T., Neuhuber, B., & Fischer, I. (2011). Acute administration of AMPA/Kainate blocker combined with delayed transplantation of neural precursors improves lower urinary tract function in spinal injured rats. *Brain Research*, *1418*, 23–31.
Mitsui, T., Shumsky, J. S., Lepore, A. C., Murray, M., & Fischer, I. (2005). Transplantation of neuronal and glial restricted precursors into contused spinal cord improves bladder and motor functions, decreases thermal hypersensitivity, and modifies intraspinal circuitry. *The Journal of Neuroscience: The Official Journal of the Society for Neuroscience*, *25*, 9624–9636.
Monje, P. V., Deng, L., & Xu, X. M. (2021). Human Schwann cell transplantation for spinal cord injury: Prospects and challenges in translational medicine. *Frontiers in Cellular Neuroscience*, *15*, 690894.
Moon, L. D., Leasure, J. L., Gage, F. H., & Bunge, M. B. (2006). Motor enrichment sustains hindlimb movement recovered after spinal cord injury and glial transplantation. *Restorative Neurology and Neuroscience*, *24*, 147–161.
Moreno-Manzano, V., Rodriguez-Jimenez, F. J., Garcia-Rosello, M., Lainez, S., Erceg, S., Calvo, M. T., et al. (2009). Activated spinal cord ependymal stem cells rescue neurological function. *Stem Cells*, *27*, 733–743.
Morshead, C. M., Reynolds, B. A., Craig, C. G., McBurney, M. W., Staines, W. A., Morassutti, D., et al. (1994). Neural stem cells in the adult mammalian forebrain: A relatively quiescent subpopulation of subependymal cells. *Neuron*, *13*, 1071–1082.
Moviglia, G. A., Fernandez Vina, R., Brizuela, J. A., Saslavsky, J., Vrsalovic, F., Varela, G., et al. (2006). Combined protocol of cell therapy for chronic spinal cord injury. Report on the electrical and functional recovery of two patients. *Cytotherapy*, *8*, 202–209.
Mujtaba, T., Han, S. S., Fischer, I., Sandgren, E. P., & Rao, M. S. (2002). Stable expression of the alkaline phosphatase marker gene by neural cells in culture and after transplantation into the CNS using cells derived from a transgenic rat. *Experimental Neurology*, *174*, 48–57.
Mujtaba, T., Piper, D. R., Kalyani, A., Groves, A. K., Lucero, M. T., & Rao, M. S. (1999). Lineage-restricted neural precursors can be isolated from both the mouse neural tube and cultured ES cells. *Developmental Biology*, *214*, 113–127.
Murray, M., Kim, D., Liu, Y., Tobias, C., Tessler, A., & Fischer, I. (2002). Transplantation of genetically modified cells contributes to repair and recovery from spinal injury. *Brain Research. Brain Research Reviews*, *40*, 292–300.
Nagoshi, N., Okano, H., & Nakamura, M. (2020). Regenerative therapy for spinal cord injury using iPSC technology. *Inflammation and Regeneration*, *40*, 40.
Nakahara, Y., Gage, F. H., & Tuszynski, M. H. (1996). Grafts of fibroblasts genetically modified to secrete NGF, BDNF, NT-3, or basic FGF elicit differential responses in the adult spinal cord. *Cell Transplantation*, *5*, 191–204.
Nakamura, M., Okano, H., Toyama, Y., Dai, H. N., Finn, T. P., & Bregman, B. S. (2005). Transplantation of embryonic spinal cord-derived neurospheres support growth of supraspinal projections and functional recovery after spinal cord injury in the neonatal rat. *Journal of Neuroscience Research*, *81*, 457–468.
Nakhjavan-Shahraki, B., Yousefifard, M., Rahimi-Movaghar, V., Baikpour, M., Nasirinezhad, F., Safari, S., et al. (2018). Transplantation of olfactory ensheathing cells

on functional recovery and neuropathic pain after spinal cord injury; systematic review and meta-analysis. *Scientific Reports*, *8*, 325.

Nazari, B., Namjoo, Z., Moradi, F., Kazemi, M., Ebrahimi-Barough, S., Sadroddiny, E., et al. (2021). miR-219 overexpressing oligodendrocyte progenitor cells for treating compression spinal cord injury. *Metabolic Brain Disease*, *36*, 1069–1077.

Neuhuber, B., Timothy Himes, B., Shumsky, J. S., Gallo, G., & Fischer, I. (2005). Axon growth and recovery of function supported by human bone marrow stromal cells in the injured spinal cord exhibit donor variations. *Brain Research*, *1035*, 73–85.

Nistor, G. I., Totoiu, M. O., Haque, N., Carpenter, M. K., & Keirstead, H. S. (2005). Human embryonic stem cells differentiate into oligodendrocytes in high purity and myelinate after spinal cord transplantation. *Glia*, *49*, 385–396.

Nori, S., Okada, Y., Yasuda, A., Tsuji, O., Takahashi, Y., Kobayashi, Y., et al. (2011). Grafted human-induced pluripotent stem-cell-derived neurospheres promote motor functional recovery after spinal cord injury in mice. *Proceedings of the National Academy of Sciences of the United States of America*, *108*, 16825–16830.

Nornes, H., Bjorklund, A., & Stenevi, U. (1983). Reinnervation of the denervated adult spinal cord of rats by intraspinal transplants of embryonic brain stem neurons. *Cell and Tissue Research*, *230*, 15–35.

Nornes, H. O., Buchanan, J., & Bjorklund, A. (1988). Noradrenaline-containing transplants in the adult spinal cord of mammals. *Progress in Brain Research*, *78*, 181–186.

Nornes, H. O., & Das, G. D. (1974). Temporal pattern of neurogenesis in spinal cord of rat. I. An autoradiographic study—Time and sites of origin and migration and settling patterns of neuroblasts. *Brain Research*, *73*, 121–138.

Nout, Y. S., Culp, E., Schmidt, M. H., Tovar, C. A., Proschel, C., Mayer-Proschel, M., et al. (2011). Glial restricted precursor cell transplant with cyclic adenosine monophosphate improved some autonomic functions but resulted in a reduced graft size after spinal cord contusion injury in rats. *Experimental Neurology*, *227*, 159–171.

Novikova, L. N., Brohlin, M., Kingham, P. J., Novikov, L. N., & Wiberg, M. (2011). Neuroprotective and growth-promoting effects of bone marrow stromal cells after cervical spinal cord injury in adult rats. *Cytotherapy*, *13*, 873–887.

Novikova, L. N., Novikov, L. N., & Kellerth, J. O. (2003). Biopolymers and biodegradable smart implants for tissue regeneration after spinal cord injury. *Current Opinion in Neurology*, *16*, 711–715.

NSCISC. (2021). *Spinal cord injury: Facts and figures at a glance, 2021* (p. 2). National Spinal Cord Injury Statistical Center. https://www.nscisc.uab.edu/Public/Facts%20and%20Figures%20-%202021.pdf.

Nygren, L. G., Olson, L., & Seiger, A. (1977). Monoaminergic reinnervation of the transected spinal cord by homologous fetal brain grafts. *Brain Research*, *129*, 227–235.

Obernier, K., & Alvarez-Buylla, A. (2019). Neural stem cells: Origin, heterogeneity and regulation in the adult mammalian brain. *Development*, *146*(4), dev156059. https://doi.org/10.1242/dev.15605.

Ogawa, Y., Sawamoto, K., Miyata, T., Miyao, S., Watanabe, M., Nakamura, M., et al. (2002). Transplantation of in vitro-expanded fetal neural progenitor cells results in neurogenesis and functional recovery after spinal cord contusion injury in adult rats. *Journal of Neuroscience Research*, *69*, 925–933.

Ohta, M., Suzuki, Y., Noda, T., Ejiri, Y., Dezawa, M., Kataoka, K., et al. (2004). Bone marrow stromal cells infused into the cerebrospinal fluid promote functional recovery of the injured rat spinal cord with reduced cavity formation. *Experimental Neurology*, *187*, 266–278.

Okubo, T., Iwanami, A., Kohyama, J., Itakura, G., Kawabata, S., Nishiyama, Y., et al. (2016). Pretreatment with a gamma-secretase inhibitor prevents tumor-like overgrowth in human iPSC-derived transplants for spinal cord injury. *Stem Cell Reports*, *7*, 649–663.

Oraee-Yazdani, S., Akhlaghpasand, M., Golmohammadi, M., Hafizi, M., Zomorrod, M. S., Kabir, N. M., et al. (2021). Combining cell therapy with human autologous Schwann cell and bone marrow-derived mesenchymal stem cell in patients with subacute complete spinal cord injury: Safety considerations and possible outcomes. *Stem Cell Research & Therapy*, *12*, 445.

Oraee-Yazdani, S., Hafizi, M., Atashi, A., Ashrafi, F., Seddighi, A. S., Hashemi, S. M., et al. (2016). Co-transplantation of autologous bone marrow mesenchymal stem cells and Schwann cells through cerebral spinal fluid for the treatment of patients with chronic spinal cord injury: Safety and possible outcome. *Spinal Cord*, *54*, 102–109.

Oudega, M., Gautier, S. E., Chapon, P., Fragoso, M., Bates, M. L., Parel, J. M., et al. (2001). Axonal regeneration into Schwann cell grafts within resorbable poly(alpha-hydroxyacid) guidance channels in the adult rat spinal cord. *Biomaterials*, *22*, 1125–1136.

Oudega, M., Varon, S., & Hagg, T. (1994). Regeneration of adult rat sensory axons into intraspinal nerve grafts: promoting effects of conditioning lesion and graft predegeneration. *Experimental Neurology*, *129*, 194–206.

Pakulska, M. M., Ballios, B. G., & Shoichet, M. S. (2012). Injectable hydrogels for central nervous system therapy. *Biomedical Materials*, *7*, 024101.

Pal, R., Venkataramana, N. K., Bansal, A., Balaraju, S., Jan, M., Chandra, R., et al. (2009). Ex vivo-expanded autologous bone marrow-derived mesenchymal stromal cells in human spinal cord injury/paraplegia: A pilot clinical study. *Cytotherapy*, *11*, 897–911.

Panni, P., Ferguson, I. A., Beacham, I., Mackay-Sim, A., Ekberg, J. A., & St John, J. A. (2013). Phagocytosis of bacteria by olfactory ensheathing cells and Schwann cells. *Neuroscience Letters*, *539*, 65–70.

Papazian, I., Kyrargyri, V., Evangelidou, M., Voulgari-Kokota, A., & Probert, L. (2018). Mesenchymal stem cell protection of neurons against glutamate excitotoxicity involves reduction of NMDA-triggered calcium responses and surface GluR1, and is partly mediated by TNF. *International Journal of Molecular Sciences*, *19*.

Park, J. H., Kim, D. Y., Sung, I. Y., Choi, G. H., Jeon, M. H., Kim, K. K., et al. (2012). Long-term results of spinal cord injury therapy using mesenchymal stem cells derived from bone marrow in humans. *Neurosurgery*, *70*, 1238–1247 (discussion 1247).

Parsons, X. H., Teng, Y. D., Parsons, J. F., Snyder, E. Y., Smotrich, D. B., & Moore, D. A. (2011). Efficient derivation of human neuronal progenitors and neurons from pluripotent human embryonic stem cells with small molecule induction. *Journal of Visualized Experiments*, *e3273*.

Patil, N., Walsh, P., Carrabre, K., Holmberg, E. G., Lavoie, N., Dutton, J. R., et al. (2021). Regionally specific human pre-oligodendrocyte progenitor cells produce both oligodendrocytes and neurons after transplantation in a chronically injured spinal cord rat model after glial scar ablation. *Journal of Neurotrauma*, *38*, 777–788.

Pearse, D. D., Pereira, F. C., Marcillo, A. E., Bates, M. L., Berrocal, Y. A., Filbin, M. T., et al. (2004). cAMP and Schwann cells promote axonal growth and functional recovery after spinal cord injury. *Nature Medicine*.

Pearse, D. D., Sanchez, A. R., Pereira, F. C., Andrade, C. M., Puzis, R., Pressman, Y., et al. (2007). Transplantation of Schwann cells and/or olfactory ensheathing glia into the contused spinal cord: Survival, migration, axon association, and functional recovery. *Glia*, *55*, 976–1000.

Peljto, M., & Wichterle, H. (2011). Programming embryonic stem cells to neuronal subtypes. *Current Opinion in Neurobiology*, *21*, 43–51.

Pencalet, P., Serguera, C., Corti, O., Privat, A., Mallet, J., & Gimenez y Ribotta, M. (2006). Integration of genetically modified adult astrocytes into the lesioned rat spinal cord. *Journal of Neuroscience Research*, *83*, 61–67.

Peterson, C. A., Murphy, R. J., Dupont-Versteegden, E. E., & Houle, J. D. (2000). Cycling exercise and fetal spinal cord transplantation act synergistically on atrophied muscle following chronic spinal cord injury in rats. *Neurorehabilitation and Neural Repair*, *14*, 85–91.

Planat-Benard, V., Varin, A., & Casteilla, L. (2021). MSCs and inflammatory cells crosstalk in regenerative medicine: Concerted actions for optimized resolution driven by energy metabolism. *Frontiers in Immunology*, *12*, 626755.

Plant, G. W., Christensen, C. L., Oudega, M., & Bunge, M. B. (2003). Delayed transplantation of olfactory ensheathing glia promotes sparing/regeneration of supraspinal axons in the contused adult rat spinal cord. *Journal of Neurotrauma*, *20*, 1–16.

Polentes, J., Stamegna, J. C., Nieto-Sampedro, M., & Gauthier, P. (2004). Phrenic rehabilitation and diaphragm recovery after cervical injury and transplantation of olfactory ensheathing cells. *Neurobiology of Disease*, *16*, 638–653.

Poplawski, G. H. D., Kawaguchi, R., Van Niekerk, E., Lu, P., Mehta, N., Canete, P., et al. (2020). Injured adult neurons regress to an embryonic transcriptional growth state. *Nature*, *581*, 77–82.

Poplawski, G. H. D., Lie, R., Hunt, M., Kumamaru, H., Kawaguchi, R., Lu, P., et al. (2018). Adult rat myelin enhances axonal outgrowth from neural stem cells. *Science Translational Medicine*, *10*(442), eaal2563. https://doi.org/10.1126/scitranslmed.aal2563.

Pous, L., Deshpande, S. S., Nath, S., Mezey, S., Malik, S. C., Schildge, S., et al. (2020). Fibrinogen induces neural stem cell differentiation into astrocytes in the subventricular zone via BMP signaling. *Nature Communications*, *11*, 630.

Priest, C. A., Manley, N. C., Denham, J., Wirth, E. D., 3rd, & Lebkowski, J. S. (2015). Preclinical safety of human embryonic stem cell-derived oligodendrocyte progenitors supporting clinical trials in spinal cord injury. *Regenerative Medicine*, *10*, 939–958.

Privat, A., Mansour, H., Pavy, A., Geffard, M., & Sandillon, F. (1986). Transplantation of dissociated foetal serotonin neurons into the transected spinal cord of adult rats. *Neuroscience Letters*, *66*, 61–66.

Ra, J. C., Shin, I. S., Kim, S. H., Kang, S. K., Kang, B. C., Lee, H. Y., et al. (2011). Safety of intravenous infusion of human adipose tissue-derived mesenchymal stem cells in animals and humans. *Stem Cells and Development*, *20*, 1297–1308.

Rabchevsky, A. G., & Streit, W. J. (1997). Grafting of cultured microglial cells into the lesioned spinal cord of adult rats enhances neurite outgrowth. *Journal of Neuroscience Research*, *47*, 34–48.

Radtke, C., Akiyama, Y., Brokaw, J., Lankford, K. L., Wewetzer, K., Fodor, W. L., et al. (2004). Remyelination of the nonhuman primate spinal cord by transplantation of H-transferase transgenic adult pig olfactory ensheathing cells. *The FASEB Journal*, *18*, 335–337.

Ramer, L. M., Au, E., Richter, M. W., Liu, J., Tetzlaff, W., & Roskams, A. J. (2004). Peripheral olfactory ensheathing cells reduce scar and cavity formation and promote regeneration after spinal cord injury. *The Journal of Comparative Neurology*, *473*, 1–15.

Ramon y Cajal, S. (1991). *Degeneration and regeneration of the nervous system* (3rd (1991) ed.). New York: Oxford University Press.

Ramon-Cueto, A., Cordero, M. I., Santos-Benito, F. F., & Avila, J. (2000). Functional recovery of paraplegic rats and motor axon regeneration in their spinal cords by olfactory ensheathing glia. *Neuron*, *25*, 425–435.

Ramon-Cueto, A., Plant, G. W., Avila, J., & Bunge, M. B. (1998). Long-distance axonal regeneration in the transected adult rat spinal cord is promoted by olfactory ensheathing glia transplants. *The Journal of Neuroscience: The Official Journal of the Society for Neuroscience*, *18*, 3803–3815.

Rao, M. S. (1999). Multipotent and restricted precursors in the central nervous system. *The Anatomical Record*, *257*, 137–148.

Rao, M. S., & Mayer-Proschel, M. (1997). Glial-restricted precursors are derived from multipotent neuroepithelial stem cells. *Developmental Biology*, *188*, 48–63.

Rao, M. S., Noble, M., & Mayer-Proschel, M. (1998). A tripotential glial precursor cell is present in the developing spinal cord. *Proceedings of the National Academy of Sciences of the United States of America*, *95*, 3996–4001.

Rauch, M. F., Hynes, S. R., Bertram, J., Redmond, A., Robinson, R., Williams, C., et al. (2009). Engineering angiogenesis following spinal cord injury: A coculture of neural progenitor and endothelial cells in a degradable polymer implant leads to an increase in vessel density and formation of the blood-spinal cord barrier. *The European Journal of Neuroscience, 29*, 132–145.
Reier, P. J. (1985). Neural tissue grafts and repair of the injured spinal cord. *Neuropathology and Applied Neurobiology, 11*, 81–104.
Reier, P. J. (2004). Cellular transplantation strategies for spinal cord injury and translational neurobiology. *NeuroRx, 1*, 424–451.
Reier, P. J., Anderson, D. K., Thompson, F. J., & Stokes, B. T. (1992). Neural tissue transplantation and CNS trauma: Anatomical and functional repair of the injured spinal cord. *Journal of Neurotrauma, 9*(Suppl. 1), S223–S248.
Reier, P. J., Bregman, B. S., & Wujek, J. R. (1985). Intraspinal transplants of embryonic spinal cord tissue in adult and neonatal rats: Evidence for topographical differentiation and axonal interactions with the host CNS. *Neural Grafting in the Mammalian CNS*, 257–263. Elsevier Science Publications.
Reier, P. J., Bregman, B. S., & Wujek, J. R. (1986). Intraspinal transplantation of embryonic spinal cord tissue in neonatal and adult rats. *The Journal of Comparative Neurology, 247*, 275–296.
Reier, P. J., Houle, J. D., Jakeman, L., Winialski, D., & Tessler, A. (1988). Transplantation of fetal spinal cord tissue into acute and chronic hemisection and contusion lesions of the adult rat spinal cord. *Progress in Brain Research, 78*, 173–179.
Reier, P. J., Howland, D. R., Mitchell, G., Wolpaw, J. R., Hoh, D. J., & Lane, M. A. (2017). *Spinal cord injury: Repair, plasticity and rehabilitation, encyclopedia of life science* (March 2017th ed.). Chichester: John Wiley & Sons, Ltd.
Reier, P. J., Perlow, M. J., & Guth, L. (1983). Development of embryonic spinal cord transplants in the rat. *Brain Research, 312*, 201–219.
Reier, P. J., Stokes, B. T., Thompson, F. J., & Anderson, D. K. (1992). Fetal cell grafts into resection and contusion/compression injuries of the rat and cat spinal cord. *Experimental Neurology, 115*, 177–188.
Ren, H., Chen, X., Tian, M., Zhou, J., Ouyang, H., & Zhang, Z. (2018). Regulation of inflammatory cytokines for spinal cord injury repair through local delivery of therapeutic agents. *Advance Science (Weinheim), 5*, 1800529.
Reynolds, L. F., Bren, M. C., Wilson, B. C., Gibson, G. D., Shoichet, M. S., & Murphy, R. J. (2008). Transplantation of porous tubes following spinal cord transection improves hindlimb function in the rat. *Spinal Cord, 46*, 58–64.
Reynolds, B. A., & Weiss, S. (1992). Generation of neurons and astrocytes from isolated cells of the adult mammalian central nervous system. *Science, 255*, 1707–1710.
Richardson, P. M., McGuinness, U. M., & Aguayo, A. J. (1980). Axons from CNS neurons regenerate into PNS grafts. *Nature, 284*, 264–265.
Richter, M. W., Fletcher, P. A., Liu, J., Tetzlaff, W., & Roskams, A. J. (2005). Lamina propria and olfactory bulb ensheathing cells exhibit differential integration and migration and promote differential axon sprouting in the lesioned spinal cord. *The Journal of Neuroscience: The Official Journal of the Society for Neuroscience, 25*, 10700–10711.
Riemann, L., Younsi, A., Scherer, M., Zheng, G., Skutella, T., Unterberg, A. W., et al. (2018). Transplantation of neural precursor cells attenuates chronic immune environment in cervical spinal cord injury. *Frontiers in Neurology, 9*, 428.
Riley, J., Federici, T., Park, J., Suzuki, M., Franz, C. K., Tork, C., et al. (2009). Cervical spinal cord therapeutics delivery: Preclinical safety validation of a stabilized microinjection platform. *Neurosurgery, 65*, 754–761 (discussion 761–752).
Robertson, M. J., Gip, P., & Schaffer, D. V. (2008). Neural stem cell engineering: Directed differentiation of adult and embryonic stem cells into neurons. *Frontiers in Bioscience, 13*, 21–50.

Robinson, J., & Lu, P. (2017). Optimization of trophic support for neural stem cell grafts in sites of spinal cord injury. *Experimental Neurology, 291*, 87–97.

Rodriguez-Jimenez, F. J., Vilches, A., Perez-Arago, M. A., Clemente, E., Roman, R., Leal, J., et al. (2021). Activation of neurogenesis in multipotent stem cells cultured in vitro and in the spinal cord tissue after severe injury by inhibition of glycogen synthase kinase-3. *Neurotherapeutics, 18*, 515–533.

Roet, K. C., & Verhaagen, J. (2014). Understanding the neural repair-promoting properties of olfactory ensheathing cells. *Experimental Neurology, 261*, 594–609.

Romanyuk, N., Amemori, T., Turnovcova, K., Prochazka, P., Onteniente, B., Sykova, E., et al. (2015). Beneficial effect of human induced pluripotent stem cell-derived neural precursors in spinal cord injury repair. *Cell Transplantation, 24*, 1781–1797.

Rosenzweig, E. S., Brock, J. H., Lu, P., Kumamaru, H., Salegio, E. A., Kadoya, K., et al. (2018). Restorative effects of human neural stem cell grafts on the primate spinal cord. *Nature Medicine, 24*, 484–490.

Ruitenberg, M. J., Levison, D. B., Lee, S. V., Verhaagen, J., Harvey, A. R., & Plant, G. W. (2005). NT-3 expression from engineered olfactory ensheathing glia promotes spinal sparing and regeneration. *Brain, 128*, 839–853.

Ruitenberg, M. J., Plant, G. W., Christensen, C. L., Blits, B., Niclou, S. P., Harvey, A. R., et al. (2002). Viral vector-mediated gene expression in olfactory ensheathing glia implants in the lesioned rat spinal cord. *Gene Therapy, 9*, 135–146.

Ruitenberg, M. J., Plant, G. W., Hamers, F. P., Wortel, J., Blits, B., Dijkhuizen, P. A., et al. (2003). Ex vivo adenoviral vector-mediated neurotrophin gene transfer to olfactory ensheathing glia: Effects on rubrospinal tract regeneration, lesion size, and functional recovery after implantation in the injured rat spinal cord. *The Journal of Neuroscience: The Official Journal of the Society for Neuroscience, 23*, 7045–7058.

Ruzicka, J., Machova-Urdzikova, L., Gillick, J., Amemori, T., Romanyuk, N., Karova, K., et al. (2017). A comparative study of three different types of stem cells for treatment of rat spinal cord injury. *Cell Transplantation, 26*, 585–603.

Ruzicka, J., Romanyuk, N., Hejcl, A., Vetrik, M., Hruby, M., Cocks, G., et al. (2013). Treating spinal cord injury in rats with a combination of human fetal neural stem cells and hydrogels modified with serotonin. *Acta Neurobiologiae Experimentalis (Wars), 73*, 102–115.

Ryu, H. H., Kang, B. J., Park, S. S., Kim, Y., Sung, G. J., Woo, H. M., et al. (2012). Comparison of mesenchymal stem cells derived from fat, bone marrow, Wharton's jelly, and umbilical cord blood for treating spinal cord injuries in dogs. *The Journal of Veterinary Medical Science, 74*, 1617–1630.

Saito, F., Nakatani, T., Iwase, M., Maeda, Y., Murao, Y., Suzuki, Y., et al. (2012). Administration of cultured autologous bone marrow stromal cells into cerebrospinal fluid in spinal injury patients: A pilot study. *Restorative Neurology and Neuroscience, 30*, 127–136.

Sakiyama-Elbert, S., Johnson, P. J., Hodgetts, S. I., Plant, G. W., & Harvey, A. R. (2012). Scaffolds to promote spinal cord regeneration. *Handbook of Clinical Neurology, 109*, 575–594.

Salazar, D. L., Uchida, N., Hamers, F. P., Cummings, B. J., & Anderson, A. J. (2010). Human neural stem cells differentiate and promote locomotor recovery in an early chronic spinal cord injury NOD-scid mouse model. *PLoS One, 5*, e12272.

Salewski, R. P., Mitchell, R. A., Li, L., Shen, C., Milekovskaia, M., Nagy, A., et al. (2015). Transplantation of induced pluripotent stem cell-derived neural stem cells mediate functional recovery following thoracic spinal cord injury through remyelination of axons. *Stem Cells Translational Medicine, 4*, 743–754.

Sandrock, R. W., Wheatley, W., Levinthal, C., Lawson, J., Hashimoto, B., Rao, M., et al. (2010). Isolation, characterization and preclinical development of human glial-restricted progenitor cells for treatment of neurological disorders. *Regenerative Medicine, 5*, 381–394.

Sarugaser, R., Hanoun, L., Keating, A., Stanford, W. L., & Davies, J. E. (2009). Human mesenchymal stem cells self-renew and differentiate according to a deterministic hierarchy. *PLoS One*, *4*, e6498.

Sasaki, M., Lankford, K. L., Zemedkun, M., & Kocsis, J. D. (2004). Identified olfactory ensheathing cells transplanted into the transected dorsal funiculus bridge the lesion and form myelin. *The Journal of Neuroscience*, *24*, 8485–8493.

Satti, H. S., Waheed, A., Ahmed, P., Ahmed, K., Akram, Z., Aziz, T., et al. (2016). Autologous mesenchymal stromal cell transplantation for spinal cord injury: A Phase I pilot study. *Cytotherapy*, *18*, 518–522.

Saunders, N. R., Fry, E. J., Lane, M., & Dziegielewska, K. M. (2002). Spinal cord injuries: Making the re-connection. *Biologist (London, England)*, *49*, 107–112.

Schuldiner, M., Yanuka, O., Itskovitz-Eldor, J., Melton, D. A., & Benvenisty, N. (2000). Effects of eight growth factors on the differentiation of cells derived from human embryonic stem cells. *Proceedings of the National Academy of Sciences of the United States of America*, *97*, 11307–11312.

Sharp, J., Frame, J., Siegenthaler, M., Nistor, G., & Keirstead, H. S. (2010). Human embryonic stem cell-derived oligodendrocyte progenitor cell transplants improve recovery after cervical spinal cord injury. *Stem Cells*, *28*, 152–163.

Shibata, R., Nagoshi, N., Kajikawa, K., Ito, S., Shibata, S., Shindo, T., et al. (2022). Administration of C5a receptor antagonist improves the efficacy of human induced pluripotent stem cell-derived neural stem/progenitor cell transplantation in the acute phase of spinal cord injury. *Journal of Neurotrauma*, *39*, 667–682.

Shih, C. H., Lacagnina, M., Leuer-Bisciotti, K., & Proschel, C. (2014). Astroglial-derived periostin promotes axonal regeneration after spinal cord injury. *The Journal of Neuroscience*, *34*, 2438–2443.

Shin, J. E., Jung, K., Kim, M., Hwang, K., Lee, H., Kim, I. S., et al. (2018). Brain and spinal cord injury repair by implantation of human neural progenitor cells seeded onto polymer scaffolds. *Experimental & Molecular Medicine*, *50*, 39.

Shin, J. C., Kim, K. N., Yoo, J., Kim, I. S., Yun, S., Lee, H., et al. (2015). Clinical trial of human fetal brain-derived neural stem/progenitor cell transplantation in patients with traumatic cervical spinal cord injury. *Neural Plasticity*, *2015*, 630932.

Shoichet, M. S., Tate, C. C., Baumann, M. D., & LaPlaca, M. C. (2008). Strategies for regeneration and repair in the injured central nervous system. In W. M. Reichert (Ed.), *Frontiers in neuroengineering. Indwelling neural implants: Strategies for contending with the in vivo environment*. Boca Raton (FL): CRC Press.

Shumsky, J. S., Tobias, C. A., Tumolo, M., Long, W. D., Giszter, S. F., & Murray, M. (2003). Delayed transplantation of fibroblasts genetically modified to secrete BDNF and NT-3 into a spinal cord injury site is associated with limited recovery of function. *Experimental Neurology*, *184*, 114–130.

Silva, N. A., Gimble, J. M., Sousa, N., Reis, R. L., & Salgado, A. J. (2013). Combining adult stem cells and olfactory ensheathing cells: the secretome effect. *Stem Cells and Development*, *22*, 1232–1240.

Slotkin, J. R., Pritchard, C. D., Luque, B., Ye, J., Layer, R. T., Lawrence, M. S., et al. (2017). Biodegradable scaffolds promote tissue remodeling and functional improvement in non-human primates with acute spinal cord injury. *Biomaterials*, *123*, 63–76.

Soubannier, V., Maussion, G., Chaineau, M., Sigutova, V., Rouleau, G., Durcan, T. M., et al. (2020). Characterization of human iPSC-derived astrocytes with potential for disease modeling and drug discovery. *Neuroscience Letters*, *731*, 135028.

Speicher, A. M., Wiendl, H., Meuth, S. G., & Pawlowski, M. (2019). Generating microglia from human pluripotent stem cells: Novel in vitro models for the study of neurodegeneration. *Molecular Neurodegeneration*, *14*, 46.

Spruance, V. M., Zholudeva, L. V., Hormigo, K. M., Randelman, M. L., Bezdudnaya, T., Marchenko, V., et al. (2018). Integration of transplanted neural precursors with the injured cervical spinal cord. *Journal of Neurotrauma*, *35*, 1781–1799.

Stamegna, J. C., Felix, M. S., Roux-Peyronnet, J., Rossi, V., Feron, F., Gauthier, P., et al. (2011). Nasal OEC transplantation promotes respiratory recovery in a subchronic rat model of cervical spinal cord contusion. *Experimental Neurology*, *229*, 120–131.

Steeves, J., & Blight, A. (2012). Spinal cord injury clinical trials translational process, review of past and proposed acute trials with reference to recommended trial guidelines. *Handbook of Clinical Neurology*, *109*, 386–398.

Steward, O., Sharp, K., Selvan, G., Hadden, A., Hofstadter, M., Au, E., et al. (2006). A re-assessment of the consequences of delayed transplantation of olfactory lamina propria following complete spinal cord transection in rats. *Experimental Neurology*.

Stokes, B. T., & Reier, P. J. (1991). Oxygen transport in intraspinal fetal grafts: Graft-host relations. *Experimental Neurology*, *111*, 312–323.

Stokes, B. T., & Reier, P. J. (1992). Fetal grafts alter chronic behavioral outcome after contusion damage to the adult rat spinal cord. *Experimental Neurology*, *116*, 1–12.

Stokols, S., & Tuszynski, M. H. (2004). The fabrication and characterization of linearly oriented nerve guidance scaffolds for spinal cord injury. *Biomaterials*, *25*, 5839–5846.

Stokols, S., & Tuszynski, M. H. (2006). Freeze-dried agarose scaffolds with uniaxial channels stimulate and guide linear axonal growth following spinal cord injury. *Biomaterials*, *27*, 443–451.

Strauss, S. (2010). Geron trial resumes, but standards for stem cell trials remain elusive. *Nature Biotechnology*, *28*, 989–990.

Strnadel, J., Carromeu, C., Bardy, C., Navarro, M., Platoshyn, O., Glud, A. N., et al. (2018). Survival of syngeneic and allogeneic iPSC-derived neural precursors after spinal grafting in minipigs. *Science Translational Medicine*, *10*(440), eaam6651. https://doi.org/10.1126/scitranslmed.aam6651.

Sugai, K., Fukuzawa, R., Shofuda, T., Fukusumi, H., Kawabata, S., Nishiyama, Y., et al. (2016). Pathological classification of human iPSC-derived neural stem/progenitor cells towards safety assessment of transplantation therapy for CNS diseases. *Molecular Brain*, *9*, 85.

Sugai, K., Sumida, M., Shofuda, T., Yamaguchi, R., Tamura, T., Kohzuki, T., et al. (2021). First-in-human clinical trial of transplantation of iPSC-derived NS/PCs in subacute complete spinal cord injury: Study protocol. *Regenerative Therapy*, *18*, 321–333.

Sugar, O., & Gerard, R. W. (1940). Spinal cord regeneration in the rat. *Journal of Neurophysiology*, *3*, 1–19.

Sullivan, P. Z., AlBayar, A., Burrell, J. C., Browne, K. D., Arena, J., Johnson, V., et al. (2021). Implantation of engineered axon tracts to bridge spinal cord injury beyond the glial scar in rats. *Tissue Engineering. Part A*, *27*, 1264–1274.

Sun, T., Ye, C., Zhang, Z., Wu, J., & Huang, H. (2013). Cotransplantation of olfactory ensheathing cells and Schwann cells combined with treadmill training promotes functional recovery in rats with contused spinal cords. *Cell Transplantation*, *22*(Suppl. 1), S27–S38.

Suzuki, H., Ahuja, C. S., Salewski, R. P., Li, L., Satkunendrarajah, K., Nagoshi, N., et al. (2017). Neural stem cell mediated recovery is enhanced by chondroitinase ABC pretreatment in chronic cervical spinal cord injury. *PLoS One*, *12*, e0182339.

Sykova, E., Jendelova, P., Urdzikova, L., Lesny, P., & Hejcl, A. (2006). Bone marrow stem cells and polymer hydrogels- -two strategies for spinal cord injury repair. *Cellular and Molecular Neurobiology*, *26*, 1113–1129.

Tabakow, P., Jarmundowicz, W., Czapiga, B., Fortuna, W., Miedzybrodzki, R., Czyz, M., et al. (2013). Transplantation of autologous olfactory ensheathing cells in complete human spinal cord injury. *Cell Transplantation*, *22*, 1591–1612.

Takeoka, A., Jindrich, D. L., Munoz-Quiles, C., Zhong, H., van den Brand, R., Pham, D. L., et al. (2011). Axon regeneration can facilitate or suppress hindlimb function after olfactory ensheathing glia transplantation. *The Journal of Neuroscience: The Official Journal of the Society for Neuroscience*, *31*, 4298–4310.

Takeoka, A., Kubasak, M. D., Zhong, H., Kaplan, J., Roy, R. R., & Phelps, P. E. (2010). Noradrenergic innervation of the rat spinal cord caudal to a complete spinal cord transection: Effects of olfactory ensheathing glia. *Experimental Neurology*, *222*, 59–69.

Takeoka, A., Kubasak, M. D., Zhong, H., Roy, R. R., & Phelps, P. E. (2009). Serotonergic innervation of the caudal spinal stump in rats after complete spinal transection: effect of olfactory ensheathing glia. *The Journal of Comparative Neurology*, *515*, 664–676.

Tam, R. Y., Fuehrmann, T., Mitrousis, N., & Shoichet, M. S. (2014). Regenerative therapies for central nervous system diseases: A biomaterials approach. *Neuropsychopharmacology*, *39*, 169–188.

Teng, Y. D., Lavik, E. B., Qu, X., Park, K. I., Ourednik, J., Zurakowski, D., et al. (2002). Functional recovery following traumatic spinal cord injury mediated by a unique polymer scaffold seeded with neural stem cells. *Proc Natl Acad Sci USA*, *99*, 3024–3029.

Tessler, A., Himes, B. T., Houle, J., & Reier, P. J. (1988). Regeneration of adult dorsal root axons into transplants of embryonic spinal cord. *The Journal of Comparative Neurology*, *270*, 537–548.

Tessler, A., Himes, B. T., Rogahn, C., Houle, J., & Reier, P. J. (1988). Enhancement of adult dorsal root regeneration by embryonic spinal cord transplants. *Progress in Brain Research*, *78*, 213–218.

Theele, D. P., Schrimsher, G. W., & Reier, P. J. (1996). Comparison of the growth and fate of fetal spinal iso- and allografts in the adult rat injured spinal cord. *Experimental Neurology*, *142*, 128–143.

Thompson, R. E., Pardieck, J., Smith, L., Kenny, P., Crawford, L., Shoichet, M., et al. (2018). Effect of hyaluronic acid hydrogels containing astrocyte-derived extracellular matrix and/or V2a interneurons on histologic outcomes following spinal cord injury. *Biomaterials*, *162*, 208–223.

Thompson, F. J., Reier, P. J., Parmer, R., & Lucas, C. C. (1993). Inhibitory control of reflex excitability following contusion injury and neural tissue transplantation. *Advances in Neurology*, *59*, 175–184.

Thompson, F. J., Reier, P. J., Uthman, B., Mott, S., Fessler, R. G., Behrman, A., et al. (2001). Neurophysiological assessment of the feasibility and safety of neural tissue transplantation in patients with syringomyelia. *Journal of Neurotrauma*, *18*, 931–945.

Thompson, R., & Sakiyama-Elbert, S. (2018). Using biomaterials to promote pro-regenerative glial phenotypes after nervous system injuries. *Biomedical Materials*, *13*, 024104.

Thomson, J. A., Itskovitz-Eldor, J., Shapiro, S. S., Waknitz, M. A., Swiergiel, J. J., Marshall, V. S., et al. (1998). Embryonic stem cell lines derived from human blastocysts. *Science*, *282*, 1145–1147.

Tobias, C. A., Dhoot, N. O., Wheatley, M. A., Tessler, A., Murray, M., & Fischer, I. (2001). Grafting of encapsulated BDNF-producing fibroblasts into the injured spinal cord without immune suppression in adult rats. *Journal of Neurotrauma*, *18*, 287–301.

Tobias, C. A., Han, S. S., Shumsky, J. S., Kim, D., Tumolo, M., Dhoot, N. O., et al. (2005). Alginate encapsulated BDNF-producing fibroblast grafts permit recovery of function after spinal cord injury in the absence of immune suppression. *Journal of Neurotrauma*, *22*, 138–156.

Tobias, C. A., Shumsky, J. S., Shibata, M., Tuszynski, M. H., Fischer, I., Tessler, A., et al. (2003). Delayed grafting of BDNF and NT-3 producing fibroblasts into the injured spinal cord stimulates sprouting, partially rescues axotomized red nucleus neurons from loss and atrophy, and provides limited regeneration. *Experimental Neurology*, *184*, 97–113.

Tom, V. J., & Houle, J. D. (2008). Intraspinal microinjection of chondroitinase ABC following injury promotes axonal regeneration out of a peripheral nerve graft bridge. *Experimental Neurology, 211*, 315–319.

Tom, V. J., Kadakia, R., Santi, L., & Houle, J. D. (2009). Administration of chondroitinase ABC rostral or caudal to a spinal cord injury site promotes anatomical but not functional plasticity. *Journal of Neurotrauma, 26*, 2323–2333.

Tom, V. J., Sandrow-Feinberg, H. R., Miller, K., Santi, L., Connors, T., Lemay, M. A., et al. (2009). Combining peripheral nerve grafts and chondroitinase promotes functional axonal regeneration in the chronically injured spinal cord. *The Journal of Neuroscience: The Official Journal of the Society for Neuroscience, 29*, 14881–14890.

Torres-Espin, A., Corona-Quintanilla, D. L., Fores, J., Allodi, I., Gonzalez, F., Udina, E., et al. (2013). Neuroprotection and axonal regeneration after lumbar ventral root avulsion by re-implantation and mesenchymal stem cells transplant combined therapy. *Neurotherapeutics, 10*, 354–368.

Totoiu, M. O., Nistor, G. I., Lane, T. E., & Keirstead, H. S. (2004). Remyelination, axonal sparing, and locomotor recovery following transplantation of glial-committed progenitor cells into the MHV model of multiple sclerosis. *Experimental Neurology, 187*, 254–265.

Tsuji, O., Miura, K., Okada, Y., Fujiyoshi, K., Mukaino, M., Nagoshi, N., et al. (2010). Therapeutic potential of appropriately evaluated safe-induced pluripotent stem cells for spinal cord injury. *Proceedings of the National Academy of Sciences of the United States of America, 107*, 12704–12709.

Tsuji, O., Sugai, K., Yamaguchi, R., Tashiro, S., Nagoshi, N., Kohyama, J., et al. (2019). Concise review: Laying the groundwork for a first-in-human study of an induced pluripotent stem cell-based intervention for spinal cord injury. *Stem Cells, 37*, 6–13.

Tuszynski, M. H., Peterson, D. A., Ray, J., Baird, A., Nakahara, Y., & Gage, F. H. (1994). Fibroblasts genetically modified to produce nerve growth factor induce robust neuritic ingrowth after grafting to the spinal cord. *Experimental Neurology, 126*, 1–14.

Tuszynski, M. H., Weidner, N., McCormack, M., Miller, I., Powell, H., & Conner, J. (1998). Grafts of genetically modified Schwann cells to the spinal cord: survival, axon growth, and myelination. *Cell Transplantation*, 7, 187–196.

Uchida, N., Buck, D. W., He, D., Reitsma, M. J., Masek, M., Phan, T. V., et al. (2000). Direct isolation of human central nervous system stem cells. *Proceedings of the National Academy of Sciences of the United States of America, 97*, 14720–14725.

Ullah, I., Subbarao, R. B., & Rho, G. J. (2015). Human mesenchymal stem cells—Current trends and future prospective. *Bioscience Reports*, 35.

Urdzikova, L. M., Ruzicka, J., LaBagnara, M., Karova, K., Kubinova, S., Jirakova, K., et al. (2014). Human mesenchymal stem cells modulate inflammatory cytokines after spinal cord injury in rat. *International Journal of Molecular Sciences, 15*, 11275–11293.

Vanecek, V., Zablotskii, V., Forostyak, S., Ruzicka, J., Herynek, V., Babic, M., et al. (2012). Highly efficient magnetic targeting of mesenchymal stem cells in spinal cord injury. *International Journal of Nanomedicine*, 7, 3719–3730.

Vawda, R., & Fehlings, M. G. (2013). Mesenchymal cells in the treatment of spinal cord injury: Current & future perspectives. *Current Stem Cell Research & Therapy, 8*, 25–38.

Vieira, M. S., Santos, A. K., Vasconcellos, R., Goulart, V. A. M., Parreira, R. C., Kihara, A. H., et al. (2018). Neural stem cell differentiation into mature neurons: Mechanisms of regulation and biotechnological applications. *Biotechnology Advances, 36*, 1946–1970.

Vodyanik, M. A., Yu, J., Zhang, X., Tian, S., Stewart, R., Thomson, J. A., et al. (2010). A mesoderm-derived precursor for mesenchymal stem and endothelial cells. *Cell Stem Cell*, 7, 718–729.

Wang, G., Ao, Q., Gong, K., Zuo, H., Gong, Y., & Zhang, X. (2010). Synergistic effect of neural stem cells and olfactory ensheathing cells on repair of adult rat spinal cord injury. *Cell Transplantation, 19*, 1325–1337.

Wang, J. J., Chuah, M. I., Yew, D. T., Leung, P. C., & Tsang, D. S. (1995). Effects of astrocyte implantation into the hemisected adult rat spinal cord. *Neuroscience, 65*, 973–981.

Wang, J., Jiang, P., Deng, W., Sun, Y., & Liu, Y. (2022). Grafted human ESC-derived astroglia repair spinal cord injury via activation of host anti-inflammatory microglia in the lesion area. *Theranostics, 12*, 4288–4309.

Wang, C., Luan, Z., Yang, Y., Wang, Z., Wang, Q., Lu, Y., et al. (2015). High purity of human oligodendrocyte progenitor cells obtained from neural stem cells: Suitable for clinical application. *Journal of Neuroscience Methods, 240*, 61–66.

Wang, C., Sun, C., Hu, Z., Huo, X., Yang, Y., Liu, X., et al. (2017). Improved neural regeneration with olfactory ensheathing cell inoculated plga scaffolds in spinal cord injury adult rats. *Neurosignals, 25*, 1–14.

Watanabe, K., Nakamura, M., Iwanami, A., Fujita, Y., Kanemura, Y., Toyama, Y., et al. (2004). Comparison between fetal spinal-cord- and forebrain-derived neural stem/progenitor cells as a source of transplantation for spinal cord injury. *Developmental Neuroscience, 26*, 275–287.

Watanabe, S., Uchida, K., Nakajima, H., Matsuo, H., Sugita, D., Yoshida, A., et al. (2015). Early transplantation of mesenchymal stem cells after spinal cord injury relieves pain hypersensitivity through suppression of pain-related signaling cascades and reduced inflammatory cell recruitment. *Stem Cells, 33*, 1902–1914.

Watzlawick, R., Rind, J., Sena, E. S., Brommer, B., Zhang, T., Kopp, M. A., et al. (2016). Olfactory ensheathing cell transplantation in experimental spinal cord injury: Effect size and reporting bias of 62 experimental treatments: A systematic review and meta-analysis. *PLoS Biology, 14*, e1002468.

Wei, Y. T., He, Y., Xu, C. L., Wang, Y., Liu, B. F., Wang, X. M., et al. (2010). Hyaluronic acid hydrogel modified with nogo-66 receptor antibody and poly-L-lysine to promote axon regrowth after spinal cord injury. *Journal of Biomedical Materials Research. Part B, Applied Biomaterials, 95*, 110–117.

Weiss, S., Dunne, C., Hewson, J., Wohl, C., Wheatley, M., Peterson, A. C., et al. (1996). Multipotent CNS stem cells are present in the adult mammalian spinal cord and ventricular neuroaxis. *The Journal of Neuroscience: The Official Journal of the Society for Neuroscience, 16*, 7599–7609.

White, T. E., Lane, M. A., Sandhu, M. S., O'Steen, B. E., Fuller, D. D., & Reier, P. J. (2010). Neuronal progenitor transplantation and respiratory outcomes following upper cervical spinal cord injury in adult rats. *Experimental Neurology, 225*, 231–236.

Wictorin, K., & Bjorklund, A. (1992). Axon outgrowth from grafts of human embryonic spinal cord in the lesioned adult rat spinal cord. *Neuroreport, 3*, 1045–1048.

Wilcox, J. T., Satkunendrarajah, K., Zuccato, J. A., Nassiri, F., & Fehlings, M. G. (2014). Neural precursor cell transplantation enhances functional recovery and reduces astrogliosis in bilateral compressive/contusive cervical spinal cord injury. *Stem Cells Translational Medicine, 3*, 1148–1159.

Willerth, S. M., Arendas, K. J., Gottlieb, D. I., & Sakiyama-Elbert, S. E. (2006). Optimization of fibrin scaffolds for differentiation of murine embryonic stem cells into neural lineage cells. *Biomaterials, 27*, 5990–6003.

Willerth, S. M., Rader, A., & Sakiyama-Elbert, S. E. (2008). The effect of controlled growth factor delivery on embryonic stem cell differentiation inside fibrin scaffolds. *Stem Cell Research, 1*, 205–218.

Willerth, S. M., & Sakiyama-Elbert, S. E. (2008). *Combining stem cells and biomaterial scaffolds for constructing tissues and cell delivery*. Cambridge (MA): StemBook.

Williams, R. R., Henao, M., Pearse, D. D., & Bunge, M. B. (2015). Permissive Schwann cell graft/spinal cord interfaces for axon regeneration. *Cell Transplantation, 24*, 115–131.
Windrem, M. S., Osipovitch, M., Liu, Z., Bates, J., Chandler-Militello, D., Zou, L., et al. (2017). Human iPSC glial mouse chimeras reveal glial contributions to schizophrenia. *Cell Stem Cell, 21*(195–208), e196.
Windus, L. C., Chehrehasa, F., Lineburg, K. E., Claxton, C., Mackay-Sim, A., Key, B., et al. (2011). Stimulation of olfactory ensheathing cell motility enhances olfactory axon growth. *Cellular and Molecular Life Sciences, 68*, 3233–3247.
Windus, L. C., Lineburg, K. E., Scott, S. E., Claxton, C., Mackay-Sim, A., Key, B., et al. (2010). Lamellipodia mediate the heterogeneity of central olfactory ensheathing cell interactions. *Cellular and Molecular Life Sciences, 67*, 1735–1750.
Wirth, E., 3rd, Lebkowski, J. S., & Lebacqz, K. (2011). Response to Frederic Bretzner et al. "Target populations for first-in-human embryonic stem cell research in spinal cord injury". *Cell Stem Cell, 8*, 476–478.
Wirth, E. D., 3rd, Theele, D. P., Mareci, T. H., Anderson, D. K., Brown, S. A., & Reier, P. J. (1992). In vivo magnetic resonance imaging of fetal cat neural tissue transplants in the adult cat spinal cord. *Journal of Neurosurgery, 76*, 261–274.
Wirth, E. D., 3rd, Theele, D. P., Mareci, T. H., Anderson, D. K., & Reier, P. J. (1995). Dynamic assessment of intraspinal neural graft survival using magnetic resonance imaging. *Experimental Neurology, 136*, 64–72.
Wirth, E. D., Reier, P. J., Fessler, R. G., Thompson, F. J., Uthman, B., Behrman, A., et al. (2001). Feasibility and safety of neural tissue transplantation in patients with syringomyelia. *Journal of Neurotrauma, 18*, 911–929.
Wright, K. T., El Masri, W., Osman, A., Roberts, S., Chamberlain, G., Ashton, B. A., et al. (2007). Bone marrow stromal cells stimulate neurite outgrowth over neural proteoglycans (CSPG), myelin associated glycoprotein and Nogo-A. *Biochemical and Biophysical Research Communications, 354*, 559–566.
Wu, L., Li, J., Chen, L., Zhang, H., Yuan, L., & Davies, S. J. (2013). Combined transplantation of GDAs(BMP) and hr-decorin in spinal cord contusion repair. *Neural Regeneration Research, 8*, 2236–2248.
Wu, Y. Y., Mujtaba, T., Han, S. S., Fischer, I., & Rao, M. S. (2002). Isolation of a glial-restricted tripotential cell line from embryonic spinal cord cultures. *Glia, 38*, 65–79.
Xu, J. C., Bernreuther, C., Cui, Y. F., Jakovcevski, I., Hargus, G., Xiao, M. F., et al. (2011). Transplanted L1 expressing radial glia and astrocytes enhance recovery after spinal cord injury. *Journal of Neurotrauma, 28*, 1921–1937.
Xu, X. M., Chen, A., Guenard, V., Kleitman, N., & Bunge, M. B. (1997). Bridging Schwann cell transplants promote axonal regeneration from both the rostral and caudal stumps of transected adult rat spinal cord. *Journal of Neurocytology, 26*, 1–16.
Xu, X. M., Guenard, V., Kleitman, N., Aebischer, P., & Bunge, M. B. (1995). A combination of BDNF and NT-3 promotes supraspinal axonal regeneration into Schwann cell grafts in adult rat thoracic spinal cord. *Experimental Neurology, 134*, 261–272.
Xu, X. M., Guenard, V., Kleitman, N., & Bunge, M. B. (1995). Axonal regeneration into Schwann cell-seeded guidance channels grafted into transected adult rat spinal cord. *The Journal of Comparative Neurology, 351*, 145–160.
Yang, Y., Pang, M., Chen, Y. Y., Zhang, L. M., Liu, H., Tan, J., et al. (2020). Human umbilical cord mesenchymal stem cells to treat spinal cord injury in the early chronic phase: Study protocol for a prospective, multicenter, randomized, placebo-controlled, single-blinded clinical trial. *Neural Regeneration Research, 15*, 1532–1538.
Yang, C. C., Shih, Y. H., Ko, M. H., Hsu, S. Y., Cheng, H., & Fu, Y. S. (2008). Transplantation of human umbilical mesenchymal stem cells from Wharton's jelly after complete transection of the rat spinal cord. *PLoS One, 3*, e3336.

Yang, J., Xiong, L. L., Wang, Y. C., He, X., Jiang, L., Fu, S. J., et al. (2018). Oligodendrocyte precursor cell transplantation promotes functional recovery following contusive spinal cord injury in rats and is associated with altered microRNA expression. *Molecular Medicine Reports*, *17*, 771–782.

Yao, R., Murtaza, M., Velasquez, J. T., Todorovic, M., Rayfield, A., Ekberg, J., et al. (2018). Olfactory ensheathing cells for spinal cord injury: Sniffing out the issues. *Cell Transplantation*, *27*, 879–889.

Yao, Z. F., Wang, Y., Lin, Y. H., Wu, Y., Zhu, A. Y., Wang, R., et al. (2017). Transplantation of PDGF-AA-overexpressing oligodendrocyte precursor cells promotes recovery in rat following spinal cord injury. *Frontiers in Cellular Neuroscience*, *11*, 79.

Yasuda, A., Tsuji, O., Shibata, S., Nori, S., Takano, M., Kobayashi, Y., et al. (2011). Significance of remyelination by neural stem/progenitor cells transplanted into the injured spinal cord. *Stem Cells*, *29*, 1983–1994.

Yeh, J. Z., Wang, D. H., Cherng, J. H., Wang, Y. W., Fan, G. Y., Liou, N. H., et al. (2020). A collagen-based scaffold for promoting neural plasticity in a rat model of spinal cord injury. *Polymers (Basel)*, *12*.

Yokota, K., Kobayakawa, K., Kubota, K., Miyawaki, A., Okano, H., Ohkawa, Y., et al. (2015). Engrafted neural stem/progenitor cells promote functional recovery through synapse reorganization with spared host neurons after spinal cord injury. *Stem Cell Reports*, *5*, 264–277.

Young, W. (2015). Electrical stimulation and motor recovery. *Cell Transplantation*, *24*, 429–446.

Yu, S., Yao, S., Wen, Y., Wang, Y., Wang, H., & Xu, Q. (2016). Angiogenic microspheres promote neural regeneration and motor function recovery after spinal cord injury in rats. *Scientific Reports*, *6*, 33428.

Yuan, X. B., Jin, Y., Haas, C., Yao, L., Hayakawa, K., Wang, Y., et al. (2016). Guiding migration of transplanted glial progenitor cells in the injured spinal cord. *Scientific Reports*, *6*, 22576.

Zachar, L., Bacenkova, D., & Rosocha, J. (2016). Activation, homing, and role of the mesenchymal stem cells in the inflammatory environment. *Journal of Inflammation Research*, *9*, 231–240.

Zhai, W., Gao, L., Qu, L., Li, Y., Zeng, Y., Li, Q., et al. (2020). Combined transplantation of olfactory ensheathing cells with rat neural stem cells enhanced the therapeutic effect in the retina of RCS rats. *Frontiers in Cellular Neuroscience*, *14*, 52.

Zhang, J., Chen, H., Duan, Z., Chen, K., Liu, Z., Zhang, L., et al. (2017). The effects of co-transplantation of olfactory ensheathing cells and Schwann cells on local inflammation environment in the contused spinal cord of rats. *Molecular Neurobiology*, *54*, 943–953.

Zholudeva, L. V., Abraira, V. E., Satkunendrarajah, K., McDevitt, T. C., Goulding, M. D., Magnuson, D. S. K., et al. (2021). Spinal Interneurons as gatekeepers to neuroplasticity after injury or disease. *The Journal of Neuroscience: The Official Journal of the Society for Neuroscience*, *41*, 845–854.

Zholudeva, L. V., Iyer, N., Qiang, L., Spruance, V. M., Randelman, M. L., White, N. W., et al. (2018). Transplantation of neural progenitors and V2a interneurons after spinal cord injury. *Journal of Neurotrauma*, *35*, 2883–2903.

Zholudeva, L. V., Jin, Y., Qiang, L., Lane, M. A., & Fischer, I. (2021). Preparation of neural stem cells and progenitors: Neuronal production and grafting applications. *Methods in Molecular Biology*, *2311*, 73–108.

Zholudeva, L. V., & Lane, M. A. (2019a). Choosing the right cell for spinal cord repair. *Journal of Neuroscience Research*, *97*, 109–111.

Zholudeva, L. V., & Lane, M. A. (2019b). Transplanting cells for spinal cord repair: Who, what, when, where and why? *Cell Transplantation*, *28*(4), 388–399. https://doi.org/10.1177/0963689718824097. Epub 2019 Jan 18.2.

Zholudeva, L. V., & Lane, M. A. (2022). *Spinal interneurons: Plasticity after spinal cord injury* (1st ed.). Academic Press.

Zholudeva, L. V., Qiang, L., Marchenko, V., Dougherty, K. J., Sakiyama-Elbert, S. E., & Lane, M. A. (2018). The neuroplastic and therapeutic potential of spinal interneurons in the injured spinal cord. *Trends in Neurosciences, 41*, 625–639.

Ziegler, M. D., Hsu, D., Takeoka, A., Zhong, H., Ramon-Cueto, A., Phelps, P. E., et al. (2011). Further evidence of olfactory ensheathing glia facilitating axonal regeneration after a complete spinal cord transection. *Experimental Neurology, 229*, 109–119.

Zurita, M., & Vaquero, J. (2006). Bone marrow stromal cells can achieve cure of chronic paraplegic rats: Functional and morphological outcome one year after transplantation. *Neuroscience Letters, 402*, 51–56.

Zweckberger, K., Ahuja, C. S., Liu, Y., Wang, J., & Fehlings, M. G. (2016). Self-assembling peptides optimize the post-traumatic milieu and synergistically enhance the effects of neural stem cell therapy after cervical spinal cord injury. *Acta Biomaterialia, 42*, 77–89.

Zweckberger, K., Liu, Y., Wang, J., Forgione, N., & Fehlings, M. G. (2015). Synergetic use of neural precursor cells and self-assembling peptides in experimental cervical spinal cord injury. *Journal of Visualized Experiments, e52105*.

Further reading

Anderson, A. J., Piltti, K. M., Hooshmand, M. J., Nishi, R. A., & Cummings, B. J. (2017). Preclinical efficacy failure of human cns-derived stem cells for use in the pathway study of cervical spinal cord injury. *Stem Cell Reports, 8*, 249–263.

Assinck, P., Duncan, G. J., Hilton, B. J., Plemel, J. R., & Tetzlaff, W. (2017). Cell transplantation therapy for spinal cord injury. *Nature Neuroscience, 20*(5), 637–647. https://doi.org/10.1038/nn.4541.4.

Lu, P. (2017). Stem cell transplantation for spinal cord injury repair. *Progress in Brain Research, 231*, 1–32. https://doi.org/10.1016/bs.pbr.2016.11.012. Epub 2017 Jan 2. PMID: 285543936.

Nazareth, L., Shelper, T. B., Chacko, A., Basu, S., Delbaz, A., Lee, J. Y. P., et al. (2020). Key differences between olfactory ensheathing cells and Schwann cells regarding phagocytosis of necrotic cells: Implications for transplantation therapies. *Scientific Reports, 10*, 18936.

Pearse, D. D., Marcillo, A. E., Oudega, M., Lynch, M. P., Wood, P. M., & Bunge, M. B. (2004). Transplantation of Schwann cells and olfactory ensheathing glia after spinal cord injury: does pretreatment with methylprednisolone and interleukin-10 enhance recovery? *Journal of Neurotrauma, 21*, 1223–1239.

Reier, P. J., Bregman, B. S., Wujek, J. R., & Tessler, A. (1986). Intraspinal transplantation of fetal spinal cord tissue: An approach toward functional repair of the injured spinal cord. In M. E. Goldberger, A. Gorio, & M. Murray (Eds.), *Development and plasticity of the mammalian spinal cord* (pp. 251–269). Padova: Liviana Press.

Thornton, M. A., Mehta, M. D., Morad, T. T., Ingraham, K. L., Khankan, R. R., Griffis, K. G., et al. (2018). Evidence of axon connectivity across a spinal cord transection in rats treated with epidural stimulation and motor training combined with olfactory ensheathing cell transplantation. *Experimental Neurology, 309*, 119–133.

Zipser, C. M., Cragg, J. J., Guest, J. D., Fehlings, M. G., Jutzeler, C. R., Anderson, A. J., et al. (2022). Cell-based and stem-cell-based treatments for spinal cord injury: Evidence from clinical trials. *Lancet Neurology, 21*(7), 659–670. https://doi.org/10.1016/S1474-4422(21)00464-6. Epub 2022 May 12.

CHAPTER FOUR

Investigating cell therapies in animal models of Parkinson's and Huntington's disease: Current challenges and considerations

Mariah J. Lelos*
Brain Repair Group, School of Biosciences, Cardiff University, Cardiff, United Kingdom
*Corresponding author: e-mail address: lelosmj@cardiff.ac.uk

Contents

International Review of Neurobiology, Volume 166
ISSN 0074-7742
https://doi.org/10.1016/bs.irn.2022.09.005

Abstract

Cell therapeutics have entered into an exciting era, with first-in-person clinical trials underway for Parkinson's disease and novel cell therapies in development for other neurodegenerative diseases. In the hope of ensuring successful translation of these novel cell products to the clinic, a significant amount of preclinical work continues to be undertaken. Rodent models of neural transplantation are required to thoroughly assess the survival, safety and efficacy of novel therapeutics. It is critical to produce robust and reliable preclinical data, in order to increase the likelihood of clinical success. As a result, significant effort has been driven into generating ever more relevant model systems, from genetically modified disease models to mice with humanized immune systems. Despite this, several challenges remain in the quest to assess human cells in the rodent brain long-term. Here, with a focus on models of Parkinson's and Huntington's disease, we discuss key considerations for choosing an appropriate rodent model for neural transplantation. We also consider the challenges associated with long-term survival and assessment of functional efficacy in these models, as well as the need to consider the clinical relevance of the model. While the choice of model will be dependent on the scientific question, by considering the caveats associated with each model, we identify opportunities to optimize the preclinical assessment and generate reliable data on our novel cell therapeutics.

1. Introduction

It is an exciting era for human stem cell replacement therapies, with an abundance of preclinical data suggesting that cell transplants may provide a valid means of improving disease outcomes and with first-in-person clinical trials commencing around the world (Barker, Parmar, Studer, & Takahashi, 2017). The concept of cell therapies is predicated on the idea that transplantation of live cells into the brain can replace those cells lost to the disease. And, indeed, for Parkinson's disease (PD), it is now well established that human fetal and stem cell-derived neurons can survive long term in rodent models and in people with PD (Lane et al., 2022; Li et al., 2016), they can integrate into the host circuitry (Grealish et al., 2015) and they can alleviate functional deficits (Kirkeby et al., 2012; Kriks et al., 2011; Lelos, Morgan, et al., 2016; Lelos, Robertson, et al., 2016). While cell therapies for PD are perhaps the most clinically advanced, considerable work is being undertaken in other neurological diseases as well, such as Huntington's (HD) and stroke, demonstrating that cell therapies may be broadly applicable to a wide range of neurodegenerative conditions.

In order to progress cell therapies from an in vitro concept to a clinical treatment, it is fundamentally necessary to validate the safety and efficacy of a novel cell product in rodent models. Only after rigorous preclinical assessment, can clinical translation commence. As a result, a wide variety of preclinical models have been established, each one designed to address a particular scientific question, but equally each one associated with caveats and compromises.

Significant advances in the development of animal models of disease have seen the move from more crude neurotoxin models, which discretely damage specific neurons, to more sophisticated genetically modified rodent models and viral vector technologies, which more closely mimic the neuropathological features of the disease. Despite this advance in modeling disease, there are several specific challenges associated with the use of animal models for testing cell therapies. This chapter will explore some of these key challenges and will use PD and HD as prototypic conditions to highlight some disease-specific considerations. The primary focus will be on the use of mouse and rat models, given that these are the most readily accessible to the majority of researchers and they are used for most of preclinical validation assays (safety, efficacy, toxicology), although areas where large animal models would be of considerable benefit will be considered. In particular, this chapter will explore the challenges of (1) finding an optimal disease model, (2) ensuring cell survival in the context of a xenograft, (3) creating a model to test functional efficacy and (4) ensuring the complex interaction between the cell therapy and the clinical context is taken into consideration (Fig. 1).

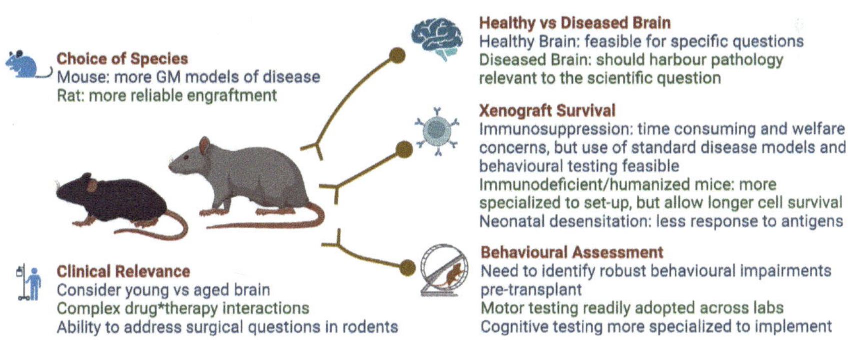

Fig. 1 Overview of key challenges and considerations associated with the use of rodent models for assessing novel cell therapies. Ultimately, the scientific question will directly influence the choice of model, but there are a number of caveats to consider with each model. Created with BioRender.com.

2. Challenge 1: Identifying a good model for cell transplantation studies

2.1 The healthy vs diseased brain

The key concept underpinning the development of cell therapies is the idea that replacement of cells that have degenerated over the course of the disease with healthy cells will improve neural function and alleviate functional impairments. However, the first consideration is whether a healthy brain or a disease model is required to address the scientific question at hand. The transplantation of cells into a healthy, intact brain is a useful model for answering basic questions associated with a novel therapeutic, such whether the cells can survive or whether there is likely to be overgrowth or extensive proliferation associated with a cell therapy product. In this circumstance, no interactions with disease mechanisms are in question, and as a consequence a simple model will provide a faster and cheaper method of generating the relevant data.

Transplantation into the intact brain is perfectly feasible from an experimental perspective, but it is worth noting that this approach may produce grafts that are structurally different to those that are situated in a degenerated brain. It has been demonstrated, using both models of HD and PD, that the size of the graft and the propensity to integrate into the host brain is affected by the extent of neurodegeneration. Indeed, while a neurotoxic lesion encourages larger grafts to survive and integrate, relative to the intact brain, it has also been observed that poor graft survival can occur if the lesion is too extensive. For example, it has been reported that fetal tissue graft volume is significantly larger when the tissue has been transplanted into a lesion model of HD, relative to an intact, healthy brain (Watts & Dunnett, 1998). Moreover, the same researchers demonstrated that graft volume decreases in a brain that received a high dose (90 nM) of the neurotoxin quinolinic acid, as compared to brains that received more moderate doses (20 and 40 nM). Thus, the existence of a lesion, as well as the extent of lesion, will influence graft survival and integration into the host brain.

As scientists work toward clinical translation of cell therapy products, it is likely there will be a requirement to investigate the novel therapeutic in the context of the diseased brain. In this case, it is important to first consider the extent to which the animal model displays features of the neurodegenerative disease under investigation. Ideally, the rodent model would present with a similar degenerative phenotype to the PD or HD patient. For example, the rodent would display progressive loss of the relevant cell type and this would be associated with disease-relevant behavioral impairments that manifest and progressively worsen across the course of the animal's life.

Furthermore, if the assumed mechanism of recovery of the cell therapy is brain repair and reconnection of the damaged neural circuitry, then some lesion or degeneration is required to test this hypothesis. Without degenerated neurons in the host brain, it would not be possible to effectively model the structural or functional impact of synaptically repairing the brain. This assumption presumes that neural repair is the goal of the cell therapy, but if reduced neuroinflammation or the release of neurotrophic factors is the predominate strategy, then cell loss may not be a key consideration in choosing the most appropriate model. Specific considerations associated with models of PD and HD are discussed below in Sections 2.3 and 2.4, respectively.

2.2 Choosing the appropriate rodent species

A second consideration is the use of mice vs rats for the experimental program. Original proof-of-concept transplantation studies were predominantly conducted in rat models, such as the Sprague-Dawley line. With the advances in genetic sequencing resulting in the identification of key genes that either cause, or are risk factors for, neurodegenerative diseases, use of genetically modified models has become an attractive prospect. However, most genetically modified models have been generated using mice and, although transplantation into the mouse brain is possible, it has been shown to be less robust than transplantation into the rat brain (Roberton et al., 2013). Transplants in the mouse brain have been reported to be substantially smaller than those surviving in the rat, even allowing for the differences in host brain size (El-Akabawy et al., 2012; Kelly et al., 2007). While the exact reason for this remains unclear, it has been suggested that the metabolism of immunosuppressants with short half-lives may be faster in mice or the immune system may be more sensitive. Immunodeficient mice, such as SCID or Rag2$^{-/-}$ host mice can be used as an alternative, but consideration of how the disease will be modeled in these mice must be undertaken, as well there being recognition that regular behavioral testing is not feasible in these models due to their propensity to infection (see Section 3 for further discussion of immunocompromised hosts). Finally, although both mice and rats can be used to test motor and cognitive function, performance in specific tasks is notably affected by the species. For example, differences in addictive behavior, impulsive behavior, social interaction and cognitive function have been reported (Ellenbroek & Youn, 2016).

2.3 Choosing the appropriate strain

The optimal choice of rat or mouse strain is dependent on several factors, and can be more readily controlled when using neurotoxin models of disease. One of the most prominent reasons for choosing a particular strain is to undertake behavioral testing to assess the functional efficacy of a cell therapy product. Animals that are more curious and active, such as Lister-hooded rats and C57Bl6 mice, are more likely to sustain a behavioral response long-term, in contrast to more passive strains, such as Sprague-Dawley rats and CD1 mice (Bohlen et al., 2014; Clemens, Jansson, Portal, Riess, & Nguyen, 2014; de Castro Gomes et al., 2013; Hasanpour, Mitricheva, Logothetis, & Noori, 2021; Moy et al., 2008).

Genetically modified mice or rats will be developed on a particular background and modification of this background is likely to affect the expression of the manipulated gene. For example, it has been reported that expression of the same Alzheimer's-relevant mutations on a DBA/2J background is both more lethal and produced more mild pathology than expression of these genes on the C57Bl6 background (Jackson et al., 2015). Thus, it may be pertinent to adapt behavioral tests to the background strain in these models, or systematically assess the impact of changing the background strain on survival, disease pathology and profiles of behavioral impairment. Indeed, certain strains, such as immunodeficient mice or rats, are not optimal for behavioral testing due to the need to maintain relatively secure barriers for protection, making use of standard behavioral equipment on a daily basis unfeasible. Thus, choosing the appropriate strain to test the efficacy of a treatment will require some consideration of the profiles of each strain and their likelihood to respond appropriately in a given task.

2.4 Parkinson's disease models for cell transplantation studies

In PD, loss of substantia nigra pars compacta dopamine neurons is a primary pathological feature of the disease and it results in dopaminergic denervation of the striatal nucleus. The neural transplant strategy that is being explored clinically is the heterotopic transplantation of cells into striatum. The rationale for this approach is that the nigral-striatal synapse will more readily be re-established by placing the grafted cells directly next to their target cells. Homotopic transplantation of cells back into the substantia nigra may be advantageous from a neural circuitry repair perspective, but the long distance that cells would need to project to reach their post-synaptic targets remains a challenge. It has been reported in rodent models that, using current strategies, only partial striatal integration and recovery is achieved with homotopic transplantation (Kauhausen, Thompson, & Parish, 2015; Thompson, Grealish, Kirik, & Björklund, 2009). New strategies are being considered, such as combining cell therapies with viral infusions that overexpress growth factors in order to encourage longer axonal growth and better reconstruction of the nigral striatal pathway (Moriarty et al., 2022), which have thus far proven fairly successful and may modify our approach to cell transplantation in the future. However, the more recent use of larger human cells in the rodent brain is also likely to enhance the ability of these cells to project along the nigrostriatal, which emphasizes the need to test novel homotopic transplantation approaches in larger brains, such as non-human primate models. Thus, at present, an optimal transplant model for PD would mimic the nigrostriatal loss and current clinical approaches favor the heterotopic transplantation of cells into the striatum.

The next important feature of a PD disease model would be for it to be either stable, or demonstrating progressive degeneration over the course of the experiment. Human stem cell-derived precursors take ~6 months to mature into dopamine neurons in vivo (Björklund & Parmar, 2020), which therefore requires a model to be stable for at least this period of time to allow assessment of the final cell therapy product. This is not an insignificant consideration given that the dopamine system has been shown to be highly plastic, with remaining, endogenous fibers compensating after a degenerative event and arborising to re-innervate the striatal nucleus (Frohna, Neal-Beliveau, & Joyce, 1997). Indeed, high levels of dopamine loss (~80% or more) are required to observe a sustained functional impairment in rodent models (Björklund & Dunnett, 2019), which is consistent with the observation that people with PD have already lost 50–80% of their dopamine neurons before they present with the first indications of the classic motor changes, such as tremor or bradykinesia (Chaudhuri & Titova, 2019). Thus, a model of PD would need to present with stable dopamine loss over the course of the experiment, to allow for the innervation and functional efficacy of the transplant to be effectively assessed.

Ideally, the model would present with other key features of the disease, such as the alpha-synuclein propagation and aggregation, neuroinflammation and loss of other cell types/neurotransmitter systems. A wide range of neurotoxin models, genetic models and viral vector/preformed fibril models have been developed to mimic aspects of PD (Breger & Fuzzati Armentero, 2019; Thirugnanam & Santhakumar, 2022). Neurotoxin models, such as intracerebral infusion of 6-hydroxydopamine (6-OHDA) or systemic injection of

MPTP in the mouse, result in loss of the nigral dopamine neurons and subsequent denervation of the striatal nucleus. However, although these approaches achieve a key requirement of the model (i.e., loss of the nigrostriatal pathway), intrastriatal infusion of 6-OHDA and systemic MPTP both result in partial loss of the nigral neurons, and plasticity and reinnervation are well-documented in these models. Only unilateral infusion of 6-OHDA along the medial forebrain bundle (MFB) results in near complete destruction of the nigral neurons and has been shown to present with stable striatal denervation long-term. Importantly, however, this 6-OHDA MFB model can only be used as a unilateral model, since significant bilateral damage to the dopamine system will result in dysphagia, weight loss and significant welfare concerns (Berridge, Venier, & Robinson, 1989). It is also worth noting that no, or insignificant, alpha-synuclein pathology manifests in the 6-OHDA and MPTP neurotoxin models, which means a primary pathological feature of the disease is lacking in this context.

A multitude of genetic models of PD have been developed (covered comprehensively by Breger & Fuzzati Armentero, 2019), which express mutations associated with increased risk of developing PD, such as on the *SNCA* and *LRRK2* genes. These models have the potential advantages of harboring bilateral pathology and may be more molecularly similar to PD, but they fail to robustly mimic the extent of dopamine cell loss and behavioral impairment required to test cell therapies long-term. Indeed, there is also evidence that exposure to immunosuppressants, which are required to maintain xenografted human cells in the brains of immunocompetent rodents, can in themselves reduce the disease-associated pathology and improve behavioral performance (Tamburrino et al., 2015). This introduces the risk that a more mild model would experience amelioration of the pathology over the course of the experiment, which may negate the ability to test the efficacy of the cell therapy, or the interaction with disease, in such a model.

The desire to more accurately model the impact of alpha-synuclein pathology on the brain has led to the development of viral vector-based models, which overexpress either human wild-type or mutant alpha-synuclein pathology (Cenci & Björklund, 2020; Ulusoy, Decressac, Kirik, & Björklund, 2010; Van der Perren, Macchi, et al., 2015; Van der Perren, Toelen, et al., 2015). These more slowly progressing models typically result in ~50% loss of nigral DA neurons by ~8 weeks post-lesion, with the subset of animals harboring 50–80% loss most likely to demonstrate robust motor phenotypes (Decressac, Mattsson, & Björklund, 2012).

Interestingly, while motor impairments correlate well with overt cell loss in the 6-OHDA lesion model, they are associated with the combined impact of cell loss and cellular dysfunction in the remaining cells in the alpha-synuclein viral vector model (Decressac et al., 2012). While these models are partially effective at degenerating the dopaminergic neurons of the substantia nigra and behavioral impairments have been reported, the loss of neurons can be variable between animals. Indeed, animals with only mild cell loss may not manifest behavioral impairments, and even in animals with greater cell loss, the remaining fibers may arborise longer term, resulting in some behavioral recovery after several months (Kirik et al., 2002).

The direct infusion of preformed fibrils can also be used as a model of PD on its own, or in combination with the viral-induced alpha-synuclein pathology. The addition of preformed fibrils to the viral vector preparations has been shown to result in the robust formation of inclusions that resemble Lewy bodies and Lewy neurites, as well as greater loss of dopamine neurons ($\sim$70% loss by 16 weeks) with behavioral impairments manifesting as early as $\sim$4 weeks post-infusion (Hoban et al., 2020). While variability is still inherent to this model, the recapitulation of the more progressive dopamine loss, the alpha-synuclein pathology and associated neuroinflammatory changes, and the manifestation of behavioral deficits, makes this one of the more representative models of PD. However, in the context of using this model for cell transplantation studies, it has again been reported that that use of immunosuppressants can alleviate the disease pathology and improve behavioral deficits, independently of the neural transplant, making robust assessment of the efficacy of the graft challenging in this model (Tamburrino et al., 2015).

Thus, overall, a wide range of models has been developed to mimic PD, but naturally there is no one ideal model. Furthermore, when considering the use of these models to explore cell therapies, it is critical to consider the precise nature of the scientific enquiry, to ensure the model presents with the key features required to address the question. Despite this, all models should be used with the appropriate caveats and limitations in mind. For example, it would be ideal to use a rat (rather than mouse) strain to ensure more consistent survival of the graft. Indeed, for PD, a particular caveat is the propensity for plasticity and neurorecovery over time and after exposure to immunosuppressants. It is for these reasons that many experiments continue to use the unilateral 6-OHDA model to test functional efficacy. The alpha-synuclein viral vector plus preformed fibril model, however, may be preferred if the relationship between the transplant and the disease itself is under consideration. Thus, some consideration of the purpose of the experiment is required to determine whether a robust model is preferred for long-term behavioral analysis or if graft-disease interactions are key to the investigation.

2.5 Huntington's disease models for cell transplantation studies

Unlike PD, HD has a clear genetic cause, with an abnormal CAG expansion on the huntingtin gene on chromosome 4 identified as the culprit. Inheritance of this mutation leads to a wide range of cellular dysfunctions, including the aggregation of mutant huntingtin (mHTT), and early loss of striatal medium spiny neurons, followed by subsequent atrophy of cortical and other brain areas. These changes are associated with the manifestation of a wide range of psychiatric, motor and cognitive dysfunctions.

Given the nature of the disease, many genetically modified rodent models have been developed that harbor the mHTT mutation (Farshim & Bates, 2018; Kosior & Leavitt, 2018; Plank et al., 2018), with the vast majority of them using mouse strains. Transgenic and knock-in models tend to have long CAG repeats (100+), which differ to the clinical presentation of HD in humans (40–60 for most people with HD), but several aspects of neuropathology and behavioral deficits have been successfully modeled in these lines.

For HD, it may be ideal to test any therapeutic, including cell therapies, in a genetically modified rodent. However, this is not necessarily straightforward and it is pertinent to consider several caveats of the models in order to choose the optimal model. First, the ability of cells to survive in these models must be explored. As discussed above, while transplantation into the mouse brain is technically feasible, the survival of the cells has been reported to be less than in rat models (Roberton et al., 2013). Additionally, it has been observed that transplantation into the diseased HD brain may yield even smaller, less healthy transplants than the wild-type brain (Adil et al., 2018). While this is undoubtedly of interest if the disease-graft interaction in the HD brain were under investigation, it is worthy of consideration if cell maturation, toxicity, safety or the potential to alleviate functional deficits are instead being studied.

Second, since the goal of some cell therapy approaches is to replace the cells that have degenerated in the diseased brain, it is important to consider the extent to which this this can be modeled effectively in a genetically modified mouse model. Most mutant mice harbor progressive pathology in the striatum

(Bayram-Weston, Jones, Dunnett, & Brooks, 2012; Bayram-Weston, Jones, Dunnett, & Brooks, 2012; Bayram-Weston, Jones, Dunnett, & Brooks, 2012; Bayram-Weston, Torres, Jones, Dunnett, & Brooks, 2012), but the pattern of early, significant atrophy in the striatum observed in patients is not replicated in most genetic models of HD. Thus, it is important to consider whether a model lacking striatal atrophy is ideal for evaluation of a cell therapy. If the goal is to repair striatal projections and reinstate neural networks, it may be that the therapeutic potential of this approach cannot be sufficiently judged in the absence of cell loss. However, it is possible that introduction of healthy, viable neurons into the diseased environment is sufficient to improve neural function in the absence of overt cell loss, if the proposed mechanism is trophic or anti-inflammatory in nature. These important considerations will influence both the model chosen to test the cell therapy, and the interpretation of the results.

Depending on the question being addressed in the experiment, there are alternative models that can be considered. Rat models of HD with more clinically relevant CAG repeat lengths have been generated and have the potential to be used for neural transplantation studies. The pathology is progressive, but relatively mild until the older ages, and behavioral deficits tend to be more neuropsychiatric in nature, with relatively little motor dysfunction manifesting (Plank et al., 2018). Nevertheless, this may present a more permissive neural environment to study the graft in the presence of HD pathology and with some relevant behavioral deficits. However, to date, no cell therapy studies have been conducted in a rat genetically modified HD model.

Alternatively, the quinolinic acid neurotoxic lesion model has been a classic model of striatal loss, which has been used extensively in the cell therapy field. Intrastriatal infusion of quinolinic acid selectively degenerates the medium spiny neurons, leaving interneuron populations relatively unaffected (Lelos & Dunnett, 2018; Lelos, Harrison, Rosser, & Dunnett, 2013). The advantage of this model is in the ability to model the striatal atrophy observed in HD and to induce relevant behavioral impairments, including both motor and cognitive deficits (Brasted et al., 1998; Lelos, Harrison, & Dunnett, 2012). Neural transplantation studies are then able to assess the ability of the cells to survive, mature, integrate, forms functional synapses and alleviate cognitive and motor impairments (Döbrössy & Dunnett, 2005; Lelos, Roberton, et al., 2016; Mazzocchi-Jones, Döbrössy, & Dunnett, 2009). Importantly, however, cell therapies are assessed in a fairly permissive environment, which lacks the key HD pathology, and which may therefore provide less clinically translational data.

Ultimately, for HD, the choice of model will depend on the scientific questions under investigation. If the ability of a therapy to replace the degenerated cells and achieve neural repair is under investigation, the quinolinic acid model may be most appropriate. However, questions pertaining to the transfer of mHTT pathology, the inflamed environment, blood–brain barrier permeability or neurotrophic effects of the graft, for example, may require a genetically modified rodent model. Assessment of the functional efficacy of the transplant, the safety of the cell therapy product, or the maturation of the cells long-term, may be explored in either neurotoxin or genetically modified models, as long as the caveats associated with graft viability are considered.

3. Challenge 2: Survival of xenografts in rodent models

The original proof-of-concept research exploring the feasibility of using cell therapies for PD and HD was conducted using a variety of cell types, but was heavily dependent on allogeneic rodent fetal tissues. As we move toward clinical translation of these therapies, however, cell therapy products are typically derived from human tissues. This raises a number of challenges for studying these cells in rodent models. A first, and critical, consideration is the relatively slow maturation of human cells. In PD, allogeneic grafts would mature into functional, dopamine-releasing grafts after ~4 weeks in situ, whereas human fetal or stem cell-derived neurons reach maturity at 20–24 weeks post-graft (Björklund & Parmar, 2020). This long maturation time requires prolonged manipulation of the immune system to allow for the xenograft to survive and mature. Therefore, it is important to consider the relative merits of the available models and identify the optimal model to address the question under investigation.

3.1 The use of immunosuppressants in immunocompetent hosts

Immunocompetent rodents can be treated pharmacologically to dampen the immune system. A popular model in rodents has been the use of the immunosuppressant Cyclosporine A, which is a calcineurin inhibitor that reduces the production of inflammatory cytokines by T-lymphocytes, and which is also sometimes used clinically for human neural transplant recipients. Tacrolimus (FK506) is an alternative calcineurin inhibitor that has been explored preclinically and used in human graft recipients.

The primary advantages of using this type of model are (1) the ability to undertake repeated tests or procedures on the animals, and (2) the relevance to the clinical context, where many graft recipients will receive some immunosuppression post-transplant surgery. In terms of undertaking repeated procedures, it is possible using this model to undertake daily behavioral tests to assess the functional efficacy of the graft in tests of motor and/or cognitive function. It is also relatively safe to undertake repeated surgeries, for example, in the context of studying synaptic integration using the rabies monosynaptic tracing virus (Grealish et al., 2015). The disadvantages lie in the welfare considerations and the requirement for a significant time investment in running the experiment. Daily intraperitoneal injections are required to ensure continuous protection of the graft from rejection, which requires commitment from researchers potentially for up to 6 months, to attend daily, at the same time of day, to administer the drug. Immunosuppressants can cause welfare issues after long-term exposure, including nephrotoxicity, growth of teeth and nails and weight loss. Therefore, close monitoring of the animal, in conjunction with a veterinarian, is required.

In terms of the clinical relevance of these pharmaceutical immunosuppression models, it is useful that the drugs used preclinically align with those used in both fetal and the first-in-person stem cell clinical trials for PD and HD. TransEURO, a trial of fetal dopamine grafts for PD, used a triple immunotherapy, which included cyclosporine (Barker et al., 2019), while the TRIDENT trial, which is assessing fetal tissue transplants for HD is using tacrolimus (Drew et al., 2021). The European Stem-PD trial will use Cyclosporine A immunosuppression (as the triple immunotherapy, in combination with azathioprine and steroids) (Kirkeby, Parmar, & Barker, 2017) and the American NYSTEM-PD and Japanese CiRA trials will use tacrolimus (Barker et al., 2017). It is also noteworthy that where an autologous approach is under investigation, patients may receive no immunosuppression post-surgery (Barker et al., 2017; Loring, 2018).

A final, notable, consideration is the potential neurotrophic impact of the immunosuppressant on the graft itself and on the disease pathology. It has been reported that the use of Cyclosporine A results in the survival of more xenografted cells and larger transplants at post-mortem, which has been attributed directly to the reduction in neuroinflammation in the brain. Specifically, using three different models of PD (the Thy1-α-synuclein transgenic mouse, a novel AAV-α-synuclein mouse model, and the MPTP mouse model), researchers reported reductions in reactive astrocytes (GFAP expression) and reduced alpha-synuclein burden (Tamburrino et al., 2015).

Similarly, tacrolimus has been reported to significantly lower the infiltration of both T helper and cytotoxic T cells and reduce the number of microglia and macrophages in an alpha-synuclein model of PD (Van der Perren, Macchi, et al., 2015). Interestingly, in vitro assays have also revealed an effect of Cyclosporine A and tacrolimus on neurogenesis, astrogenesis and proliferation (Ashwini et al., 2018; Guo et al., 2007), which suggests that exposure to these drugs may influence the maturation of stem cell derived precursor cells post-transplantation. Overall, immunosuppressants have been reported to improve graft survival, to reduce disease burden and they may also influence the graft maturation process, all of which are important to consider when using this approach to model the clinical application of a cell therapy.

3.2 The use of immunodeficient rodents and rodents with humanized immune systems

The majority of cell transplant trials in patients will use an allogeneic approach, with graft recipients receiving short-term immunosuppressants to enhance graft survival. However, as discussed above, implementing this model preclinically is both time-consuming and labor intensive. Thus, the use of immunodeficient mouse or rat models provides an opportunity to transplant cells without having to contend with the host immune response. The primary advantage of these models is the ease with which rodents can be left long-term (e.g., months) while the grafts mature and relatively little intervention is required from the researchers. This also eliminates variability due to practical differences in administering injections. However, the limitations of such models lie in the inability to undertake complex interventions or run multiple behavioral assays. The susceptibility to infection, housing in barrier facilities or individually ventilated cages, and the risk associated with invasive neural surgeries makes daily handling or repeated interventions more challenging in these models. Thus, the use of these models is determined in part by the capabilities of the animal facility itself, as well as more often by the complexity and requirements of the experimental design.

A drawback of the preclinical xenotransplantation approach is the inability to model the human immune response to a transplant. In the approaches described above, the immunocompetent rodent is likely to produce a different immune response to human patients, and immunodeficient hosts will be even less translational in their immune profile. To address these challenges, humanized rodent models have been generated, which allow some investigation of the graft-immune interaction in preclinical models. To achieve this, immunodeficient mice are engrafted with human immune cells and their progenitors, such as peripheral blood mononuclear cells or hematopoietic stem cells and subsequently receive intracerebral transplants of human cells.

As an example of this, in an attempt to determine whether an immune response to autologous cells would differ to allogeneic cells, Schweitzer et al. (2020) created humanized autologous and allogeneic mice. They used non-obese diabetic mice with severe combined immunodeficiency and depletion of the interleukin-2 receptor γ (NOD SCID gamma mice) and created patient-humanized (or allogeneic volunteer) NOD SCID gamma mice. To achieve this, they harvested peripheral-blood mononuclear cells from a Parkinson's transplant recipient at 18 and 24 months post-surgery and from a volunteer, and injected these into the NOD SCID gamma mice. The results revealed long-term survival of the autologous cells when matched to the autologous humanized mouse, but poor survival in the allogenic humanized mouse and the infiltration of CD4+ t-lymphocytes, which supported their hypothesis that autologous transplantation would confer greater protection than allogeneic transplants.

Despite the advances in developing continuously more sophisticated humanized models, challenges in their production and implementation remain. First, they are more costly and specialized than the use of more standardized immunocompetent or immunodeficient rodents. Second, in some humanized models, a robust human T cell-mediated immune response could not be established (Traggiai et al., 2004), although improved models have also been generated (Brehm, Wiles, Greiner, & Shultz, 2014; Kooreman et al., 2017). And finally, in some models, natural killer cell activity from the host rodent has interfered with efficient survival of human neural transplants (for a comprehensive review, see Petrus-Reurer et al., 2021).

3.3 Neonatal desensitization to support graft survival

An alternative approach has been developed, in which neonatal desensitization is used to reduce graft rejection. Here, researchers exposed post-natal pups (P1–5) to human fetal-derived cells via in intraperitoneal injection and hypothesized that this exposure during a critical window of immune system maturation, would allow the rodents to develop a tolerance to human neural cells. Indeed, it was shown, using different combinations of cell types, that neural transplants could survive for 12 weeks in adult Sprague-Dawley rats after neonatal desensitization, and to a comparable level as rats exposed to Cyclosporine A (Kelly et al., 2009). A modified version of the neonatal desensitization protocol was developed using stem cell derived preparations, in which heterogenous populations of cells (progenitors, plus mature dopamine neurons) were injected, based on the hypothesize that exposure to a wider range of antigens across the developmental range would ensure survival throughout all stages of graft maturation (Heuer, Jönsson, Pfisterer, Kirkeby, & Parmar, 2016). Rats were maintained for 6 months post-graft, with evidence of some loss of grafts over the longer time periods. Indeed, grafts have been shown to survive up to 1 year post-transplant with this method, although the percentage of surviving grafts is quite low by the point (~25%) (Lane et al., 2022). Despite only conferring partial protection, this model has the potential to avoid the welfare and time challenges associated with pharmacological manipulation of the immune system, while allowing the immunocompetent hosts to undergo multiple surgical or behavioral procedures. However, there have also been studies suggesting that this approach is best used in rat models, with little success evident in mice (Janowski et al., 2012).

Given the complexities of modulating the immune system to support allogenic grafts, and the significant health side effects associated with immunosuppressant drugs, an alternative approach under investigation is the use of "universal" or non-immunogenic cell lines. The goal is to evade the immune system by eliminating HLA and other immune signatures from the lines, but preclinical validation of this approach has not yet been reported (Frederiksen, Doehn, Tveden-Nyborg, & Freude, 2020).

In conclusion, there are a wide range of rodent models available that will support the long-term survival of xenografts, but the choice of model is dependent on the scientific question under investigation, as well as the complexity and requirements of the experimental design.

4. Challenge 3: Assessing the functional efficacy of the graft

4.1 General considerations for assessing the efficacy of a cell therapy

In anticipation of bringing a novel cell therapy to the clinic, it is imperative to demonstrate that the product is actually efficacious. While evidence of survival, safety and toxicity are relatively straightforward data to obtain using the available rodent models, assessing functional efficacy remains a greater challenge. Many neurodegenerative diseases present with both motor and non-motor/cognitive dysfunctions and these are prominent in PD and HD.

Therefore, in assessing the efficacy of a novel cell therapy product, it is important to consider not just the ability to improve motor function, but also the impact of the cell therapy on other symptoms. Here, general considerations associated with testing functional efficacy will be discussed first, followed by more specific consideration of behavioral testing in models of PD and HD in Section 4.2.

A wide range of simple motor tests have been described that are sensitive to basal ganglia function and measure changes in forelimb speed and accuracy, gait, spontaneous reaching behaviors and sensorimotor function. The nature of the model will, to some extent, determine which tests are applicable. For example, in PD and HD, the unilateral 6-OHDA or quinolinic acid models are regularly used for stable, long-term assessment of human neuron grafts, but the unilateral nature of these models means that the researcher should preferentially identify tests that delineate ipsi- and contralateral paw function (Fig. 2). In this context, one would anticipate that the rodent would be impaired on measurements of contralateral paw function, while the ipsilateral paw would remain fully functioning and, instead, would act as a within-subjects control. Tests such as the adjusting steps task, the vibrissae-evoked touch test, the corridor task, balance beam and the cylinder

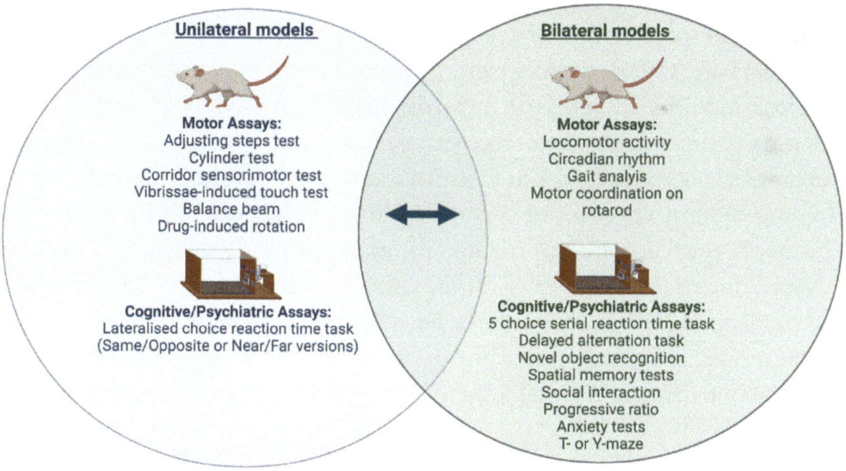

Fig. 2 Overview of behavioral tests optimized for models with either unilateral or bilateral damage to the basal ganglia circuitry. While it is possible for any of these tests to be used in any model, the tests listed under "Unilateral models" are designed to isolate one limb or hemisphere for sensitive analysis of contralateral versus ipsilateral function. Similarly, the tests listed under "Bilateral models" engage both limbs/hemispheres of the brain, making them less sensitive in unilateral models. Created with BioRender.com.

test, among others, all reliably measure this imbalance in paw function or sensorimotor function (Brooks, Trueman, & Dunnett, 2012; Carter, Morton, & Dunnett, 2001; Dowd, Monville, Torres, & Dunnett, 2005).

Indeed, one of the most reliable tests in the field is the use of drug-induced rotation, where the rodent is exposed to a drug that agonizes or affects aspects of the striatal nucleus that are imbalanced in the brain (e.g., amphetamine for PD, or the D2/D2 receptor agonist apomorphine for HD). Activation of one hemisphere drives locomotion ipsilaterally and the number of rotations per minute during the course of the drugs' active period can discriminate between a disease model vs a intact brain. The modulation of this motor bias has also been shown to be sensitive to the impact of therapeutics in the disease model (Björklund & Dunnett, 2019).

Assessing motor dysfunction in a bilateral model of PD or HD can be conducted using either the same tests as described above (but anticipating bilateral impairment, in comparison to healthy control rodents), or using alternative tests that assess more general gait or locomotor activity. More specialized tasks have also been established, which require more time and effort to run and quantify, such as the single-pellet reaching test (Klein & Dunnett, 2012), but these also provide some of the most sensitive and translational data on motor (dys)function.

Non-motor symptoms are prominent features of neurodegenerative diseases that can span a range of symptoms including, but not limited to, autonomic nervous system dysfunction, disrupted sleep, executive dysfunctions, dementia and neuropsychiatric impairments such as anxiety and apathy. The non-motor impairments most likely to be impacted by a neural transplant are neurological symptoms such as cognitive and neuropsychiatric dysfunctions. Assessing the impact of a cell therapy on these behaviors can be challenging because (1) specialized equipment is often required, (2) bilateral models are preferred due to the propensity for the intact hemisphere to influence behavioral performance and (3) the mechanism of recovery should be appropriate for the model and therapeutic. For example, in PD, a dopamine-dependent cognitive impairment is most likely to be impacted by a dopamine-rich neural transplant.

Some maze tests, such as the elevated plus or circle maze and the light/dark arena, are able to effectively measure aspects of anxiety. The classic T-maze can be used to measure alternation behavior, while problem solving, olfaction and spatial memory can be assessed using puzzle box and object-in-arena tests. Social deficits can be measured using well-established social interaction tests and some basic associative tests, such as place preference, can be conducted in relatively simple arenas. While these maze tests are able to cover many aspects of non-motor function, there are much more complex cognitive and neuropsychiatric changes that manifest in these diseases that can only be captured by more sensitive cognitive assessment using automated operant equipment, such as two-lever Skinner boxes and 9-hole operant boxes. For example, evaluating incentive motivation, visuospatial dysfunction, rule learning, implicit learning, habit formation, fear learning, attention, associative learning, contextual discrimination or impulsivity would best be measured in programable operant boxes, which are capable of measuring tens to hundreds of trials per session in an unbiased fashion. Implementation of these types of tests is costly in terms of purchasing the hardware/software, requires basic knowledge of coding and requires investment in time to establish the training and test protocols.

The preference for bilateral models of disease is based on the fact that an intact neural system in one hemisphere will likely support learning and performance on tasks of non-motor dysfunction. Thus, bilateral models, such as viral vector mediated alpha-synuclein overexpression or mutant models of HD are well suited to cognitive testing. Use of these models for testing the efficacy of a cell therapy specifically is only caveated by the considerations discussed above, such as rationalizing the extent to which cell loss occurs in mutant models of HD and whether neural circuitry repair is likely to improve the behavioral deficits that manifest in these models.

4.2 Assessing the efficacy of a cell therapy product on cognitive function in models of PD and HD

For PD, recovery of motor function post-transplantation of human-derived cells has been reported, typically using the classic unilateral 6-OHDA lesion model, due to the stability of the model long-term (for example, Kirkeby et al., 2012; Kriks et al., 2011). However, assessing the impact of the therapies on cognitive dysfunctions has proven more challenging.

The intrastriatal 6-OHDA model is also prone to plasticity and compensation to support post-transplant assessment long-term. Relatively mild deficits in genetically modified rodents also makes evaluation of cognitive recovery somewhat challenging. Indeed, inherent variability between animals that are infused with the alpha-synuclein viral vectors and the neurorecovery induced by systemic immunosuppressants, makes assessment of functional efficacy in these models challenging as well. For this reason, the neurotoxic models, such as 6-OHDA and quinolinic acid, are often preferred. It is well-documented that bilateral 6-OHDA lesions of the medial forebrain bundle will have a severe impact on animal welfare and therefore are not viable to maintain long-term. To overcome these issues, scientists have developed a limited number of cognitive tasks that can be undertaken in the unilateral model, such as the lateralized choice reaction time task, and there has been some success in identifying positive effects of fetal dopamine grafts on attention, motivation and visuospatial function (Dowd & Dunnett, 2004; Lelos, Morgan, et al., 2016) (Fig. 2).

In a similar vein for HD, post-transplantation recovery of motor function has been reported in both the quinolinic acid lesion model (Besusso et al., 2020; Lelos, Roberton, et al., 2016) and in genetically modified mouse models of HD (Adil et al., 2018; Reidling et al., 2018). However, assessment of cognitive or neuropsychiatric function remains more challenging. Using the quinolinic acid lesion model, bilateral lesions are possible if they avoid complete destruction of the lateral neostriatum, otherwise lack of appetite and weight loss manifest and welfare issues arise. Some studies have reported improvements in cognitive function after transplantation of rodent fetal medium spiny neurons into the unilateral and bilateral quinolinic acid lesion models (Brasted, Watts, Robbins, & Dunnett, 1999). Indeed, an improvement in object recognition was also observed in the Q140 mutant mouse model of HD after transplantation of human neural stem cells (Reidling et al., 2018). Thus, while there are models available for assessing the functional efficacy of a transplant, evaluation of the cognitive/neuropsychiatric dysfunctions remains challenging, with relatively few studies successfully reporting on this aspect of behavioral recovery.

5. Challenge 4: Creating clinically relevant models

In order to safely move a cell therapy from the bench to the clinic, it is important to consider the extent to which preclinical models are clinically relevant—as well as where they fail to translate to the clinical context. Only by recognizing these caveats can we work toward more relevant models and not allow unforeseen factors to affect the clinical success of a cell therapy product. In pursuit of the perfect preclinical host, scientists have developed models that mimic the disease state more accurately, with many mutant HD lines created, including rat models harboring 51 CAG repeats, and both genetically modified and alpha-synuclein viral vector models of PD. They have created rodent lines with humanized immune systems. They have developed ever more sensitive and translational behavioral assays. However, despite all these efforts, the models used to test the safety and efficacy of cell therapies are often not optimal and not clinically relevant. As discussed above, this may be due to the particular requirements for the model, based on the scientific question at hand. These requirements may lead the researcher to choose a neurotoxic model and forsake the more disease-relevant genetically modified model, for example; or to use a healthy intact rat, if only cell survival is under investigation. If these choices are rationale and well-considered, taking into account the experimental question and the facilities available, then scientists should be capable of interpreting the results with an acknowledgement of the caveats.

However, after the more basic characterization of the cell therapy product is achieved, it is important to then consider the extent to which the preclinical assessments are likely to support robust clinical translation. At this stage, it is pertinent to consider the aspects of our preclinical models that are not representative of the clinical scenario. Some of these areas include (1) the age of the animals (i.e., the frequent use of young animals to test therapeutic, although the therapeutics are intended for an aged population), (2) the lack of medical complexity (i.e., these therapeutics will be used by patients with complex medical needs and likely to require additional medicines to manage symptoms or co-morbid conditions) and (3) the use of surgical implantation devices that differ from the human surgical theaters.

5.1 Using young animals to model the aged brain

The primary risk factor for most neurodegenerative diseases is age, with the prevalence of Parkinson's, Huntington's, Alzheimer's and stroke all increasing dramatically with increasing age. Despite this, the vast majority of studies conducted to assess cell therapies use either relatively young mice or rats (typically 6–8 weeks old at the point of transplantation), or many studies even transplant into the neonatal brain. These are choices likely influenced by very pragmatic considerations: the purchase and/or cumulative maintenance costs of young animals will be significantly less than aged animals; aged animals will require greater time to maintain good welfare post-surgery and post-immunosuppressant exposure; human cells take 5–6 months to mature in vivo, which is already makes for long, expensive and labor intensive experiments and aging animals would add to this timescale. However, despite this, it is important to recognize that this choice to use neonatal or young rodents is likely skewing our expectations of the cell therapy products. It is, without doubt, the best experimental approach to use aged rodents, from a purely translational perspective.

In studies that have directly compared the use of young and aged rats, it is clear that neural transplantation into young rats yields more efficacious transplants. For example, use of fetal dopamine tissues in aged (compared to young) rats revealed reduced survival of cells, inferior behavioral recovery, and poorer neurite outgrowth and host brain integration (Collier et al., 2015; Collier, Sortwell, & Daley, 1999). Even in the context of human cells, it has been reported that transplantation of human stem-cell derived neurons into the brains of 2-year-old rats resulted in significantly less fiber outgrowth and graft-to-host innervation than the same cells engrafted into the neonatal brain (Grinand & Takahashi, 2022). Consistent with this, where clinical trials of human fetal cells have been undertaken in PD patients, it has been reported that cell survival and functional improvements were evident to a higher degree in younger patients (<60 years old) than in older patients (Freed et al., 2001). Thus, this raises the possibility that the data produced in most rodent preclinical studies is not fully representative of what we might observe in the clinic. It also suggests that younger brains may better support the survival and integration of transplanted cells, but unless this age effect is more thoroughly researched and a strategy of using younger patients for clinical trials is more widely rationalized and supported by ethics boards, it will remain that the preclinical and clinical data are produced in populations of differing ages, making the preclinical data therefore insufficient for robust clinical translation.

5.2 Acknowledging complex transplant-drug interactions

A recurring concern with preclinical models is the extent to which they truly recapitulate the clinical situation and can be predictive of clinical outcome. People with PD or HD are likely to have complex medical needs. They will almost certainly require medications to alleviate the motor symptoms (anti-Parkinsonian drugs for PD, anti-chorea drugs for HD), medications to treat symptoms of non-motor aspects of the disease (e.g., depression, anxiety, sleep disorders) and potential drugs for unrelated conditions. Thus, a key consideration is the extent to which pharmaceuticals taken for co-morbid medical conditions, or to treat the symptoms of the diseases, may directly impact upon the survival, integration or maturation of the transplant.

To address this, there has been some consideration of the effect of L-DOPA exposure on the survival and maturation of neural grafts. These studies have shown that in rats given post-transplant administration of L-DOPA, xenograft survival, volume and functional recovery were similar to rats not exposed to L-DOPA, but the immune response to the xenograft was greater in the presence of L-DOPA (Breger, Kienle, Smith, Dunnett, & Lane, 2017). Indeed, the impact of other drugs, such as Glucagon Like Peptide-1 receptor (GLP-1R) agonists exendin-4 or liraglutide, on the survival and maturation of a neural graft has also been explored in combination with L-DOPA therapy (Elabi, Davies, & Lane, 2021). While some beneficial effects of particular drug combinations were identified (larger graft volumes, improved functional efficacy), there was also evidence of increased leukocyte infiltration in some conditions. Additionally, a recent study also identified greater expression of GIRK2 in hESC-derived grafts after exposure to L-DOPA, suggesting an influence of the drug on neuronal differentiation (Elabi et al., 2022). These data reveal the need to consider these complex interactions between the novel cell therapy and existing medical conditions/treatments in order to more robustly predict the survival or efficacy of the cell therapy in the patient brain.

5.3 Identifying when brain size matters

Rodents can be used to recapitulate aspects of disease and to test the safety and efficacy of a novel cell therapy. However, there remain several key questions that cannot readily be addressed using rodent models, due to the significant differences between the size of the human brain and mouse/rat brains. For example, critical questions related to the performance of the surgical implantation device require a larger brain in order to test the exact instrument that will be used in the surgical theater. Preclinical surgical practices in rodent neurosurgery broadly resemble the process that occurs within the human theater, however, the likelihood of cell survival after expulsion from the instrument in the clinic, or the expected diffusion, cannot be estimated based on the preclinical surgical data.

The larger sizes of the pig, sheep and primate brains, as well as the greater inherent complexity of the cortical-basal ganglia networks, allows for more reliable assessment of the integration of neural transplants into both distal and biologically relevant regions. Conversely, the placement of large human cells into the small rodent brain makes it challenging to develop a convincing profile of graft-host integration over large distances. For example, as discussed in Section 2.4, success in reconstructing the nigrostriatal pathway after homotopic transplantation in models of PD is likely, in part, related to the larger size of human cells and the greater ability of these cells to project longer distances in the relatively small rat or mouse brains. In a similar vein, it is challenging to make precise assumptions about the biodistribution of a population of cells or the required dosage for functional efficacy, based on the xenograft data obtained from the transplantation of human cells into the mouse or rat brain.

It is, therefore, important when choosing a model to consider the extent to which the scientific question can be addressed convincingly, or whether it is time to look beyond rodent models, to larger animals to address specific outstanding questions.

6. Discussion

Preclinical rodent models are an essential tool to test novel therapeutics prior to clinical translation. There have been a number of exciting advances in the field, with the development of more relevant models of molecular neuropathology for PD and HD, the establishment of ever more sensitive behavioral tasks and a better recognition of how to model the complex clinical environment in which a patient exists. For testing cell therapeutics, specifically, there have been considerable attempts to create "humanized" models, which harbor human alpha-synuclein (Hoban et al., 2020), humanized immune systems (Petrus-Reurer et al., 2021) or tolerance to human antigens (Kelly et al., 2009). Thus, with these scientific advances, examining a xenograft in a rodent brain allows a wide range of scientific questions to be explored.

Ultimately, the specific scientific question drives the choice of model to be used. As researchers adopt currently available models, it is critical to think carefully about the question being addressed to choose the most appropriate model. The use healthy or a disease model could be debated in the context of considering the survival, maturation or toxicity of a novel cell therapeutic. And while neurotoxin or protein overexpression models may only mimic certain features of the disease in humans, this might be pertinent to address specific questions such as neural network repair, rejection mechanisms, protein propagation or the effect of neuroinflammation on cell survival. Where long-term maturation and behavioral analysis are key components of the experiment, daily immunosuppression may be used and a robust model insensitive to drug-induced neuroprotection or endogenous compensation is required. Thus, it is important to recognize that no single model is optimal for testing cell therapeutics, but the choice of the model must be well rationalized for the scientific question.

The aim of this chapter, however, was to consider the specific challenges associated with using rodents to explore novel cell therapies. Significant challenges do exist is using rodent models to effectively assess human tissues long-term, for example, with poorer survival identified in the mouse brain, and neurorecovery and compensation evident both in mild disease models and after exposure to immunosuppressant drugs. The challenge of maintaining human xenografts in the rodent brain continues, but with a range of models being developed, including daily immunosuppression, neonatal tolerization, immunodeficient rodents and humanized mice, it is feasible to identify an immune manipulation strategy most suitable to the particular question. Only by being cognizant of some of these challenges and the caveats associated with the models, however, can we recognize the limitations in our data and hope to develop more robust systems to test cell therapeutics. Recognizing the need to assess therapeutics in the aged brain and in the context of complex medical regimes, for example, opens up the opportunity to collect more reliable data that has a greater chance of translating effectively to the clinic.

As we look toward the future, we would expect further development of preclinical models of disease, in pursuit of better understanding of the mechanisms of molecular dysfunction. Since cell therapies for neurodegenerative diseases rely on the use of human cells for transplantation, we are likely to see newer and more refined models developed, as well as move toward larger animals for the final stages of clinical translation.

References

Adil, M. M., Gaj, T., Rao, A. T., Kulkarni, R. U., Fuentes, C. M., Ramadoss, G. N., et al. (2018). hPSC-derived striatal cells generated using a scalable 3D hydrogel promote recovery in a Huntington disease mouse model. *Stem Cell Reports*, *10*(5), 1481–1491. https://doi.org/10.1016/J.STEMCR.2018.03.007.

Ashwini, A., Naganur, S. S., Smitha, B., Sheshadri, P., Prasanna, J., & Kumar, A. (2018). Cyclosporine A-mediated IL-6 expression promotes neural induction in pluripotent stem cells. *Molecular Neurobiology*, *55*(5), 4267–4279. https://doi.org/10.1007/S12035-017-0633-0.

Barker, R. A., Farrell, K., Guzman, N. V., He, X., Lazic, S. E., Moore, S., et al. (2019). Designing stem-cell-based dopamine cell replacement trials for Parkinson's disease. *Nature Medicine*, *25*(7). https://doi.org/10.1038/s41591-019-0507-2.

Barker, R. A., Parmar, M., Studer, L., & Takahashi, J. (2017). Human trials of stem cell-derived dopamine neurons for Parkinson's disease: Dawn of a new era. *Cell Stem Cell*, *21*(5), 569–573. https://doi.org/10.1016/j.stem.2017.09.014.

Bayram-Weston, Z., Jones, L., Dunnett, S. B., & Brooks, S. P. (2012a). Light and electron microscopic characterization of the evolution of cellular pathology in Hdh Q92 Huntington's disease knock-in mice. *Brain Research Bulletin*, *88*(2–3), 171–181. https://doi.org/10.1016/j.brainresbull.2011.03.013.

Bayram-Weston, Z., Jones, L., Dunnett, S. B., & Brooks, S. P. (2012b). Light and electron microscopic characterization of the evolution of cellular pathology in the R6/1 Huntington's disease transgenic mice. *Brain Research Bulletin, 88*(2–3), 104–112. https://doi.org/10.1016/j.brainresbull.2011.07.009.

Bayram-Weston, Z., Jones, L., Dunnett, S. B., & Brooks, S. P. (2012c). Light and electron microscopic characterization of the evolution of cellular pathology in YAC128 Huntington's disease transgenic mice. *Brain Research Bulletin, 88*(2–3), 137–147. https://doi.org/10.1016/j.brainresbull.2011.05.005.

Bayram-Weston, Z., Torres, E. M., Jones, L., Dunnett, S. B., & Brooks, S. P. (2012). Light and electron microscopic characterization of the evolution of cellular pathology in the Hdh (CAG)150 Huntington's disease knock-in mouse. *Brain Research Bulletin, 88*(2–3), 189–198. https://doi.org/10.1016/j.brainresbull.2011.03.014.

Berridge, K. C., Venier, I. L., & Robinson, T. E. (1989). Taste reactivity analysis of 6-hydroxydopamine-induced aphagia: Implications for arousal and anhedonia hypotheses of dopamine function. *Behavioral Neuroscience, 103*(1), 36–45. https://doi.org/10.1037//0735-7044.103.1.36.

Besusso, D., Schellino, R., Boido, M., Belloli, S., Parolisi, R., Conforti, P., et al. (2020). Stem cell-derived human striatal progenitors innervate striatal targets and alleviate sensorimotor deficit in a rat model of Huntington disease. *Stem Cell Reports, 14*(5), 876–891. https://doi.org/10.1016/J.STEMCR.2020.03.018.

Björklund, A., & Dunnett, S. B. (2019). The amphetamine induced rotation test: A re-assessment of its use as a tool to monitor motor impairment and functional recovery in rodent models of Parkinson's disease. *Journal of Parkinson's Disease, 9*, 17–29. https://doi.org/10.3233/JPD-181525.

Björklund, A., & Parmar, M. (2020). Neuronal replacement as a tool for basal ganglia circuitry repair: 40 years in perspective. *Frontiers in Cellular Neuroscience, 14*. https://doi.org/10.3389/FNCEL.2020.00146.

Bohlen, M., Hayes, E. R., Bohlen, B., Bailoo, J. D., Crabbe, J. C., & Wahlsten, D. (2014). Experimenter effects on behavioral test scores of eight inbred mouse strains under the influence of ethanol. *Behavioural Brain Research, 272*, 46–54. https://doi.org/10.1016/J.BBR.2014.06.017.

Brasted, P. J., Döbrössy, M. D., Robbins, T. W., Dunnett, S. B., Dobrossy, M. D., Robbins, T. W., et al. (1998). Striatal lesions produce distinctive impairments in reaction time performance in two different operant chambers. *Brain Research Bulletin, 46*(6), 487–493. Retrieved from http://www.ncbi.nlm.nih.gov/pubmed/9744285.

Brasted, P. J., Watts, C., Robbins, T. W., & Dunnett, S. B. (1999). Associative plasticity in striatal transplants. *Proceedings of the National Academy of Sciences of the United States of America, 96*(18), 10524–10529. Retrieved from http://www.pubmedcentral.nih.gov/articlerender.fcgi?artid=17922&tool=pmcentrez&rendertype=abstract.

Breger, L. S., & Fuzzati Armentero, M. T. (2019). Genetically engineered animal models of Parkinson's disease: From worm to rodent. *European Journal of Neuroscience, 49*, 533–560. https://doi.org/10.1111/ejn.14300.

Breger, L. S., Kienle, K., Smith, G. A., Dunnett, S. B., & Lane, E. L. (2017). Influence of chronic L-DOPA treatment on immune response following allogeneic and xenogeneic graft in a rat model of Parkinson's disease. *Brain, Behavior, and Immunity, 61*, 155–164. https://doi.org/10.1016/J.BBI.2016.11.014.

Brehm, M. A., Wiles, M. V., Greiner, D. L., & Shultz, L. D. (2014). Generation of improved humanized mouse models for human infectious diseases. *Journal of Immunological Methods, 410*, 3. https://doi.org/10.1016/J.JIM.2014.02.011.

Brooks, S. P., Trueman, R. C., & Dunnett, S. B. (2012). Assessment of motor coordination and balance in mice using the Rotarod, elevated bridge, and footprint tests. *Current Protocols in Mouse Biology, 2*(1). https://doi.org/10.1002/9780470942390.MO110165.

Carter, R. J., Morton, J., & Dunnett, S. B. (2001). Motor coordination and balance in rodents. *Current Protocols in Neuroscience, 15*(1). https://doi.org/10.1002/0471142301.ns0812s15.

Cenci, M. A., & Björklund, A. (2020). Animal models for preclinical Parkinson's research: An update and critical appraisal. *Progress in Brain Research, 252*, 27–59. https://doi.org/10.1016/BS.PBR.2020.02.003.

Chaudhuri, K. R., & Titova, N. (2019). Societal burden and persisting unmet needs of Parkinson's disease. *European Neurological Review, 14*(1), 28–35. https://doi.org/10.17925/ENR.2019.14.1.28.

Clemens, L. E., Jansson, E. K. H., Portal, E., Riess, O., & Nguyen, H. P. (2014). A behavioral comparison of the common laboratory rat strains Lister hooded, Lewis, Fischer 344 and Wistar in an automated homecage system. *Genes, Brain, and Behavior, 13*(3), 305–321. https://doi.org/10.1111/GBB.12093.

Collier, T. J., O'Malley, J., Rademacher, D. J., Stancati, J. A., Sisson, K. A., Sortwell, C. E., et al. (2015). Interrogating the aged striatum: Robust survival of grafted dopamine neurons in aging rats produces inferior behavioral recovery and evidence of impaired integration. *Neurobiology of Disease*, 77, 191–203. https://doi.org/10.1016/J.NBD.2015.03.005.

Collier, T. J., Sortwell, C. E., & Daley, B. F. (1999). Diminished viability, growth, and behavioral efficacy of fetal dopamine neuron grafts in aging rats with long-term dopamine depletion: An argument for neurotrophic supplementation. *Journal of Neuroscience, 19*(13), 5563–5573. https://doi.org/10.1523/JNEUROSCI.19-13-05563.1999.

de Castro Gomes, V., Hassan, W., Maisonnette, S., Johnson, L. R., Ramos, A., & Landeira-Fernandez, J. (2013). Behavioral evaluation of eight rat lines selected for high and low anxiety-related responses. *Behavioural Brain Research, 257*, 39–48. https://doi.org/10.1016/J.BBR.2013.09.028.

Decressac, M., Mattsson, B., & Björklund, A. (2012). Comparison of the behavioural and histological characteristics of the 6-OHDA and α-synuclein rat models of Parkinson's disease. *Experimental Neurology, 235*(1), 306–315. https://doi.org/10.1016/J.EXPNEUROL.2012.02.012.

Döbrössy, M. D., & Dunnett, S. B. (2005). Training specificity, graft development and graft-mediated functional recovery in a rodent model of Huntington's disease. *Neuroscience, 132*(3), 543–552. https://doi.org/10.1016/j.neuroscience.2005.01.016.

Dowd, E., & Dunnett, S. B. (2004). Deficits in a lateralized associative learning task in dopamine-depleted rats with functional recovery by dopamine-rich transplants. *The European Journal of Neuroscience, 20*(7), 1953–1959. https://doi.org/10.1111/j.1460-9568.2004.03637.x.

Dowd, E., Monville, C., Torres, E. M., & Dunnett, S. B. (2005). The corridor task: A simple test of lateralised response selection sensitive to unilateral dopamine deafferentation and graft-derived dopamine replacement in the striatum. *Brain Research Bulletin, 68*(1–2), 24–30. https://doi.org/10.1016/j.brainresbull.2005.08.009.

Drew, C. J. G., Sharouf, F., Randell, E., Brookes-Howell, L., Smallman, K., Sewell, B., et al. (2021). Protocol for an open label: Phase I trial within a cohort of foetal cell transplants in people with Huntington's disease. *Brain. Communications, 3*(1). https://doi.org/10.1093/BRAINCOMMS/FCAA230.

Elabi, O. F., Davies, J. S., & Lane, E. L. (2021). L-dopa-dependent effects of GLP-1R agonists on the survival of dopaminergic cells transplanted into a rat model of Parkinson disease. *International Journal of Molecular Sciences, 22*(22). https://doi.org/10.3390/IJMS222212346.

Elabi, O. F., Pass, R., Sormonta, I., Nolbrant, S., Drummond, N., Kirkeby, A., et al. (2022). Human embryonic stem cell-derived dopaminergic grafts alleviate L-DOPA induced dyskinesia. *Journal of Parkinson's Disease, 1–16.* https://doi.org/10.3233/JPD-212920.

El-Akabawy, G., Rattray, I., Johansson, S. M., Gale, R., Bates, G., & Modo, M. (2012). Implantation of undifferentiated and pre-differentiated human neural stem cells in the R6/2 transgenic mouse model of Huntington's disease. *BMC Neuroscience*, *13*(1). https://doi.org/10.1186/1471-2202-13-97.

Ellenbroek, B., & Youn, J. (2016). Rodent models in neuroscience research: Is it a rat race? *Disease Models & Mechanisms*, *9*(10), 1079–1087. https://doi.org/10.1242/DMM.026120.

Farshim, P. P., & Bates, G. P. (2018). Mouse models of Huntington's disease. *Methods in Molecular Biology (Clifton, N.J.)*, *1780*, 97–120. https://doi.org/10.1007/978-1-4939-7825-0_6.

Frederiksen, H. R., Doehn, U., Tveden-Nyborg, P., & Freude, K. K. (2020). Non-immunogenic induced pluripotent stem cells, a promising way forward for allogenic transplantations for neurological disorders. *Frontiers in Genome Editing*, *2*, 623717. https://doi.org/10.3389/FGEED.2020.623717.

Freed, C. R., Greene, P. E., Breeze, R. E., Tsai, W.-Y., DuMouchel, W., Kao, R., et al. (2001). Transplantation of embryonic dopamine neurons for severe Parkinson's disease. *New England Journal of Medicine*, *344*(10), 710–719. https://doi.org/10.1056/NEJM200103083441002.

Frohna, P. A., Neal-Beliveau, B. S., & Joyce, J. N. (1997). Delayed plasticity of the mesolimbic dopamine system following neonatal 6-OHDA lesions. *Synapse (New York, N.Y.)*, *25*(3), 293–305. https://doi.org/10.1002/(sici)1098-2396(199703)25:3<293::aid-syn9>3.0.co;2-6.

Grealish, S., Heuer, A., Cardoso, T., Kirkeby, A., Jönsson, M., Johansson, J., et al. (2015). Monosynaptic tracing using modified rabies virus reveals early and extensive circuit integration of human embryonic stem cell-derived neurons. *Stem Cell Reports*, *4*(6), 975–983. https://doi.org/10.1016/j.stemcr.2015.04.011.

Grinand, L., & Takahashi, J. (2022). Automated measurement of fluorescence signals reveals a significant increase of the graft-derived neurite extension in neonates compared to aged rats. *Regenerative Therapy*, *19*, 97–106. https://doi.org/10.1016/J.RETH.2022.01.002.

Guo, J., Zeng, Y., Liang, Y., Wang, L., Su, H., & Wu, W. (2007). Cyclosporine affects the proliferation and differentiation of neural stem cells in culture. *Neuroreport*, *18*(9), 863–868. https://doi.org/10.1097/WNR.0B013E32811D6D36.

Hasanpour, M., Mitricheva, E., Logothetis, N., & Noori, H. R. (2021). Intensive longitudinal characterization of multidimensional biobehavioral dynamics in laboratory rats. *Cell Reports*, *35*(2), 108987. https://doi.org/10.1016/J.CELREP.2021.108987.

Heuer, A., Jönsson, M. E., Pfisterer, U., Kirkeby, A., & Parmar, M. (2016). hESC-derived neural progenitors prevent xenograft rejection through neonatal desensitisation. *Experimental Neurology*, *282*, 78–85. https://doi.org/10.1016/J.EXPNEUROL.2016.05.027.

Hoban, D. B., Shrigley, S., Mattsson, B., Breger, L. S., Jarl, U., Cardoso, T., et al. (2020). Impact of α-synuclein pathology on transplanted hESC-derived dopaminergic neurons in a humanized α-synuclein rat model of PD. *Proceedings of the National Academy of Sciences of the United States of America*, *117*(26), 15209–15220. https://doi.org/10.1073/PNAS.2001305117/-/DCSUPPLEMENTAL.

Jackson, H. M., Onos, K. D., Pepper, K. W., Graham, L. C., Akeson, E. C., Byers, C., et al. (2015). DBA/2J genetic background exacerbates spontaneous lethal seizures but lessens amyloid deposition in a mouse model of Alzheimer's disease. *PLoS One*, *10*(5). https://doi.org/10.1371/JOURNAL.PONE.0125897.

Janowski, M., Jablonska, A., Kozlowska, H., Orukari, I., Bernard, S., Bulte, J. W., et al. (2012). Neonatal desensitization does not universally prevent xenograft rejection. *Nature Methods*, *9*, 856–858. https://doi.org/10.1038/nmeth.2146.

Kauhausen, J. A., Thompson, L. H., & Parish, C. L. (2015). Chondroitinase improves midbrain pathway reconstruction by transplanted dopamine progenitors in parkinsonian mice. *Molecular and Cellular Neurosciences*, *69*, 22–29. https://doi.org/10.1016/J.MCN.2015.10.002.

Kelly, C. M., Precious, S. V., Penketh, R., Amso, N., Dunnett, S. B., & Rosser, A. E. (2007). Striatal graft projections are influenced by donor cell type and not the immunogenic background. *Brain: A Journal of Neurology*, *130*(Pt 5), 1317–1329. https://doi.org/10.1093/brain/awm053.

Kelly, C. M., Precious, S. V., Scherf, C., Penketh, R., Amso, N. N., Battersby, A., et al. (2009). Neonatal desensitization allows long-term survival of neural xenotransplants without immunosuppression. *Nature Methods*, *6*(4), 271–273. https://doi.org/10.1038/nmeth.1308.

Kirik, D., Rosenblad, C., Burger, C., Lundberg, C., Johansen, T. E., Muzyczka, N., et al. (2002). Parkinson-like neurodegeneration induced by targeted overexpression of alpha-synuclein in the nigrostriatal system. *The Journal of Neuroscience*, *22*(7), 2780–2791. https://doi.org/20026246.

Kirkeby, A., Grealish, S., Wolf, D. A., Nelander, J., Wood, J., Lundblad, M., et al. (2012). Generation of regionally specified neural progenitors and functional neurons from human embryonic stem cells under defined conditions. *Cell Reports*, *1*(6), 703–714. https://doi.org/10.1016/j.celrep.2012.04.009.

Kirkeby, A., Parmar, M., & Barker, R. A. (2017). Strategies for bringing stem cell-derived dopamine neurons to the clinic: A European approach (STEM-PD). *Progress in Brain Research*, *230*, 165–190. https://doi.org/10.1016/bs.pbr.2016.11.011.

Klein, A., & Dunnett, S. B. (2012). Analysis of skilled forelimb movement in rats: The single pellet reaching test and staircase test. *Current Protocols in Neuroscience*. Chapter 8(Suppl. 58) https://doi.org/10.1002/0471142301.NS0828S58.

Kooreman, N. G., de Almeida, P. E., Stack, J. P., Nelakanti, R. V., Diecke, S., Shao, N. Y., et al. (2017). Alloimmune responses of humanized mice to human pluripotent stem cell therapeutics. *Cell Reports*, *20*(8), 1978. https://doi.org/10.1016/J.CELREP.2017.08.003.

Kosior, N., & Leavitt, B. R. (2018). Murine models of Huntington's disease for evaluating therapeutics. *Methods in Molecular Biology (Clifton, N.J.)*, *1780*, 179–207. https://doi.org/10.1007/978-1-4939-7825-0_10.

Kriks, S., Shim, J.-W., Piao, J., Ganat, Y. M., Wakeman, D. R., Xie, Z., et al. (2011). Dopamine neurons derived from human ES cells efficiently engraft in animal models of Parkinson's disease. *Nature*, *480*(7378), 547–551. https://doi.org/10.1038/nature10648.

Lane, E. L., Harrison, D. J., Ramos-Varas, E., Hills, R., Turner, S., & Lelos, M. J. (2022). Spontaneous graft-induced Dyskinesias are independent of 5-HT neurons and levodopa priming in a model of Parkinson's disease. *Movement Disorders: Official Journal of the Movement Disorder Society*, *37*(3), 613–619. https://doi.org/10.1002/MDS.28856.

Lelos, M. J., & Dunnett, S. B. (2018). Generating excitotoxic lesion models of Huntington's disease. *Methods in Molecular Biology (Clifton, N.J.)*, *1780*, 209–220. https://doi.org/10.1007/978-1-4939-7825-0_11.

Lelos, M. J., Harrison, D. J., & Dunnett, S. B. (2012). Intrastriatal excitotoxic lesion or dopamine depletion of the neostriatum differentially impairs response execution extrapersonal space. *European Journal of Neuroscience*, *36*(10), 3420–3428. https://doi.org/10.1111/J.1460-9568.2012.08256.X.

Lelos, M. J., Harrison, D. J., Rosser, A. E., & Dunnett, S. B. (2013). The lateral neostriatum is necessary for compensatory ingestive behaviour after intravascular dehydration in female rats. *Appetite*, *71*, 287–294. https://doi.org/10.1016/J.APPET.2013.09.003.

Lelos, M. J., Morgan, R. J., Kelly, C. M., Torres, E. M., Rosser, A. E., & Dunnett, S. B. (2016). Amelioration of non-motor dysfunctions after transplantation of human dopamine neurons in a model of Parkinson's disease. *Experimental Neurology*, *278*. https://doi.org/10.1016/j.expneurol.2016.02.003.

Lelos, M. J., Roberton, V. H., Vinh, N. N., Harrison, C., Eriksen, P., Torres, E. M., et al. (2016). Direct comparison of rat- and human-derived ganglionic Eminence tissue grafts on motor function. *Cell Transplantation, 25*(4), 665–675. https://doi.org/10.3727/096368915X690297.

Li, W., Englund, E., Widner, H., Mattsson, B., van Westen, D., Lätt, J., et al. (2016). Extensive graft-derived dopaminergic innervation is maintained 24 years after transplantation in the degenerating parkinsonian brain. *Proceedings of the National Academy of Sciences, 113*(23), 6544–6549. https://doi.org/10.1073/pnas.1605245113.

Loring, J. F. (2018). Autologous induced pluripotent stem cell-derived neurons to treat Parkinson's disease. *Stem Cells and Development, 27*(14), 958–959. https://doi.org/10.1089/SCD.2018.0107.

Mazzocchi-Jones, D., Döbrössy, M., & Dunnett, S. B. (2009). Embryonic striatal grafts restore bi-directional synaptic plasticity in a rodent model of Huntington's disease. *The European Journal of Neuroscience, 30*(11), 2134–2142. https://doi.org/10.1111/j.1460-9568.2009.07006.x.

Moriarty, N., Gantner, C. W., Hunt, C. P. J., Ermine, C. M., Frausin, S., Viventi, S., et al. (2022). A combined cell and gene therapy approach for homotopic reconstruction of midbrain dopamine pathways using human pluripotent stem cells. *Cell Stem Cell, 29*(3), 434–448. https://doi.org/10.1016/J.STEM.2022.01.013.

Moy, S. S., Nadler, J. J., Young, N. B., Nonneman, R. J., Segall, S. K., Andrade, G. M., et al. (2008). Social approach and repetitive behavior in eleven inbred mouse strains. *Behavioural Brain Research, 191*(1), 118–129. https://doi.org/10.1016/J.BBR.2008.03.015.

Petrus-Reurer, S., Romano, M., Howlett, S., Jones, J. L., Lombardi, G., & Saeb-Parsy, K. (2021). Immunological considerations and challenges for regenerative cellular therapies. *Communications Biology, 4*(1), 798. https://doi.org/10.1038/S42003-021-02237-4.

Plank, A. C., Canneva, F., Raber, K. A., Urbach, Y. K., Dobner, J., Puchades, M., et al. (2018). Early alterations in operant performance and prominent huntingtin aggregation in a congenic F344 rat line of the classical CAG n51trunc model of Huntington disease. *Frontiers in Neuroscience, 12*(JAN). https://doi.org/10.3389/FNINS.2018.00011.

Reidling, J. C., Relaño-Ginés, A., Holley, S. M., Ochaba, J., Moore, C., Fury, B., et al. (2018). Human neural stem cell transplantation rescues functional deficits in R6/2 and Q140 Huntington's disease mice. *Stem Cell Reports, 10*(1), 58–72. https://doi.org/10.1016/J.STEMCR.2017.11.005.

Roberton, V. H., Evans, A. E., Harrison, D. J., Precious, S. V., Dunnett, S. B., Kelly, C. M., et al. (2013). Is the adult mouse striatum a hostile host for neural transplant survival? *Neuroreport, 24*(18), 1010–1015. https://doi.org/10.1097/WNR.0000000000000066.

Schweitzer, J. S., Song, B., Herrington, T. M., Park, T.-Y., Lee, N., Ko, S., et al. (2020). Personalized iPSC-derived dopamine progenitor cells for Parkinson's disease. *The New England Journal of Medicine, 382*(20), 1926. https://doi.org/10.1056/NEJMOA1915872.

Tamburrino, A., Churchill, M., Wan, O., Colino-Sanguino, Y., Ippolito, R., Bergstrand, S., et al. (2015). Cyclosporin promotes neurorestoration and cell replacement therapy in pre-clinical models of Parkinson's disease. *Acta Neuropathologica Communications, 3*, 84. https://doi.org/10.1186/S40478-015-0263-6.

Thirugnanam, T., & Santhakumar, K. (2022). Chemically induced models of Parkinson's disease. *Comparative Biochemistry and Physiology. Toxicology & Pharmacology: CBP, 252*. https://doi.org/10.1016/J.CBPC.2021.109213.

Thompson, L. H., Grealish, S., Kirik, D., & Björklund, A. (2009). Reconstruction of the nigrostriatal dopamine pathway in the adult mouse brain. *The European Journal of Neuroscience, 30*(4), 625–638. https://doi.org/10.1111/j.1460-9568.2009.06878.x.

Traggiai, E., Chicha, L., Mazzucchelli, L., Bronz, L., Piffaretti, J. C., Lanzavecchia, A., et al. (2004). Development of a human adaptive immune system in cord blood cell-transplanted mice. *Science (New York, N.Y.), 304*(5667), 104–107. https://doi.org/10.1126/SCIENCE.1093933.

Ulusoy, A., Decressac, M., Kirik, D., & Björklund, A. (2010). Viral vector-mediated overexpression of α-synuclein as a progressive model of Parkinson's disease. *Progress in Brain Research*, *184*(C), 89–111. https://doi.org/10.1016/S0079-6123(10)84005-1.

Van der Perren, A., Macchi, F., Toelen, J., Carlon, M. S., Maris, M., de Loor, H., et al. (2015). FK506 reduces neuroinflammation and dopaminergic neurodegeneration in an α-synuclein-based rat model for Parkinson's disease. *Neurobiology of Aging*, *36*(3), 1559–1568. https://doi.org/10.1016/J.NEUROBIOLAGING.2015.01.014.

Van der Perren, A., Toelen, J., Casteels, C., Macchi, F., Van Rompuy, A. S., Sarre, S., et al. (2015). Longitudinal follow-up and characterization of a robust rat model for Parkinson's disease based on overexpression of alpha-synuclein with adeno-associated viral vectors. *Neurobiology of Aging*, *36*(3), 1543–1558. https://doi.org/10.1016/J.NEUROBIOLAGING.2014.11.015.

Watts, C., & Dunnett, S. B. (1998). Effects of severity of host striatal damage on the morphological development of intrastriatal transplants in a rodent model of Huntington's disease: Implications for timing of surgical intervention. *Journal of Neurosurgery*, *89*(2), 267–274. https://doi.org/10.3171/JNS.1998.89.2.0267.

CHAPTER FIVE

Considerations for the use of biomaterials to support cell therapy in neurodegenerative disease

Victoria H. Roberton[a,b,*] and James B. Phillips[a,b]
[a]UCL School of Pharmacy, University College London, London, United Kingdom
[b]UCL Centre for Nerve Engineering, London, United Kingdom
*Corresponding author: e-mail address: v.roberton@ucl.ac.uk

Contents

Abstract

Using biomaterials to complement cell therapies for a range of neurodegenerative disorders provides a potential opportunity to improve cell survival, integration and regeneration. Materials can be developed to serve as cell or drug delivery systems, temporary scaffolds or longer-term encapsulation structures. However, as yet clinical translation has been limited, with much work still required to achieve this. This chapter discusses a number of considerations for using biomaterials to support cell therapies, which are likely to be essential to achieve successful translation.

Biomaterials offer a sophisticated therapeutic cell delivery system which can provide an opportunity to overcome issues associated with conventional injection approaches, including poor cell survival and integration after transplantation. Cell therapies are under investigation as a novel treatment option for a range of neurodegenerative disorders, including Parkinson's disease, Huntington's disease, stroke, and progressive degeneration following

International Review of Neurobiology, Volume 166
ISSN 0074-7742
https://doi.org/10.1016/bs.irn.2022.09.009

trauma. Biomaterials may take many forms, from gels and sponges to fibers and particles, which can enable protection of implanted cells, modulation of host responses, and controlled release of soluble factors. In this chapter we will summarize some of the challenges associated with using biomaterials in the context of cell therapies for the treatment of neurodegenerative disorders.

1. Using biomaterials to support cell therapies

The treatment of neurodegenerative conditions often requires slow injection of small volumes of therapeutic cells into deep anatomical structures via long, thin cannulae in multiple injection tracts to target specific areas. As cells pass through the delivery device they are exposed to complex mechanical forces which have an impact on their viability and subsequent functionality following transplantation (Amer, Rose, Shakesheff, Modo, & White, 2017). The local forces experienced by cells can be modulated through combination with biomaterials, in which viscosity can be controlled, rather than standard cell suspension media. This can reduce the detrimental impact of sheer stress during the injection process on the survival of implanted cells. Furthermore, after loading cells into the delivery device and during the slow process of delivery, cells may settle in long cannulae or needles leading to uneven injection deposits, or leaving cell suspension remaining (Torres, Trigano, & Dunnett, 2015). Increasing the viscosity of the delivery material can also alleviate these issues and reduce the potential for reflux of cells after injection.

In addition to protection of cells during the delivery itself, incorporating the cells within a biomaterial matrix can provide support post-transplantation. During delivery and immediately post-transplantation may be the period of greatest cell loss (Brundin et al., 2000). A range of factors are responsible for this including; apoptosis due in part to a lack of oxygen, the lack of substrate for adherence ("anoikis"), inflammation, and the transfer from a growth factor rich in vitro environment to the diseased/damaged adult brain (Dunnett & Rosser, 2014; Newland et al., 2015; Praet et al., 2015). Using biomaterials can not only provide a substrate for adherence of cells, but also contribute to the rebuilding of tissue architecture where required, for example, in the treatment of stroke or traumatic injury. In addition, biomaterials offer the potential for the incorporation of growth factors or small molecules to promote survival and regeneration of transplanted and host cells (reviewed in Orive, Anitua, Pedraz, & Emerich, 2009). This approach can potentially improve the therapeutic effect of molecules which have short half-lives and are quickly degraded, and as such have not been successfully applied clinically due to variable outcomes when delivered via injection.

2. Considerations for the use of biomaterials

The following suggestions are not exhaustive, but are important for developing cell based therapies which may be combined with biomaterials and are summarized in Table 1. A lack of consideration for these early on can be a barrier to clinical translation.

2.1 Material selection

There are an extensive and increasing number of biomaterial options available and one of the key challenges for researchers interested in selecting a biomaterial for central nervous system (CNS) cell therapy is how to make an appropriate choice (Tam, Fuehrmann, Mitrousis, & Shoichet, 2014). Biomaterials can be categorized in various ways but one of the traditional subdivisions is based on whether they occur naturally or are synthetic (Gregory & Phillips, 2021). Natural biomaterials include proteins such as collagen, fibrin and silk and polysaccharides such as hyaluronan, chitosan and alginate, whereas synthetic materials tend to be polyesters such as poly-ε-caprolactone, poly-l-lactic acid, poly-glycolic acid and poly lactic-co-glycolic acid, although there are many other variations on these themes. There are advantages and disadvantages to both synthetic and natural materials. Natural biomaterials are inherently biocompatible and often bioactive (Joyce, Fabra, Bozkurt, & Pandit, 2021) and there is a long history of their effective use in medical applications, however there can be challenges associated with their complexity, poor mechanical properties in some situations, and risk of infection/variability when sourced from living organisms. Synthetic materials are generally easier to manufacture reliably at scale and can have more controllable chemical and mechanical properties, but are often poor at integration with host tissue. Synthetic materials can degrade to yield undesirable products whereas natural materials tend not to cause toxicity but have more potential to trigger immune responses (Lim & Spector, 2017). Combining synthetic and natural biomaterials can be an

Table 1 A summary of important considerations for the use of biomaterials to support cell therapy in neurodegenerative disease.

Material Selection	
Natural/synthetic	Biocompatibility and good integration of natural materials vs. ease of manufacture at scale and more control of chemical and mechanical properties of synthetic. Will depend on required function, delivery route, and integration requirements.
Formulation	Can be tailored to requirements for specific indication and delivery method. Also allows for combination with small molecules/neurotrophic factors and gene therapy tools. Examples include hydrogels, nano/microparticles, tissue scaffolds/grafts.
Delivery Method	Dependent on therapeutic target, e.g., via injection or implantation, and application, e.g., cell encapsulation or tissue replacement. Need for different delivery methods will affect selection for material selection and formulation
Biocompatibility	Biocompatibility essential to support survival of donor cells and allow integration into host.
Degradation	Complete degradation not essential if biocompatible, though more commonly will degrade and be cleared after therapeutic purpose is served. Consideration of degradation products and duration of degradation is important.
Mechanical Properties	Matching stiffness to native tissue as closely as possible to improve integration and reduce tissue response.
Host Response	Characterization required to determine effect of material on host and graft function.
Clinical Translation	
Manufacturing	Therapeutics must be suitable for manufacturing to a large enough scale under GMP conditions for distribution to the clinic.
Reproducibility	Manufacturing processes must produce consistent therapeutics whose quality can be routinely assessed to ensure safety and efficacy prior to implantation.
Regulatory Considerations	Comprehensive non-clinical and clinical trial data will be required to demonstrate safety and efficacy of therapeutics for assessment via relevant regulatory agencies to allow routine clinical use.

effective approach for overcoming limitations and maximizing benefits, but the increased complexity can cause additional challenges.

Choosing an appropriate biomaterial should theoretically involve (1) deciding on the function required (e.g., temporary protection of cells during delivery, long-term encapsulation of cells, local release of small molecules or growth factors etc.), (2) considering the delivery route and (3) understanding the requirements for integration at the intended host destination (e.g., appropriate stiffness and dimensions, positive modulation of host tissue response, nature and timescale of degradation). Having considered these aspects, an appropriate biomaterial would ideally be selected from the substantial array of available options, or a new version developed and refined as required. However the reality of the situation is that numerous other considerations tend to encroach on this rational approach and researchers find their choice is significantly limited by factors such as cost and availability of materials, as well as access to biomaterials science knowledge and expertise. Innovation in this area is frequently hampered by the assumption that selecting a commonly used material or repurposing one established for use in another indication will be more acceptable to the scientific community and de-risk subsequent regulatory requirements to demonstrate safety. A pragmatic approach therefore tends to prevail, in which a relatively small number of biomaterial options are used in the majority of studies. Because the majority of previously reported biomaterials have been developed for use in non-nervous system tissues and organs, often the properties are not ideal when deployed in the CNS. In tissue engineering it can be useful to consider tissues as being either matrix-dominated or cell-dominated, depending on their main function (Brown & Phillips, 2007). The skin and most of the musculoskeletal system are matrix-dominated because key functions are for example, being stiff, strong, and resilient, whereas the nervous system (along with organs such as the liver and glands) is much more cell-dominated. In recent decades much biomaterials development has focused predominantly on the matrix-dominated tissues, a consequence of which is that many popular and available biomaterials are not necessarily well suited to application in the cell-dominated environment of nervous system cell transplantation.

A common approach in tissue engineering is to try and select the biomaterial based on similarity to the natural extracellular matrix, either by matching it directly (e.g., using fibrillar collagen as a biomaterial to repair tissues where that is the dominant matrix component) or adapting synthetic polymers to mimic the key structural features of the natural extracellular matrix. While the CNS may be considered to be a cell-dominated tissue, it possesses a sophisticated and dynamic extracellular matrix containing a rich diversity of proteins and polysaccharides which differs according to brain region, compartment (Lau, Cua, Keough, Haylock-Jacobs, & Yong, 2013) and in response to damage and disease. It is possible that future developments in biomaterials science will produce materials which either contain the appropriate molecules to reconstitute the complex natural brain extracellular matrix or to mimic it closely. It is appealing to think that once such materials are available they will be useful for cell transplantation, but it is also worth considering that the microenvironment which is most appropriate for improving cell transplantation may well be different to that found in healthy brain tissue.

2.2 Material formulation and delivery

In addition to considering the source of biomaterial, i.e., the molecular components that will be chosen, the formulation most suitable for organizing those components to best serve the requirements of the chosen application must be optimized. For cell delivery to the CNS, hydrogels are the standard approach since they closely mimic the soft tissue environment and are highly permeable to oxygen and nutrients (Oliveira, Silva-Correia, Reis, & Oliveira, 2018; Tam et al., 2014). Hydrogels are 3-dimensional networks of hydrophilic polymers swollen with water which set to form highly porous soft material. Cells can be mixed with the precursor solution prior to gelation, then become trapped by the biomaterial structure as it sets. The process of gelation is often triggered by a thermal change, a change in the ionic environment or pH, enzyme action, or in some cases chemical cross-linking initiated by light (Gasperini, Mano, & Reis, 2014). One consideration for using hydrogels in cell delivery to the CNS via injection is whether to inject a cellular hydrogel that has been formed in advance, or to inject the precursor solution and rely on subsequent in situ gelation. Another important consideration is to ensure the method used in the gelation process does not harm the encapsulated cells, e.g., avoid subjecting them to low pH or UV light. The range of hydrogels which have been explored for cell delivery have been reviewed in detail elsewhere (Chen, Harn, & Chiou, 2018; Gasperini et al., 2014; Jarrin, Cabré, & Dowd, 2021; Lim & Spector, 2017; Oliveira et al., 2018; Tam et al., 2014) and these include hydrogels formed using natural and synthetic polymers. The porosity of hydrogels can be controlled through manipulating the concentration of polymer and the extent of crosslinking, providing opportunities to control ingress of immune cells and large molecules in order to limit exposure of transplanted cells to components of the host immune response (Eleftheriadou et al., 2022; Orive et al., 2015). Hydrogels can be patterned and organized using techniques such as electrospinning and bioprinting as well as through natural cellular self-alignment to provide topographical cues that can help integration and direct neuronal regeneration (Cangellaris & Gillette, 2018; East, de Oliveira, Golding, & Phillips, 2010). In combination with cell encapsulation, hydrogels can be used to deliver small molecules, neurotrophic factors and gene therapy tools either through incorporating them directly into the material or trapping carrier systems such as nanoparticles or exosomes (Jarrin, Cabré, & Dowd, 2021; Katoh, Yokota, & Fehlings, 2019; Tam et al., 2014; Yuan et al., 2017).

Nano and micro particles are another useful biomaterial for supporting cell therapy in treating CNS disorders. They can be formed from synthetic or natural polymers using various manufacturing approaches which allow drugs and biologics to be incorporated for subsequent release, and surface chemistry and can be modified to help targeting and movement across barriers such as the BBB and the plasma membrane of cells (Lauzon, Daviau, Marcos, & Faucheux, 2015). Size, drug-loading, biomaterial structure and chemical composition can be varied to provide control over the location, dosage and duration of therapeutic delivery. Examples of relevant applications might include incorporation of growth factors to improve survival and phenotype of transplanted cells or elicit positive host responses such as myelination and angiogenesis (Chen et al., 2015; Metcalfe, Strom, Williams, & Fahmy, 2015; Samal & Segura, 2021), chondroitinase ABC to increase plasticity (Muir, De Winter, Verhaagen, & Fawcett, 2019), small molecules to modulate host glial responses and local immunosuppression to improve engraftment (Eleftheriadou et al., 2022; Song, Agrawal, Griffin, & Schmidt, 2019).

Selection of an appropriate material formulation will depend on the indication for cell therapy, and therefore the mechanism of delivery required. In some instances, for example, delivery to deep brain structures where there is some cell loss but tissue structure is largely intact, the most appropriate delivery method may be injection of cells via a long cannula,. In these cases the incorporation of biomaterials into the delivery system demands that the material can be passed through a narrow cannula, to reduce damage to the brain parenchyma on injection. Thus for these applications combining therapeutic cells with hydrogels which form a gel in situ may be most appropriate. While hydrogels and nano/micro particles are frequently the most relevant biomaterials for use in CNS cell therapy, it is worth noting that other biomaterial formulations such as scaffolds, decellularized tissue grafts and fibrous mats have been developed for nervous system tissue engineering and regenerative medicine (Oliveira et al., 2018). These are most often designed with tissue replacement in mind, e.g., filling cavities and bridging gaps in the injured spinal cord. An advantage is that more structured scaffold materials can provide control over topography and stiffness, providing additional ways to control cell and tissue responses, but a major limitation is these approaches tend to require more invasive approaches to transplantation which makes them less appealing in treating neurodegenerative diseases (Chen et al., 2018).

2.3 Biocompatibility and degradation

There is a long history of materials being used in the CNS for applications such as shunt systems for treating hydrocephalus, neural prosthetics for electrical stimulation and recording, and various local drug delivery systems (Zhong & Bellamkonda, 2008). For these approaches materials need to be biocompatible but they do not necessarily need to be biodegradable, i.e., they can be relatively inert structures which remain as prosthetics for the duration of their use. In regenerative medicine it is more common to seek degradable materials, with the intention that they serve their therapeutic purpose and then are cleared from the body. In cases where natural extracellular matrix components are used as biomaterials this is relatively straightforward, since these can integrate with host tissue and be remodeled and cleared by natural cell-matrix interactions (Brown & Phillips, 2007). For other materials, careful consideration needs to be given to the mechanism and timing of degradation. If a material such as a synthetic polymer degrades rapidly by hydrolysis this can cause changes in local pH as well as the release of high concentrations of soluble products which can then diffuse to other CNS areas as well as other tissues and organs via the circulation before excretion (Sun, Mei, Song, Cui, & Wang, 2006). Slower degradation can help mitigate detrimental consequences associated with this, but prolonging the presence of the material may have other disadvantages. When selecting a biomaterial formulation it is therefore important to consider the duration over which it should degrade and the effect on both the local environment and distal tissues and organs of the products which will be released as the material breaks down.

2.4 Mechanical properties and the host response

A critical consideration in the use of biomaterials for CNS applications is their mechanical properties. Unlike musculoskeletal tissue engineering applications where materials might need to exhibit certain minimum levels of tensile strength (e.g., in tendon repair), hardness (e.g., in bone or tooth repair) or elasticity (e.g., in connective tissue repair), the predominant concern for biomaterials to be deployed within the CNS is that they can integrate appropriately with the host tissue. Arguably the most important mechanical property of an implantable biomaterial for the CNS is stiffness, defined as the extent to which an object can resist external deformation. The stiffness of brain tissue varies according to location and health but in general it is one of the least stiff (most compliant) tissues in the body (Budday et al., 2017; Moeendarbary et al., 2017). In addition, CNS cells and in particular astrocytes are exquisitely sensitive to local changes in stiffness, which means that implanting or injecting a material with a stiffness mismatch compared to the host tissue is likely to elicit a reactive response similar to that associated with traumatic injury (Moshayedi et al., 2010, 2014; Pekny & Pekna, 2014). Regenerating host neurons may also respond to stiffness cues in terms of rate and direction of growth (Balgude, Yu, Szymanski, & Bellamkonda, 2001; Kayal, Moeendarbary, Shipley, & Phillips, 2020), and the significant effect of the local stiffness environment on differentiation and behavior of transplanted therapeutic cells is well known (Engler, Sen, Sweeney, & Discher, 2006).

Despite the critical importance of this topic, mechanical properties of biomaterials for CNS application, and indeed of CNS tissue itself, are poorly characterized. There are challenges with being able to measure stiffness in very soft materials, additionally complicated by hydrogels (and tissues) tending to exhibit viscoelastic properties. Viscoelasticity is the term used to describe materials with both viscous (fluid) and elastic (solid) deformation characteristics, where the stiffness is dependent on the rate at which the force is applied. This means that there is not a single value of stiffness that can be assigned, making it challenging to try and match material and tissue stiffnesses without using techniques such as rheometry or dynamic mechanical analysis (Bartlett, Eleftheriadou, Evans, Choi, & Phillips, 2020). Dynamic mechanical analysis has been used to establish the stiffness properties of brain and spinal cord tissues from rats and these values have been used as a benchmark to identify specific biomaterial formulations (of collagen, fibrin, and alginate) that have similar stiffness profiles (Bartlett, Eleftheriadou, et al., 2020), however many challenges remain in trying to match biomaterial stiffness to that of host brain tissue in living humans and animals. Many biomaterials can be "tuned" in terms of their stiffness, e.g., by altering their density, concentration, cross-linking etc., so as better ways to quantify the mechanical properties of native CNS tissue emerge it should be possible to tailor the biomaterials of the future to integrate mechanically after implantation.

The host response to biomaterials and their degradation products must be considered and characterized to determine the effect this will have on both the host and graft function. The factors discussed above including; material type and formulation, stiffness, biocompatibility, degradation, and mechanical properties, may all have an effect on the host response. This may be further mitigated through careful material selection, and incorporation of factors which may mediate the host response.

3. Promising biomaterial and cell based therapies for neurodegenerative disease

To date, few clinical trials for cell therapies have included biomaterials despite the development of numerous potential biomaterial formulations, and a great deal of preclinical research in relevant models of disease (reviewed in Jarrin, Cabré, & Dowd, 2021). However, some biomaterial based approaches show particular promise for use in combination with cell therapies, including; structural support for cell delivery, targeted growth factor and small molecule delivery, and encapsulation of cells for release of therapeutics at the target.

As for neurodegenerative disorders, cell therapies for spinal cord injuries face problems with low cell survival rates, poor retention of cells at the lesion site, and challenges filling the lesion cavity (Liu, Schackel, Weidner, & Puttagunta, 2018). Therefore, a potential role for biomaterials is to provide a structure for cell transplantation to help generate living replacement tissue. It was reported that, in 2019, around a quarter of 49 reviewed clinical trials used cells in combination with biomaterial matrices where cells are delivered directly into the spinal cord lesion site (Bartlett, Burley, Ip, Phillips, & Choi, 2020). Similarly, neural repair strategies for neurodegenerative disease or stroke could benefit from the support provided from biomaterials in the delivery of therapeutic cells, using hydrogels or electrospun scaffolds (reviewed by Bruggeman, Moriarty, Dowd, Nisbet, & Parish, 2019).

Delivery of growth factors or small molecules alongside therapeutic cells can promote survival and differentiation of cells following transplantation, as well as being pro-regenerative for protection of remaining endogenous cells. Delivery of growth factors, in particular GDNF, has been of great interest for the treatment of PD. Pre-clinical data from GDNF delivery to the brain parenchyma in rodent and primate models prompted clinical trials which produced less striking results, partly in relation to issues with growth factor delivery (Barker et al., 2020; Sidorova & Saarma, 2020). Achieving a sustained effective dose of GDNF at the target site is a challenge, prompting investigation of alternative delivery methods, including via biomaterials (Jarrin, Hakami, Newland, & Dowd, 2021). Using growth factors or small molecules in combination with cell therapies offers the potential to address issues with cell survival, and improve axon growth and integration of cells, allowing homotopic transplantation to the substantia nigra to achieve connectivity to the striatum (Moriarty et al., 2022; Moriarty, Pandit, & Dowd, 2017; Xiao, Zhang, & Tan, 2022).

Finally, as an alternative to transplantation of cells in combination with biomaterials to improve survival while allowing integration of cells, semi-permeable materials can be used to encapsulate cells and isolate them from the host environment while allowing diffusion of nutrients and therapeutic factors. In an approach successfully implemented for Type I diabetes, encapsulation of islets in semi-permeable materials prevents rejection without the need for continuous immunosuppression, while allowing passage of nutrients and secretion of insulin (Zhang et al., 2022). Likewise, cells can potentially be used as delivery vehicles for therapeutics in treating neurodegenerative disease, allowing release of neurotransmitters such as dopamine, or growth factors such as GDNF. Encapsulation of these cells in semi-permeable membranes can allow their delivery to the nervous system with protection from rejection, while allowing release of therapeutics (Eleftheriadou et al., 2022).

4. Clinical translation

Combinations of the factors discussed above will have an impact on whether cell therapies incorporating biomaterials may be successfully translated to the clinic, and how long this may take. Much research is carried out into the use of novel biomaterials in combination with cell therapies, but whether these approaches can be successfully translated must be considered, including the feasibility of manufacturing a reproducible product under

GMP which can be delivered to a patient by a clinician. Although some approaches show great promise in preclinical studies, these practicalities are frequently overlooked and can be a hindrance to translation. Complex manufacturing requirements which require much manual input are less likely to be translated into a therapeutic product, as these will be a challenge to manufacture to a larger scale and under GMP conditions. Material formulations whose production can be transferred to scaled up, automated processes are much more feasible, and therefore clinical translation is more likely.

An important consideration, as mentioned briefly, is the process of regulatory approval. Complexity and novelty of biomaterial formulations combined with cell/gene/pharmacological therapies must be weighed against these considerations. Materials with approval for use in other indications will have already undergone rigorous testing, and there will be a lot of existing information on factors such as biocompatibility and degradation. For novel material formulations this work will all need to be undertaken, increasing the package of preclinical data required, and extending timelines for clinical translation. However, as discussed, these considerations must be weighed against the appropriate material selection for the indication in question. These factors should be considered early on in research into the use of biomaterials for cell therapies. It may well be much easier to develop material formulations with these factors in mind in the first place, than to try to adjust protocols further down the line.

5. Conclusion

Biomaterial technologies have the potential to help address some of the challenges associated with the transplantation of cells into the CNS. For this to be effective, many factors must be taken into account, including some of the considerations discussed here. Biomaterial selection and formulation requires an understanding of chemical, biological and mechanical features of the materials and of the host tissue, in addition to the already formidable complexity of CNS cell transplantation. The effective use of biomaterials in supporting this type of therapy therefore requires a collaborative interdisciplinary approach which draws upon the wealth of possibilities available in the biomaterials field but balances this carefully with the practicalities required to progress treatments to safe and effective clinical application.

References

Amer, M. H., Rose, F. R. A. J., Shakesheff, K. M., Modo, M., & White, L. J. (2017). Translational considerations in injectable cell-based therapeutics for neurological applications: Concepts, progress and challenges. *NPJ Regenerative Medicine*, *2*, 23. https://doi.org/10.1038/s41536-017-0028-x.

Balgude, A. P., Yu, X., Szymanski, A., & Bellamkonda, R. V. (2001). Agarose gel stiffness determines rate of DRG neurite extension in 3D cultures. *Biomaterials*, *22*(10), 1077–1084. https://doi.org/10.1016/S0142-9612(00)00350-1.

Barker, R. A., Björklund, A., Gash, D. M., Whone, A., Van Laar, A., Kordower, J. H., et al. (2020). GDNF and Parkinson's disease: Where next? A summary from a recent workshop. *Journal of Parkinson's Disease*, *10*, 875–891. IOS Press.

Bartlett, R. D., Burley, S., Ip, M., Phillips, J. B., & Choi, D. (2020). Cell therapies for spinal cord injury: Trends and challenges of current clinical trials. *Neurosurgery*, *87*, E456–E472. https://doi.org/10.1093/NEUROS/NYAA149.

Bartlett, R. D., Eleftheriadou, D., Evans, R., Choi, D., & Phillips, J. B. (2020). Mechanical properties of the spinal cord and brain: Comparison with clinical-grade biomaterials for tissue engineering and regenerative medicine. *Biomaterials*, *258*, 120303. https://doi.org/10.1016/j.biomaterials.2020.120303.

Brown, R. A., & Phillips, J. B. (2007). Cell responses to biomimetic protein scaffolds used in tissue repair and engineering. *International Review of Cytology*, *262*, 75–150. https://doi.org/10.1016/S0074-7696(07)62002-6.

Bruggeman, K. F., Moriarty, N., Dowd, E., Nisbet, D. R., & Parish, C. L. (2019). Harnessing stem cells and biomaterials to promote neural repair. *British Journal of Pharmacology*, *176*, 355–368. https://doi.org/10.1111/bph.14545.

Brundin, P., Karlsson, J., Emgård, M., Schierle, G. S. K., Hansson, O., Petersén, Å., et al. (2000). Improving the survival of grafted dopaminergic neurons: A review over current approaches. *Cell Transplantation*, *9*, 179–195. https://doi.org/10.1177/096368970000900205.

Budday, S., Sommer, G., Birkl, C., Langkammer, C., Haybaeck, J., Kohnert, J., et al. (2017). Mechanical characterization of human brain tissue. *Acta Biomaterialia*, *48*, 319–340. https://doi.org/10.1016/j.actbio.2016.10.036.

Cangellaris, O. V., & Gillette, M. U. (2018). Biomaterials for enhancing neuronal repair. *Frontiers in Materials*, *5*, 21. https://doi.org/10.3389/fmats.2018.00021.

Chen, Y.-S., Harn, H.-J., & Chiou, T.-W. (2018). The role of biomaterials in implantation for central nervous system injury. *Cell Transplantation*, *27*(3), 407–422. https://doi.org/10.1177/0963689717732991.

Chen, Y., Pan, C., Xuan, A., Xu, L., Bao, G., Liu, F., et al. (2015). Treatment efficacy of NGF nanoparticles combining neural stem cell transplantation on Alzheimer's disease model rats. *Medical Science Monitor*, *21*, 3608–3615. https://doi.org/10.12659/msm.894567.

Dunnett, S. B., & Rosser, A. E. (2014). Challenges for taking primary and stem cells into clinical neurotransplantation trials for neurodegenerative disease. *Neurobiology of Disease*, *61*, 79–89. https://doi.org/10.1016/j.nbd.2013.05.004.

East, E., de Oliveira, D. B., Golding, J. P., & Phillips, J. B. (2010). Alignment of astrocytes increases neuronal growth in three-dimensional collagen gels and is maintained following plastic compression to form a spinal cord repair conduit. *Tissue Engineering. Part A*, *16*(10), 3173–3184. https://doi.org/10.1089/ten.tea.2010.0017.

Eleftheriadou, D., Evans, R. E., Atkinson, E., Ahmed, A., Gavins, F. K. H., Boyd, A. S., et al. (2022). An alginate-based encapsulation system for delivery of therapeutic cells to the CNS. *RSC Advances*, *12*, 4005–4015.

Engler, A. J., Sen, S., Sweeney, H. L., & Discher, D. E. (2006). Matrix elasticity directs stem cell lineage specification. *Cell*, *126*(4), 677–689. https://doi.org/10.1016/j.cell.2006.06.044.

Gasperini, L., Mano, J. F., & Reis, R. L. (2014). Natural polymers for the microencapsulation of cells. *Journal of the Royal Society, Interface*, *11*(100), 20140817. https://doi.org/10.1098/rsif.2014.0817.

Gregory, H., & Phillips, J. B. (2021). Materials for peripheral nerve repair constructs: Natural proteins or synthetic polymers? *Neurochemistry International*, *143*, 104953. https://doi.org/10.1016/j.neuint.2020.104953.

Jarrin, S., Cabré, S., & Dowd, E. (2021). The potential of biomaterials for central nervous system cellular repair. *Neurochemistry International*, *144*, 104971. https://doi.org/10.1016/J.NEUINT.2021.104971.

Jarrin, S., Hakami, A., Newland, B., & Dowd, E. (2021). Growth factor therapy for Parkinson's disease: Alternative delivery systems. *Journal of Parkinson's Disease*, *11*, S229–S236. https://doi.org/10.3233/JPD-212662.

Joyce, K., Fabra, G. T., Bozkurt, Y., & Pandit, A. (2021). Bioactive potential of natural biomaterials: Identification, retention and assessment of biological properties. *Signal Transduction and Targeted Therapy*, *6*(1), 122. https://doi.org/10.1038/s41392-021-00512-8.

Katoh, H., Yokota, K., & Fehlings, M. G. (2019). Regeneration of spinal cord connectivity through stem cell transplantation and biomaterial scaffolds. *Frontiers in Cellular Neuroscience*, *13*, 248. https://doi.org/10.3389/fncel.2019.00248.

Kayal, C., Moeendarbary, E., Shipley, R. J., & Phillips, J. B. (2020). Mechanical response of neural cells to physiologically relevant stiffness gradients. *Advanced Healthcare Materials*, *9*(8), e1901036. https://doi.org/10.1002/adhm.201901036.

Lau, L. W., Cua, R., Keough, M. B., Haylock-Jacobs, S., & Yong, V. W. (2013). Pathophysiology of the brain extracellular matrix: A new target for remyelination. *Nature Reviews Neuroscience*, *14*(10), 722–729. https://doi.org/10.1038/nrn3550.

Lauzon, M.-A., Daviau, A., Marcos, B., & Faucheux, N. (2015). Nanoparticle-mediated growth factor delivery systems: A new way to treat Alzheimer's disease. *Journal of Controlled Release*, *206*, 187–205. https://doi.org/10.1016/j.jconrel.2015.03.024.

Lim, T. C., & Spector, M. (2017). Biomaterials for enhancing CNS repair. *Translational Stroke Research*, *8*(1), 57–64. https://doi.org/10.1007/s12975-016-0470-x.

Liu, S., Schackel, T., Weidner, N., & Puttagunta, R. (2018). Biomaterial-supported cell transplantation treatments for spinal cord injury: Challenges and perspectives. *Frontiers in Cellular Neuroscience*, *11*, 430. https://doi.org/10.3389/fncel.2017.00430.

Metcalfe, S. M., Strom, T. B., Williams, A., & Fahmy, T. M. (2015). Multiple sclerosis and the LIF/IL-6 Axis: Use of nanotechnology to harness the Tolerogenic and reparative properties of LIF. *Nanobiomedicine (Rij)*, *2*, 5. https://doi.org/10.5772/60622.

Moeendarbary, E., Weber, I. P., Sheridan, G. K., Koser, D. E., Soleman, S., Haenzi, B., et al. (2017). The soft mechanical signature of glial scars in the central nervous system. *Nature Communications*, *8*, 14787. https://doi.org/10.1038/ncomms14787.

Moriarty, N., Gantner, C. W., Hunt, C. P. J., Ermine, C. M., Frausin, S., Viventi, S., et al. (2022). A combined cell and gene therapy approach for homotopic reconstruction of midbrain dopamine pathways using human pluripotent stem cells. *Cell Stem Cell*, *29*(3), 434–448.e435. https://doi.org/10.1016/j.stem.2022.01.013.

Moriarty, N., Pandit, A., & Dowd, E. (2017). Encapsulation of primary dopaminergic neurons in a GDNF-loaded collagen hydrogel increases their survival, re-innervation and function after intra-striatal transplantation. *Scientific Reports*, 7, 16033. https://doi.org/10.1038/s41598-017-15970-w.

Moshayedi, P., da Costa, F. L., Christ, A., Lacour, S. P., Fawcett, J., Guck, J., et al. (2010). Mechanosensitivity of astrocytes on optimized polyacrylamide gels analyzed by quantitative morphometry. *Journal of Physics: Condensed Matter*, *22*(19), 194114. https://doi.org/10.1088/0953-8984/22/19/194114.

Moshayedi, P., Ng, G., Kwok, J. C. F., Yeo, G. S. H., Bryant, C. E., Fawcett, J. W., et al. (2014). The relationship between glial cell mechanosensitivity and foreign body

reactions in the central nervous system. *Biomaterials*, *35*(13), 3919–3925. https://doi.org/10.1016/j.biomaterials.2014.01.038.
Muir, E., De Winter, F., Verhaagen, J., & Fawcett, J. (2019). Recent advances in the therapeutic uses of chondroitinase ABC. *Experimental Neurology*, *321*, 113032. https://doi.org/10.1016/j.expneurol.2019.113032.
Newland, B., Welzel, P. B., Newland, H., Renneberg, C., Kolar, P., Tsurkan, M., et al. (2015). Tackling cell transplantation Anoikis: An injectable, shape memory Cryogel microcarrier platform material for stem cell and neuronal cell growth. *Small*, *11*(38), 5047–5053. https://doi.org/10.1002/smll.201500898.
Oliveira, E. P., Silva-Correia, J., Reis, R. L., & Oliveira, J. M. (2018). Biomaterials developments for brain tissue engineering. In H. J. Chun, C. H. Park, I. K. Kwon, & G. Khang (Eds.), *Cutting-edge enabling Technologies for Regenerative Medicine* (pp. 323–346). Singapore: Springer Singapore.
Orive, G., Anitua, E., Pedraz, J. L., & Emerich, D. F. (2009). Biomaterials for promoting brain protection, repair and regeneration. *Nature Reviews Neuroscience*, *10*, 682–692.
Orive, G., Santos, E., Poncelet, D., Hernandez, R. M., Pedraz, J. L., Wahlberg, L. U., et al. (2015). Cell encapsulation: Technical and clinical advances. *Trends in Pharmacological Sciences*, *36*(8), 537–546. https://doi.org/10.1016/j.tips.2015.05.003.
Pekny, M., & Pekna, M. (2014). Astrocyte reactivity and reactive Astrogliosis: Costs and benefits. *Physiological Reviews*, *94*(4), 1077–1098. https://doi.org/10.1152/physrev.00041.2013.
Praet, J., Santermans, E., Daans, J., Le Blon, D., Hoornaert, C., Goossens, H., et al. (2015). Early inflammatory responses following cell grafting in the CNS trigger activation of the subventricular zone: A proposed model of sequential cellular events. *Cell Transplantation*, *24*, 1481–1492. https://doi.org/10.3727/096368914X682800.
Samal, J., & Segura, T. (2021). Injectable biomaterial shuttles for cell therapy in stroke. *Brain Research Bulletin*, *176*, 25–42. https://doi.org/10.1016/j.brainresbull.2021.08.002.
Sidorova, Y. A., & Saarma, M. (2020). Can growth factors cure Parkinson's disease? *Trends in Pharmacological Sciences*, *41*(12), 909–922. https://doi.org/10.1016/j.tips.2020.09.010.
Song, Y. H., Agrawal, N. K., Griffin, J. M., & Schmidt, C. E. (2019). Recent advances in nanotherapeutic strategies for spinal cord injury repair. *Advanced Drug Delivery Reviews*, *148*, 38–59. https://doi.org/10.1016/j.addr.2018.12.011.
Sun, H., Mei, L., Song, C., Cui, X., & Wang, P. (2006). The in vivo degradation, absorption and excretion of PCL-based implant. *Biomaterials*, *27*(9), 1735–1740. https://doi.org/10.1016/j.biomaterials.2005.09.019.
Tam, R. Y., Fuehrmann, T., Mitrousis, N., & Shoichet, M. S. (2014). Regenerative therapies for central nervous system diseases: A biomaterials approach. *Neuropsychopharmacology*, *39*(1), 169–188. https://doi.org/10.1038/npp.2013.237.
Torres, E. M., Trigano, M., & Dunnett, S. B. (2015). Translation of cell therapies to the clinic: Characteristics of cell suspensions in large-diameter injection cannulae. *Cell Transplantation*, *24*, 737–749. https://doi.org/10.3727/096368914X685429.
Xiao, B., Zhang, S.-C., & Tan, E.-K. (2022). Combination therapy using GDNF and cell transplant in Parkinson's disease. *Molecular Neurodegeneration*, *17*(1), 49. https://doi.org/10.1186/s13024-022-00553-9.
Yuan, D., Zhao, Y., Banks, W. A., Bullock, K. M., Haney, M., Batrakova, E., et al. (2017). Macrophage exosomes as natural nanocarriers for protein delivery to inflamed brain. *Biomaterials*, *142*, 1–12. https://doi.org/10.1016/j.biomaterials.2017.07.011.
Zhang, Q., Gonelle-Gispert, C., Li, Y., Geng, Z., Gerber-Lemaire, S., Wang, Y., et al. (2022). Islet encapsulation: New developments for the treatment of type 1 diabetes. *Frontiers in Immunology*, *13*, 869984. https://doi.org/10.3389/fimmu.2022.869984.
Zhong, Y., & Bellamkonda, R. V. (2008). Biomaterials for the central nervous system. *Journal of the Royal Society, Interface*, *5*(26), 957–975. https://doi.org/10.1098/rsif.2008.0071.

CHAPTER SIX

Challenges of translating a cell therapy to GMP

Gerhard Bauer* and Brian Fury
University of CA Davis, School of Medicine, Stem Cell Program, Institute for Regenerative Cures, GMP Facility, Sacramento, CA, United States
*Corresponding author: e-mail address: gbauer@ucdavis.edu

Contents

Abstract

Over the last decade, cell and gene therapies have contributed remarkably to the array of novel therapies combating diseases that did not have any hope for an effective treatment or, let alone, a cure. This remarkable achievement was underlined by the marketing approval of CAR T cell therapies in 2017 in the United States, followed by many other countries, world-wide. Since then, thousands of patients have benefited from this autologous, gene modified cell therapy (Abou-El-Enein et al., 2021). Rare diseases, particularly

International Review of Neurobiology, Volume 166
ISSN 0074-7742
https://doi.org/10.1016/bs.irn.2022.09.002

innate neurological diseases such as Huntington's disease have also been a target for cell therapies. The notion of being able to augment or replace the function of diseased neurons with progenitor cells or neurons derived from human stem cells has been researched for the last 10 years and is finally reaching the stage of clinical translation (Holley et al., 2018; Reidling et al., 2018). With these cellular and gene therapies reaching clinical applicability, it is important to bring them to patients in a safe, efficacious and reliable way, and for this purpose, Good Manufacturing Practice (GMP) needs to be applied to the manufacturing of such novel and often life-saving therapies. In the first decade of the 21st century, gene therapies, particularly in vivo adenoviral vector gene therapy (Wilson, 2009) and hematopoietic stem cell gene therapies (Hacein-Bey-Abina et al., 2008) were associated with adverse events that were highly publicized and gave the field a bad reputation in the public eye. The last two decades, however, due to the meticulous work of dedicated researchers, and excellent progress in GMP manufacturing, cell and gene therapies have become safe and efficacious and have propelled the field to the forefront of the most promising novel therapies available for current unmet medical needs. This book chapter will discuss the historical perspective of cellular therapies and their development, will describe the currently available cell and gene therapies for different diseases and their GMP manufacturing methods and challenges, and will point out the future direction of these therapies and their envisioned manufacturing, as can be foreseen currently.

1. The history of cell and gene therapy manufacturing

1.1 The bone marrow transplant experience

The successful application of cellular therapies started in the 1950s with the pioneering work of E. Donnel Thomas, he performed the first allogeneic bone marrow transplant and reported it in the New England Journal of Medicine on September 12, 1957. The initial success was limited, six patients received chemotherapy and radiation, followed by the intravenous infusion of diluted bone marrow from a healthy donor (Henig & Zuckerman, 2014). Of the six patients, two engrafted, but all patients eventually died 3 to 4 months post marrow infusion. Looking back, this complication was expected, as little to nothing was known about donor/recipient tissue matching; the underlying factors, the histocompatibility antigens, were not yet known or defined. As could also be expected, many physicians abandoned the field of marrow transplantation at that time. Histocompatibility factors, the human leukocyte antigens (HLA) were discovered and classified during the 1960s, which explained on a very scientific level the reason why donor tissues would be rejected by the immune system of the recipient, or in the case of bone marrow transplantation, the graft would even reject the recipient tissue as it would develop a foreign immune system in the recipient. E. Donnel Thomas moved to Seattle in the late 1960s and joined the faculty of the Fred Hutchinson Cancer Research Center in 1974 to become the director of Medical Oncology. In 1977, at the "Hutch", 100 bone marrow transplants with conditioning using chemotherapy and radiation had been performed. 54 Patients had acute myeloid leukemia (AML) and 46 patients had acute lymphoblastic leukemia (ALL). After a follow up period of 1 to 4.5 years, 13 patients were in complete remission, reflecting only a 13% cure rate (Henig & Zuckerman, 2014). It was realized at that time that a treatment point earlier in the disease may be needed, and E. Donnel Thomas reported a cure of 50% in AML patients transplanted early, in their first remission. Currently, with careful disease management and treatment scheduling, an 80% cure rate in allogeneic bone marrow transplantation is achievable. In 1990, the pioneering work of E. Donnel Thomas was recognized, he won the Nobel Prize for in Physiology or Medicine (Henig & Zuckerman, 2014).

What are the most important take home points from this short history of bone marrow transplantation? Firstly, it could clearly be demonstrated that the immune system played an important role in cancer surveyance and was able to elicit a cure in AML and ALL, in the majority of transplanted patients. Secondly, only meticulous research and persistence in it, with a critical look at hypothesis versus actual results after an experiment is imperative. Thirdly, the unwavering participation of patients in clinical trials that may pose a lot of risk needs to be pointed out. Clinical trials may not necessarily benefit them, but may benefit society and medical research tremendously, in the long run.

1.2 The stem cell (gene) therapy experience

Hematopoietic stem cell gene therapy applications helped to pioneer the field of cell engineering and the GMP manufacturing of a gene modified, cellular product, for patient administration.

As discussed above, allogeneic bone marrow transplantation, which owes its efficacy to the transplantation and engraftment of allogeneic hematopoietic stem cells to generate a completely new hematopoietic system, can successfully treat blood cancers, and therefore has become the standard therapy for genetic blood cell diseases. Its greatest drawback, however, is the need for a matched hematopoietic stem cell donor; the better the match, the fewer unwanted side effects, namely graft versus host disease, will be experienced by the recipient. However, only a limited number of suitable matched donors is available for the number of patients that need to receive such a transplant, leaving many of them without the option of a potentially curative and life saving treatment. In addition, the conditioning regimen, often consisting of marrow ablative chemotherapy is, in itself, very toxic and can lead to morbidity and significant mortality. If autologous hematopoietic stem cells could be used by gene modifying them to correct such a hematopoietic disease, the field could be opened up to almost all affected patients.

This treatment option was already pioneered in the late 1980s and came into clinical applications in the early 1990s. Our group, under the leadership of Dr. Donald Kohn, developed a stem cell gene therapy treatment for adenosine deaminase deficient severe combined immunodeficiency (ADA SCID) (Morgan, Gray, Lomova, & Kohn, 2017). This disease was chosen, as it is a monogenic disease. Only one gene, coding for the enzyme adenosine deaminase, is defective. Mutations in the gene, for instance, can be single amino acid substitutions, insertions, deletions, splicing errors, and post-translational defects in the ADA protein. In all of these cases, the resulting protein either does not function or has a very much reduced enzymatic activity. This led to the hypothesis that the insertion of a functional ADA gene should result in the production of functional ADA enzyme, with the possibility of the correction of the disease. ADA SCID occurs when T cells cannot mature from their hematopoietic progenitors, due to lacking "detoxification" by ADA. Functional enzyme should allow for the restoration of a functional T cell compartment, as could be demonstrated by the application of pegylated ADA via injection into ADA SCID patients.

Initially, a retroviral vector transferring the ADA gene was developed in Dr. Donald Kohn's lab and tested under laboratory conditions. Retroviral vectors are Moloney leukemia virus based "oncoretroviral vectors," as they produce cancer in mice. However, in a re-engineered version, they are very well suited as transport vehicles for permanent gene transfer, since they naturally integrate their payload into the genome of the target cell. They are engineered not to be replication competent, and therefore, they only transport the gene of interest into the nucleus; as replication incompetent vectors, they also do not cause disease, they do not integrate any genes of the oncovirus, they only integrate the gene of interest, that was packaged into the viral particles. A drawback associated with these vectors is that they need actively replicating target cells in order to successfully integrate the gene of interest into the cellular genome; the break-down of the nuclear membrane is required for the pre-integration complex to enter the cell nucleus. For cells that easily divide in culture, this is not an issue, however, for relatively quiescent cells, this poses a problem. True HSCs are unfortunately relatively quiescent. However, it could be shown that oncoretroviruses can successfully transduce engrafting human hematopoietic stem cells, if these cells are cultured under very specific conditions. Human HSCs have a tendency to quickly differentiate into progenitors, and do not remain in the stem cell state. Only true hematopoietic stem cells, however, lead to a durable engraftment of a new hematopoietic system, and in this special case, a gene corrected hematopoietic system. It was therefore necessary to develop culture conditions that preserved the "stemness" and in vivo engrafting ability of these HSCs. Our group was able to develop such culture conditions in the early 1990s (Bauer et al., 1997).

It now became necessary to develop culture conditions that could be applied for clinical grade transductions on patient cells. Translation of these culture conditions into a GMP manufacturing scenario posed great difficulties. We needed to obtain clinical grade reagents, particularly cytokines, which at that time, were not yet routinely available GMP grade, or pharmaceutical grade. Clinical grade culture of hematopoietic stem and progenitor cells was completely novel. We therefore had to use research grade reagents for the most part, which needed to be tested thoroughly to be useable in these transduction cultures. The separation and purification of hematopoietic stem and progenitor cells was initially performed using the clinical grade CellPro CD34+ cell separation device, based on a Streptavidin binding and elution process (Hawkins, Marley, O'Brien, Gordon, & Goldman, 1997). After a patent suit, this system became unavailable, and the clinical grade Baxter Isolex magnetic bead selection system was applied, all the way through the 1990s into the early 2000s. This system also became obsolete and was replaced with the clinical grade Miltenyi CliniMacs magnetic cell selection system, which is still available to date. In order for HSCs to maintain their "stemness" and in vivo engraftment ability, they need to adhere to a matrix. In the beginning of the GMP development of the culture system, the HSC culture matrix was autologous bone marrow stroma that had to be cultured and irradiated prior to the initiation of the HSC culture. Irradiation of the stroma was required since it should be prevented from actively replicating and overgrowing the culture. Later in the 1990s, this stroma culture was replaced with fibronectin fragment CH296, subsequently receiving the trademark "Retronectin."

As can be envisioned, all of these changes, particularly the use of different CD34+ cells separation devices and the switch in matrices, also introduced major changes in the standard operating procedures and final batch records for cell manufacturing in the GMP facility. It needs to be understood that, at that time, the field was at the very beginning, and GMP manufacturing procedures for cell and gene therapy had to be developed. It took us a good number of years to implement and improve procedures for GMP manufacturing of such products, with constant communication with the regulatory agency and application of necessary changes in the day-to day-operations and in the quality system, which was also under constant development.

1.3 Adverse events in gene therapy

In 1999, the field of gene therapy experienced a setback, since an 18 year old participant in an adenoviral vector gene therapy clinical trial designed to treat ornithine transcarbamylase deficiency died after the direct in vivo administration of the gene therapy vector (Wilson, 2009). The patient developed an intense immune response to the vector, leading to a significant inflammatory reaction, followed by a blood-clotting disorder, and finally kidney, liver, and lung failure. This incidence was widely reported in the media at that time.

In vivo gene therapy differs from the previously described ex vivo stem cell gene therapy approach, where cells get transduced using retro or lentiviral vectors ex vivo, and vector does not get directly administered into the patient. In 2000, all gene therapy clinical trials were halted at the University of Pennsylvania, where this incidence occurred and many other centers that conducted gene therapy clinical trials were put on alert by the FDA or received audits. Ex vivo gene therapy clinical trials continued, however, until another tragic incidence occurred: In 2003, it was reported that a pediatric patient who had received stem cell gene therapy for X-linked SCID, a different kind of Severe Combined Immunodeficiency caused by a defective IL-2 receptor on T cells, had developed a T cell leukemia (Hacein-Bey-Abina et al., 2008). The gene therapy approach for X linked SCID consisted of the transfer of an intact IL-2 receptor gamma chain into autologous hematopoietic stem cells using an oncoretroviral vector. 9 Patients overall were treated with this type of stem cell gene therapy, and eventually 4 of these 9 patients developed a T cell leukemia between 31 and 68 months after the administration of the gene modified cells. It was found that insertional oncogenesis was the cause of the T cell leukemia. The oncoretroviral vector had integrated next to the LIM domain-only 2 (LMO2) proto-oncogene and had activated it, causing the affected cell to clonally expand in a malignant fashion. After this report, all stem cell gene therapy clinical trials in the US were halted for an extended period of time, severely impacting the field of clinical gene therapy.

This series of adverse events led to a re-thinking within the field. Much safer vectors needed to be generated for in vivo gene therapy applications, so the risk to the patients receiving such therapies would be reduced, while maintaining the benefit of gene correction. For ex vivo gene therapy applications, the utilization of integrating vectors, particularly oncoretroviral vectors in long lived cells, such as hematopoietic cells, also had to be to re-evaluated.

1.4 Advances in safer ex vivo gene therapy product manufacturing

The very concerning adverse events precipitated by insertional oncogenesis in hematopoietic stem cells while using retroviral vectors for transduction were carefully evaluated over the years following the adverse events. It was found that several important factors were involved in the oncogenic transformation of transduced cells (Gabriel, Schmidt, & von Kalle, 2012):

1. A very high multiplicity of infection (number of viral particles per target cell) leads to many integration sites.
2. It was found that oncoretroviral vectors tended to "semi-randomly" integrate, favoring transcriptional start regions in active genes.
3. It was realized that a strong "selective survival advantage" was conferred onto progeny arising from the transduced hematopoietic cells in the X-linked SCID clinical gene therapy trial, favoring cells with a proliferative potential, such as cells with an upregulated proto-oncogene.
4. It was found that lentiviral vectors had one distinct advantage over oncoretroviral vectors, they did not preferentially integrate into transcriptional start regions of genes, while they still integrated into active genes, however.

In order to make ex vivo stem cell gene therapy safer, it was therefore suggested to switch to lentiviral vectors for the gene transduction of long lived cells, such as hematopoietic stem cells (Guenechea et al., 2000), to reduce the number of integrants per genome to one or two by lowering the multiplicity of infection, and to make sure a selective survival advantage of transduced cells is carefully monitored. The FDA also initially required the life-long follow up of patients participating in gene therapy trials using blood cells (not just hematopoietic stem cells), which was over time reduced to a 15 year follow up period.

These measures contributed to much safer gene therapy clinical trials applying transduced cells, although a small number of adverse events have still been reported to date, the most publicized event occurring in February of 2021; two patients with sickle cell disease were diagnosed with leukemia 5 years after receiving the gene therapy, however, it was later found that the gene therapy vector was most likely not the cause of the cancer.

1.5 Good manufacturing practice

Good Manufacturing Practice (GMP) is a national standard for the manufacturing of safe and efficacious therapeutics. It is important to point out the "national" aspect of this guiding principle. In this chapter, we will focus on US and European regulations for GMP, with the main emphasis being put on GMP product development and manufacturing in the US.

In the US, the GMP requirements are part of the federal law, set forth in 21 Code of Federal Regulations (CFR), part 210 and 211 (CFR, 2022). The Code of Federal Regulations is legally binding on a federal level, in all States of the US. Guidance for the Industry documents are also released by the FDA for certain aspects of product manufacturing. These guidances are not binding, but will often be requested to be adhered to by the FDA when an Investigational New Drug (IND) application is filed, so product manufacturing may commence. Additional state regulations also apply, for instance, the State of California requires a "Drug Manufacturing License" issued by the CA Department of Public Health, allowing the manufacture of a drug administered into patients, including biologics.

In the European Union (EU), great efforts were made to unify GMP requirements; therefore, Regulations, Directives & Guidances were put in place in the EU member states. Regulations have binding legal force in every member state of the EU; Directives define outcomes that must be achieved, each member state, however, may interpret Directives when writing them into national laws. In Eudralex, Volume IV, the rules which govern medicinal products in the EU are defined, and it provides guidance on Good Manufacturing Practice (EudraLex, 2011).

The European Commission approves (authorizes) medicinal products after the European Medicines Agency (EMA) issues a recommendation. The EMA is responsible for scientific evaluation, supervision and safety monitoring of medicines developed for use in the European Union. The EMA is only responsible for medicines that are managed through the Centralized Authorization Procedure. The EMA not only works with the EU Member States, it also works with the European Economic Area countries (Norway, Iceland and Liechtenstein, etc.). National Competent Authorities (NCA) are responsible for the authorization of medicines available at a national level.

There is currently no mutual recognition agreement in place between the US and the EU. Therefore, many drugs approved in the EU are not approved in the US. Any medicinal (therapeutic) product that is intended to be used in the US will need to be approved by the FDA, regardless of approval status in the EU. This may mean that new FDA approved and supervised clinical trials may need to be performed, and any GMP manufacturing of a product will have to conform to US rules and regulations.

1.6 Requirements for GMP manufacturing of cell and gene therapy products in the US

To bring a therapeutic product to market in the United States and to test such a product for safety and efficacy, a hierarchy of pre-clinical and clinical trial phases has been established. Cell and gene therapy products fall into the category of biologics, and these products also need to conform to the clinical trial hierarchy that was originally developed for small molecule drugs. The regulatory agency responsible for the licensing of cell and gene therapy products is the Center for Biologics Evaluation and Research (CBER) at the FDA. It is the final authority for the approval of clinical trials involving cell and gene therapy products. To enter a cell and gene therapy product into a clinical trial, an Investigational New Drug (IND) application needs to be filed with CBER. Prior to this filing, it is recommended to have a pre-IND meeting with the agency to make sure the filing is within the regulatory guidelines set forth for such a therapeutic product.

To aid the manufacturing of safe and potent cell and gene therapy products, the FDA released several "guidance for the industry" documents. These guidance documents are updated regularly, and also encompass a specific guidance for the manufacturing of such products for Phase I clinical trials (Guidance for Industry CGMP for Phase 1 Investigational Drugs, 2008). The FDA requires increased levels of manufacturing and regulatory compliance scrutiny with the increase in Phase levels. Phase I clinical trials, for instance, focus on safety of the product as a primary endpoint, with efficacy as a secondary endpoint. Phase II clinical trials are focused on both safety and efficacy and should be designed in a way so statistically valid efficacy signals can be detected. If these clinical trials produce a favorable outcome, the product can move into a Phase III clinical trial, which is usually a large scale, multi-center clinical trial, where safety and efficacy is being tested with a large number of subjects.

"Phase appropriate" product manufacturing under GMP conditions is often the term used for products manufactured for these different phases of clinical trials. For Phase I clinical trials of cell and gene therapy products, the specific FDA guidance document defines minimum standards the product has to comply with. These standards are, as expected, safety related, and specific release tests are prescribed to assure the product is sterile, the allowable limit of endotoxin is not exceeded, mycoplasma is not detected, and if it is a cellular product, a minimum threshold of 70% viability is reached. Often, for the identity of a product, specific characterization tests need to be added. For an allogeneic cell therapy product, additional safety tests, such as adventitious virus tests may be required. For Phase I products, a potency assay is not yet required as a release test, however, it is strongly encouraged that the potency assay is being developed, as it needs to be implemented and validated by the start of a Phase III clinical trial. For Phase I manufacturing, changes in the manufacturing procedure are acceptable, however, they need to be fully documented and justified. If a major manufacturing change is introduced, it is important to communicate this to the agency, as it may need to be handled as an amendment to the IND. For Phase II manufacturing, very similar regulations as for Phase I clinical trial product manufacturing apply, however, changes in the manufacturing procedure should be minimized, and the development of a potency assay should be completed. A Phase II clinical trials serves as a good test phase for the implementation and validation of a potency assay. However, the issue of the potency assay for cell and gene therapy products is complex. Such an assay is often not a straight-forward matter, as it may involve the application of an in vivo model; a correlation of certain cellular properties to be measured in vitro, with the therapeutic outcome achieved after the administration of the cells in the in vivo model, is desired. As an example, in the application of a cellular therapy to promote the generation of blood vessels, it will be needed to show the efficacy of the therapeutic cells to generate blood vessels in the appropriate in vivo model, and then find a factor the cell releases that causes or contributes to neovascularization. The expression or concentration of this factor released by the cells can then be correlated to the in vivo therapeutic action. As can be seen, potency assay development in cell and gene therapy may be complicated and will need to be discussed with the FDA. For a Phase III clinical trial, such a potency assay needs to be validated and implemented. In addition, Phase III product manufacturing requires a locked in manufacturing process, a fully vetted and implemented quality system for product manufacturing, a validated manufacturing environment (GMP facility) and validated release assays. After the successful completion of Phase III clinical trials, discussion for a Biologics License Application (BLA) may be initiated with the FDA. This part may take quite a while, since cell and gene therapy therapeutics are complex and many additional regulatory requirements may need to be fulfilled, as requested by the agency. Often, a post-marketing clinical study is requested by the FDA to further monitor the safety and efficacy of the product.

1.7 GMP grade cell product development and manufacturing for neurological disorders

The GMP facility at UC Davis, in collaboration with Leslie Thompson's lab at UC Irvine, has been involved in the manufacturing of a novel cell therapy product for neurological disorders, namely Huntington's Disease (HD), since 2011. We would like to present the development of this cellular product as an example for the GMP manufacturing of a highly complex cellular product slated for the treatment of a currently incurable neurological disease.

HD is a neurodegenerative disease caused by the "gain of function" repeat of the glutamine coding region of the Huntingtin (HTT) gene. There is no good gene therapy strategy available for this kind of gene mutation, as the increase in "function" of the gene product, dependent on the length of the coding region, causes the disease symptoms and disease progression over time. Medium spiny neurons are mostly affected and will die over time, after exposure to the protein produced by the malfunctioning gene (Ross et al., 2014). As a therapeutic option it has been envisioned to possibly excise the malfunctioning gene, however, a certain amount of gene product is required for normal function of neurons, and the number of repeats is variable in affected patients. Therefore, simple gene correction procedures will not adequately address the cause of the disease. It was found by the group at UC Irvine, however, that a significant therapeutic effect in the appropriate in vivo models of Huntington's disease could be achieved by the implantation of pluripotent stem cell derived neuronal stem cells (NSCs). This is a different option as compared to a gene therapy approach, as the implanted cells provide neuroprotective factors, in the face of the gain of the protein produced by the malfunctioning gene. The pluripotent stem cell line used for the derivation of the neuronal stem cell line was the human embryonic stem cell line ESI-017, procured from BioTime.

In 2007, the California Institute for Regenerative Medicine (CIRM), the voter-initiated stem cell funding agency, was able to offer their first grants in the state of California, geared toward the development of stem cell-based therapies. Through this granting mechanism, funding was obtained by our collaborator, the group at UC Irvine, to develop a neuronal stem cell line from newly generated embryonic stem cell lines. We, at the UC Davis GMP facility, which was also funded by CIRM, had, at that time, tested several embryonic stem cell lines as starting products for the generation of a stable neuronal stem cell line. The differentiation of a pluripotent stem cell into the neuronal lineage follows the default neuronal differentiation pathway, however, very precise conditions need to be applied consistently, and certain reagents for such differentiation runs need to be used, without any deviations. After trying 5 different embryonic stem cell lines, we were only able to differentiate, reliably, neuronal stem cell lines from one hESC line, which was the ESI-017 line. This aspect of the project was fortunate in another way, as the ESI-017 cell line could be made available GMP grade by the company BioTime, with all pertinent documentation, so a complete derivation and initial manufacturing chain, under GMP conditions, could be established (Facts About the Current Good Manufacturing Practices CGMPs, 2021).

The first step in the manufacturing of a GMP grade cell line was the generation of a master cell bank from the original, incoming seed bank material. It was initially suggested by the company that the hESC line should be cultured on an irradiated human foreskin fibroblast (HFF) feeder cell line. However, in our hands, when the hESC line was thawed and expanded on HFFs, it did not exhibit a stable karyotype beyond a few passages and also did not exhibit consistent morphology. Karyotypic instability is a concerning safety factor. Feeder free matrices, available at that time, were also tried and showed even greater karyotypic instability with this cell line. We would like to remark that karyotypic instability of hESC lines on feeder free matrices is cell line dependent, and other lines may not display this behavior. After several unsuccessful expansion runs, we resorted to the use of irradiated mouse embryonic feeder (MEF) cells, which proved to be the only way of keeping the ESI-017 cell line karyotypically stable and morphologically consistent. Using this approach, we were able to generate a master cell bank with more than 100 vials, from which we generated a working cell bank, again consisting of more than 100 vials. Preliminary tests performed on both cell banks consisted of safety tests (sterility, endotoxin, mycoplasma) and characterization tests. Among the most difficult tasks was the reliable and reproducible flow cytometric characterization of the cell line, which we performed for the markers OCT4, SOX2, NANOG and SSEA4. The characterization panel outlined here was our best effort for a consistent characterization assay. To satisfy GMP efforts for manufacturing and characterization of the cell line, we had to make certain that the antibodies used for the assay were available consistently, with the same quality attributes and negligible lot variations. Besides the characterization of the cell bank by flow cytometry to establish an appropriate identity assay, both master and working cell banks also needed to be tested for adventitious viruses, for all the agents the cells may have come in contact with; in our case, testing was needed for human, bovine and murine adventitious agents.

Vials from the ESI-017 working cell bank were then used for the differentiation of the hESCs into neuronal stem cells. Fig. 1 shows a diagram of the differentiation process.

hESCs were plated onto MEFs, and daily media changes were performed. It is of utmost importance to keep up the daily cell feeding regimen, so the cells can be maintained under optimal conditions for the subsequent cell differentiation.

After 14 days, the hESCs were taken into embryoid body (EB) generation culture, for which ultra-low attachment plates were utilized. This allowed for

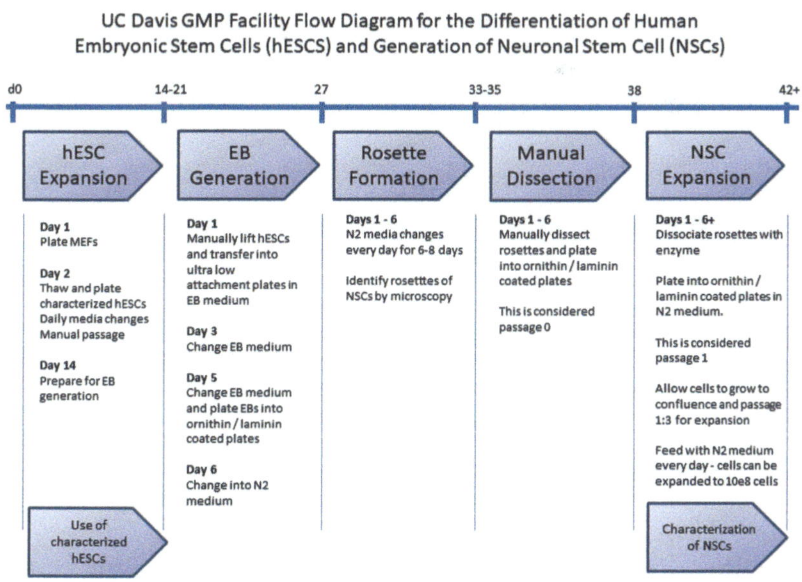

Fig. 1 Flow chart for the generation of neuronal stem cells.

the generation of floating embryoid bodies after day three of culture. The floating EBs were then harvested on day six and transferred into "N2 medium," which is a defined, serum-free medium providing the conditions required for neuronal stem cell culture. In addition, the culture vessels (usually six well plates) were coated with ornithine/laminin to provide the matrix for neuronal stem cells to adhere to and remain stable, not maturing down the neuronal lineage. After 1 to 6 days in N2 medium, attached neuronal stem cell "rosettes" became visible, and individual rosettes could be dissected manually; individual cells were picked by hand, under the microscope, in a strictly controlled clean room environment. The individual cells were then re-plated into single ornithine / laminin coated wells, which were then observed for continued cell growth and expansion in N2 medium. Cells from wells that showed robust growth, were enzymatically lifted and transferred into larger wells. This process was repeated until a sufficient number of neuronal stem cells (NSCs) was obtained. Passage counts were carefully taken and cell feeding occurred daily. Low passage, enzymatically lifted cells were cryopreserved to serve as a seed bank for the generation of an NSC master cell bank, from which a working cell bank was generated; from the working cell bank, a product bank for cells used in clinical trials was generated (Holley et al., 2018; Reidling et al., 2018).

As can be understood, this process of NSC differentiation is highly labor intensive, it is all manual and it requires an enormous amount of skill from the manufacturing technician. In addition, all of these steps need to be carried out in a clean room in a GMP facility, which will be explained in a subsequent paragraph.

2. Characterization of the NSCs

The cells needed to be tested for safety, karyotypic stability and the appropriate marker expression. We looked for the positive expression of NESTIN, SOX2, CD271, CD24, SOX and CD184 to demonstrate that these cells had the marker expression pattern of neuronal stem cells, and we measured for the absence of the marker OCT3/4, to demonstrate that these cells were not pluripotent. Additional markers were added for information only (see table below). In addition, adventitious virus testing was also required. These tests are also illustrated in the Fig. 2 and 3 below.

For the clinical use of the NSC line, it is necessary to thaw the cryopreserved product, culture it for 48 h, formulate the product in final formulation buffer, fill the cell suspension into appropriate vials that can

GMP ESI-017 NSC Lot Release Testing

- Vials from each lot were submitted for release testing including the following:

Test	Status	Specification	Pass/Fail	Test performed at
14 Day Sterility QCTL 2000	Performed	No Growth	PASS	UC Davis
Gram Stain QCTL 2400	Performed	No Organisms Seen	PASS	UC Davis
Endotoxin QCTL 2800	Performed	Less or equal 0.500 EU/mL	PASS	UC Davis
Mycoplasma PCR GMP 4436B	Performed	No Bands Seen	PASS	UC Davis
Viability	Performed	Greater or equal 70%	PASS	UC Davis
ADVENTITIOUS VIRUS TESTING Bovine Porcine Murine	In process currently	Not seen	Pending	Charles River Labs

Fig. 2 Lot release testing for GMP grade human embryonic stem cells.

Next: GMP ESI-017 NSC Lot Characterization

- Vials from each lot were thawed, plated and then harvested for the following tests:

Test	Status	Specification	Pass/Fail	Test performed at
Cell Count	Performed	±5x10^6	PASS	UC Davis
Viability	Performed	Equal or greater 70%	PASS	UC Davis
Purity - Flow Cytometry	In process currently	Equal or greater 95%	Pending	UC Davis
Markers - Flow Cytometry	In process currently	% CD24+ % CD271+ % CD184+ % SOX2+ % SOX1+ % NESTIN+ % PAX6+ % SSEA4+ % OCT3/4+	Pending	UC Davis

Fig. 3 Lot characterization of GMP grade, hESC derived neuronal stem cells.

be used at the clinical site, and transport the fresh product to the clinical site for immediate administration.

Prior to administration, safety tests such as stat viability count, gram stain, setup of a 14 day sterility culture, stat endotoxin and mycoplasma tests need to be performed. The 14 day sterility results will not be available prior to the administration of the cells, so a positive culture action plan must be in place. In case the 14 day sterility culture reads positive, the results need to be communicated immediately to the sponsor, the principal (clinical) investigator, the IRB and the FDA. It will be the responsibility of the clinical investigator to determine any necessary clinical action.

2.1 Steps to consider for later phase GMP manufacturing of an NSC line

The development of this completely novel GMP grade NSC product took approximately 10 years. Numerous talks with the regulatory agency were required to address the issue of the use of a xeno-product (irradiated mouse feeder cells) for the manufacturing of the hESC line. It has been agreed with the agency that the use of such a feeder cell line is acceptable, at least for Phase I clinical applications, after appropriate testing of the cells is conducted and the endpoints are met. It is an advantage that the initial hESC line is not the final product, which will be administered clinically, but the derived differentiated neuronal cell line is the final product. It is many passages removed from the original cell line, which was cultured on irradiated murine feeder cells. Nevertheless, appropriate safety testing needs to be conducted on the final product, and it needs to be demonstrated that murine cells are absent in the NSC product. In addition, further safety testing with respect to surviving, undifferentiated hESCs in the final NSC product also needs to be conducted. For this purpose, a sensitive flow cytometric assay needed to be developed, demonstrating that small numbers of hESCs could be detected reliably in this population, if present. This would assure enough sensitivity for this assay to serve as a release test for the absence of hESCs in the final product. This was among the most difficult and most sophisticated in vitro assays that needed to be developed for this project. A publication describing this assay is planned.

A very important assay that is still under development at this time is an in vitro potency assay. An in vivo potency does exist and is performed by our collaborator at UC Irvine. However, for routine clinical applications of the final product, it will be required to correlate the in vivo assay with an in vitro assay that could possibly be performed stat on the final product. The final, validated potency assay will be available prior to initiation of a Phase III clinical trial.

In summary, with the development of appropriate cell culture and characterization assays, and many hours of highly skilled labor in a state of the art

GMP facility, a GMP grade neuronal stem cell line with safety and efficacy demonstrated in the relevant in vivo models for Huntington's disease could be developed. The utility of this cell line for human clinical applications in patients with Huntington's disease will need to be tested in upcoming Phase I and II safety and efficacy clinical trials.

2.2 Future direction of cell and gene therapy for neurological disorders

The NSC line our group manufactured is, by default, an allogeneic cell line. Currently, it is not known if such an allogeneic cell line will persist, long term, after administration into the recipient and will have long term therapeutic effects. Allogeneic tissue, in spite of any immune-privilege of the area where the tissue is implanted, may be prone to being recognized and eliminated by the recipient's immune system, in the long run. To counteract this effect, a strategy can be envisioned that will allow to modify the pluripotent cell serving as the starting tissue, so the major tissue antigens leading to the recognition of the allogeneic cells by the host immune system will be altered or knocked out. This strategy, however, must not change the pluripotent cell in its ability to differentiate into NSCs and also must not alter any function or therapeutic benefit the derived NSCs display in the recipient. In addition, the modified cells must not pose any risk, they need to be karyotypically stable and must not lead to any increased proliferation that may possibly form a tumor. Gene modification strategies to accomplish such a Major Histocompatibilty Complex knockout may utilize CRISPR-Caspase approaches or other technologies; these strategies, however, are currently only at the very beginning of their applicability to generate cell based therapeutics for human applications, and a good amount of research at the basic and translational level is still required until such cells may become available, routinely, for safe and efficacious clinical applications.

For other therapeutic modalities, NSCs generated from pluripotent cells may be gene modified to express certain factors that may act on neurons. The over-expression of Brain Derived Neurotrophic Factor (BDNF) has been suggested in the past, as it may be neuroprotective in Huntington's disease. The gene modification of NSCs may be carried out with lentiviral vectors, transducing cells coming from a master cell bank of NSCs. This would not result in the gene modification of a single, clonal NSC, this would be a polyclonal gene modification; nevertheless, however, the establishment of a transduced working cell bank that could be thoroughly characterized, would be a possibility. This kind of gene modification, if clinically useful, could be accomplished rather quickly and would only be a relatively small addition to an already existing therapeutic cell line. The safety tests of such a cell line would encompass all tests outlined previously, but would include additionally the testing for the number of gene integrants per cell and the location of the integration sites. The number of integrants should be kept at 1 or 2 per genome, for safety reasons. As a potency assay, the measurement of the secreted BDNF could possibly suffice. The regulatory agency, however, may require extensive in vivo studies to determine toxicity and dose relation of secreted BDNF per cell administered, correlated to the therapeutic effect measured in vivo.

A final consideration should be given to automation of cell line manufacturing for any future cell therapy applications. As was pointed out earlier, the generation of the NSC line by our group demanded a lot of patience, skills and repeated attempts. This could possibly be overcome by the use of automated systems for neural cell differentiation and then clonal picking of cells from rosettes. It can be envisioned that artificial intelligence algorithms could recognize imaged neural rosettes, individual cells in these rosettes could be picked robotically and then picked cells could be robotically transferred into individual wells for clonal expansion. This may lead to more reproducible results, and could possibly make autologous therapies possible, as induced pluripotent cells could be generated from an individual; such cells could be gene corrected and then used for the differentiation into gene corrected NSCs. This would be accomplished in a more successful manner than currently available with the manual method of NSC differentiation we described previously. Such autologous cells would not get rejected by the host immune system and would offer the ability of long-term gene correction.

2.3 A state of the art GMP facility for cell and gene therapy product manufacturing

For the manufacturing of a safe and potent cell or gene therapy product for clinical application it is imperative to utilize a clean room environment suitable or specifically designed for the manufacturing of such products. Clinical applications of cell and gene therapy, and in particular, autologous cell manufacturing procedures with extensive open manipulations only arose routinely in the 1990s. Such manufacturing scenarios, when performed in an open system, which means culture flasks or cell stacks, are "high risk" manufacturing procedures, in the language of GMP manufacturers. When only few products for early phase cell and gene therapy clinical trials were manufactured, the use of smaller academic GMP facilities was appropriate, however, when cell and gene therapies were found to be very useful therapeutics to cure inherited diseases such as ADA SCID, and currently, to cure hematologic malignancies, another closer look at specific GMP facilities for the manufacturing of such products was required.

A GMP facility for such applications should have multiple manufacturing rooms, independent from each other, so multiple autologous products can be manufactured simultaneously without cross contamination risk, at the same time, the manufacturing rooms should be large enough to also handle allogeneic cell bank manufacturing. The same manufacturing facility should be able to manufacture gene therapy vector, particularly lentiviral vector, without any aerosolization or cross contamination risk, and the facility should be versatile and adaptable enough to manufacture any products that can be envisioned in future manufacturing scenarios. There should be enough segregated controlled storage areas for incoming products and materials, and the same should also be the case for outgoing materials, and particularly, final products. The facility should be of sturdy construction, have controlled temperature and humidity within prescribed limits, and controlled and stable air flow with a specific air cleanliness standard, also within prescribed limits. The air cleanliness standards are typically ISO 7 and ISO 8 for US GMP facilities, corresponding to the older US air cleanliness standard of Class 10,000 (ISO 7) and Class 100,000 (ISO 8), translating into less than 10,000 or 100,000 particles greater 0.5 μm per cubic foot per minute in the clean room air (Facts About the Current Good Manufacturing Practices CGMPs, 2021).

In Europe, the classification of air cleanliness in a clean room space is different, and only to a degree comparable to the US standards (European Medicines Agency, 2018). The particle measurements in the US are taken in non-metric units, while in Europe, measurements are taking in metric units. Also, cleanrooms in Europe are classified as A, B, C and D, and particles are measured while the room is in rest and in operation. Only under specific circumstances, therefore, are US cleanrooms for cell and gene therapy product manufacturing comparable; as an approximation one could say that a Class B cleanroom may be equivalent to an ISO 7 cleanroom, when in operation. It is worth mentioning that a Biosafety Cabinet also maintains an ISO 5 environment (Figs. 4 and 5).

In the US, there is a sliding scale of regulatory requirements that increases in scrutiny with the clinical trial phases. As long as product manufacturing occurs for only early phase clinical trials, such as Phase I and II clinical trials,

	Maximum Permitted Numbers of Particles/m³ At Rest		In Operation	
	≥ 0.5 µm	≥ 5.0 µm	≥ 0.5 µm	≥ 5.0 µm
A	3 520	20	3 520	20
B	3 520	29	352 000	2 900
C	352 000	2 900	3 520 000	29 000
D	3 520 000	29 000	Not defined	Not defined

Fig. 4 Maximum allowable numbers of particles for for Grade A to D cleanrooms.

Clean Room Classification / Air Quality Equivalence, Europe versus ISO Classification

EU Classification	Particles at rest	Particles in operation	ISO Classification
A	3 520	3 520	5/5
B	3 520	352 000	5/7
C	352 000	3 520 000	7/8
D	3 520 000	Not defined	8/-

Fig. 5 Comparison of cleanroom classifications (Grade A to D versus ISO 5 to 8).

GMP rules and regulations still apply, however, the facility will not be held to the same standards as required for Phase III and marketed product manufacturing. However, this should not be distracting the designer of a new GMP facility for cell and gene therapy applications to "under-build" such a GMP facility.

2.4 Example of a GMP facility for cell and gene therapy manufacturing at UC Davis

At UC Davis, in 2006, our group was asked to design and construct a new cell and gene therapy facility. The design encompassed all of the qualities of a high performance GMP facility just as outlined above, and it was also designed for maximum redundancy and reliability.

The GMP facility at the UC Davis Medical Center is a multi-use, six manufacturing room, Class 10,000 (ISO 7) laminar air flow clean room facility, the design phase started in in 2006, construction started in 2008, it was opened in 2010. It was specifically designed as a facility for the manufacturing of cell and gene therapy products, and to be compliant with Good Manufacturing Practice and Good Tissue Practice rules and regulations.

In addition, it is compliant with USP <797> regulations and can therefore also be used for the manufacturing of other therapeutics for clinical use, such as small molecule formulations. All six manufacturing rooms are independent of each other, no room shares any mechanical air handling equipment, each room has a dedicated air handler, redundant main air intake and exhaust air handlers are employed. Redundant power is supplied by a gas turbine generator supplying the medical campus with electricity, an automatic switch to municipal power is available, in case the gas turbine fails, and a backup diesel generator supplies a third electric power option, if the two main power sources fail.

UC Davis GMP Facility layout, with one-way personnel, product and waste flow indicated (Fig. 6).

The facility features six ISO 7 manufacturing rooms (MANUF), which are completely independent of each other. The easiest way to visualize the environment in each manufacturing room is as the interior of a large biosafety cabinet, with laminar air flow provided by spaced out terminal HEPA filters in the ceiling, no ceiling tiles, but a completely sealed ceiling, low air returns located close to the entrance and exit doors and an air-tight room design, with additional air ways from the outside areas only available through adjustable door sweeps. Adjacent to the manufacturing rooms are

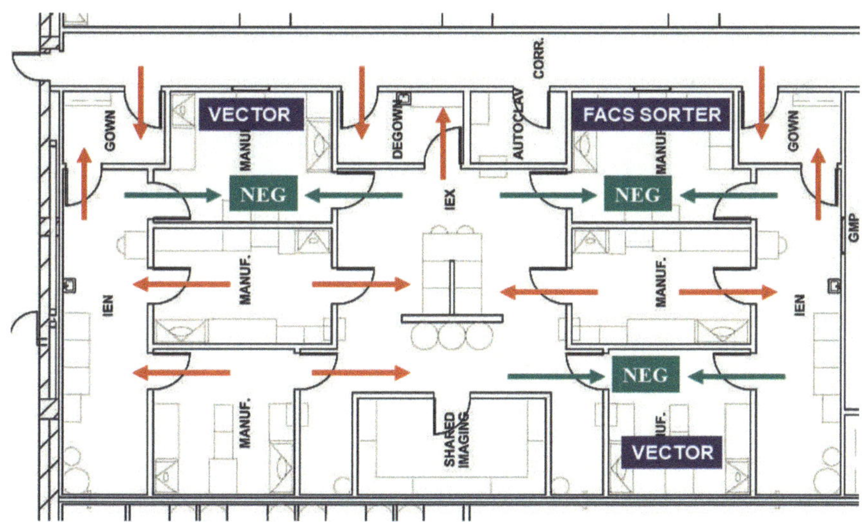

Fig. 6 GMP facility room pressurization for different manufacturing conditions. Arrows pointing inward indicate negative room pressurization, arrows pointing outward indicate positive room pressurization.

intermediate entry (IEN) and intermediate exit (IEX) areas, which are designated as ISO 8 areas. These areas are also operated using individual air handlers. Gowning and de-gowning areas (GOWN) are non-classified, but are clean spaces, they allow for proper gowning and de-gowning of personnel, and also serve as the areas where materials, reagents and products are entering and exiting the facility.

The facility maintains one-way personnel, product and waste flow for avoidance of cross contamination. All mechanical equipment is in a walkable interstitial ceiling space for easy service access and avoidance of facility contamination. Ceiling light fixtures are sealed, lamp changes are performed from the interstitial ceiling without breaching room cleanliness. A computerized facilities management and monitoring system controls and monitors room pressurization, temperature, humidity and equipment. One-way personnel, product and waste flow is depicted in the diagram: 1. Enter Gowning Room, 2. Enter IEN (Intermediate Entry Room.), 3. Enter Manufacturing Room, 4. Perform Manufacturing Procedures, 5. Enter IEX (Intermediate Exit Room.) 6. Enter De-gowning Room, 7. Exit Facility (Fig. 7).

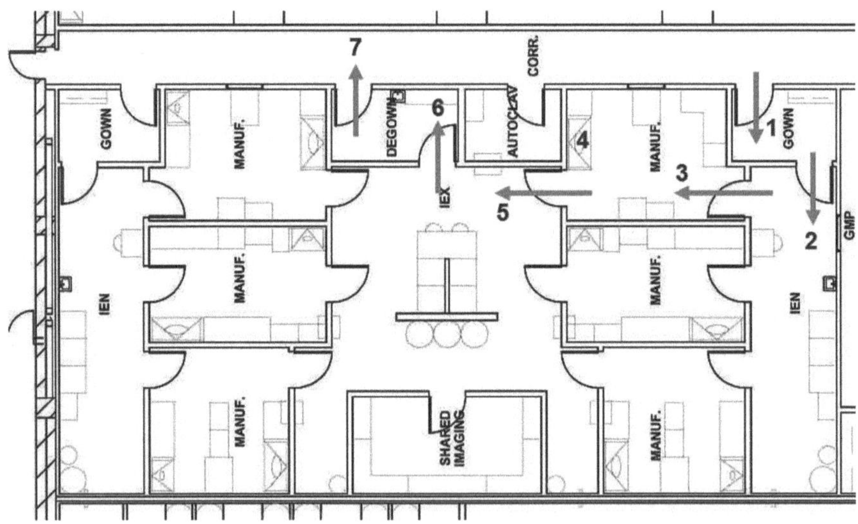

Fig. 7 One way personnel, product and waste flow. 1. Personnel gowns up, 2. personnel enters the intermediate entry area to enter the designated manufacturing suite, 3. personnel manufactures a specific product in this suite, 4. personnel exits the manufacturing suite after manufacturing is completed, 5. personnel enters the intermediate exit area, 6. personnel enters the degowning area and degowns, 7. personnel exits the facility. The intermediate entry area acts as a buffer area and as a storage area for incoming products and materials, the intermediate exit area acts as a buffer area and as storage area for outgoing/final products.

2.5 UC Davis GMP facility air pressurization

The manufacturing, entry and exit areas are separated by air pressure gradients. A special feature of this facility is switchable manufacturing room pressurization, which allows to adapt the manufacturing rooms to the required manufacturing conditions. Validated, switchable room pressurization allows for the manufacturing of aerosolizable products, such as gene therapy vectors or the use of a Fluorescent Activated Cell Sorter (FACS) in pressure sink manufacturing rooms. This generates containment conditions during manufacturing of such products, preventing the escape of potential aerosols into other areas of the facility. In 2002, this option was already pioneered by our group at another GMP facility we designed and built, with the involvement of the US FDA. This principle was then repeated in the UC Davis GMP Facility, and again vetted by the US FDA in a Type C pre-facilities meeting.

A question often asked is if the negatively pressurized areas are contaminated with dirty air and have a higher microbiological burden. This is a valid question, and the solution to such a scenario was already accounted for in the facility design in the 2002 facility. The manufacturing rooms feature 60 air changes per hour, and a 70% recirculating air HVAC system, with laminar air flow and low returns close to the manufacturing room doors. Particles in the air will be transported, by the laminar air flow, to the floor, the air will then immediately be moved into the low return grills. 70% of this air will be recirculated and scrubbed by the terminal HEPA filters in the ceiling and released back into the manufacturing room as clean air. This is the same principle as is applied in a recirculating biosafety cabinet. Every biosafety cabinet features negative pressure to contain possibly biohazardous materials, while at the same time, providing a sterile work environment, so samples can be processed using sterile technique. We have been collecting environmental monitoring data for the last 12 years in the UC Davis facility, and we cannot find a statistically valid difference in microbiological burden between negatively and positively pressurized manufacturing rooms (data not shown). However, we need to point out that our environmental cleaning procedures are stringent, and we employ daily environmental cleanings. This may also contribute to the cleanliness of these manufacturing rooms.

An elaborate interlock system controls the opening and closing of access doors, which avoids cross-contamination of products by air turbulences. Only one door at one time can be opened. The facility features an electronically controlled and automatically monitored environment, with additional 7 days per week manual monitoring by GMP facility personnel, controlling and verifying the automatic monitoring. A strict QC/QA program with environmental cleaning and monitoring procedures is in place to assure freedom from microbiological contamination. A Quality Control (QC) laboratory (not depicted in the figures) is associated with the UC Davis GMP facility, which provides QC assays such as 14-day sterility according to USP <71>, limulus amebocyte lysate (LAL) endotoxin assay and Gram stain assay for release testing.

3. History and track record

The UC Davis GMP facility within the Stem Cell Program at the School of Medicine has been in operation since March of 2010. The facility has been manufacturing GMP grade AAV, retroviral and lentiviral vectors for several projects. The GMP facility has been used for the manufacturing of clinical grade hESCs and NCS, mesenchymal and hematopoietic stem cell products, retinal progenitor cells and most recently, CAR T cells. The GMP facility also produces GMP compliant formulations of novel drugs for clinical trials. Temporal and spatial segregation is observed with all manufacturing procedures. Only one product is manufactured at a given time in one manufacturing room.

After extensive audits, the facility was deemed suitable for Phase III cell and gene therapy clinical trial product manufacturing. This is an extraordinary achievement and milestone for this facility, as only very few academic facilities are able to achieve this manufacturing standard, being able to manufacture products for all phases of clinical trials.

4. QC/QA at the UC Davis GMP facility

All products for clinical application, including cell and gene therapies need to adhere to established criteria for identity, purity, potency and other quality characteristics, ensuring safety and efficacy of such products. For this purpose, a Quality Control Unit (QCU) is in place at the UC Davis (UCD) GMP facility that creates, monitors, and implements a quality system. The responsibilities of the QCU is divided into two functions: Quality Control (QC) and Quality Assurance (QA).

Tasks assigned to QC: Assessing the suitability of incoming materials, containers, closures, labeling, in-process materials, final products, evaluating the performance of the manufacturing process to ensure adherence to proper specifications and limits, and determining the acceptability of each batch (final product) for release.

Tasks assigned to QA: Review and approval of all procedures related to manufacturing, facility performance and maintenance, generation of all SOPs in collaboration with the manufacturing team, review of associated records, auditing, producing process performance evaluations and trend analyses.

The QA unit is the final authority in the quality system. QA assures that appropriate controls are implemented, it ensures that procedures and specifications are appropriate and that they are followed. QA approves or rejects incoming materials and final products, reviews manufacturing records and investigates discrepancies.

QC is responsible for implementing all SOPs and ensures adherence to these. QC performs the quality control of the manufacturing facility, products and personnel. This includes facility monitoring, environmental monitoring, preventive maintenance, inventory management and recertification of the GMP facility. QA controls the outcome of these activities and assures compliance with GMP regulations. For each product manufactured, batch records are created. QC checks the manufacturing process and personnel during manufacturing, making sure that every step in the SOP for manufacturing of a certain product was carried out.

All products, materials and reagents are tracked and recorded throughout the whole manufacturing process. QA checks the accuracy of all these records and retains all documentation indefinitely. In case of a deviation, a deviation report is opened; the deviation is described and recorded (e.g., time of the deviation, personnel, instrument and process involved), corrective action is applied, and the deviation report is closed upon successful correction of the deviation.

Finally, release of a product is controlled by QA, generating a Certificate of Analysis (COA) after the release tests are passed.

Environmental Monitoring Procedures at the UCD GMP facility:

Routine environmental monitoring takes place at least once a week, when a manufacturing room is not in use, and daily, if a manufacturing room is in use. Environmental monitoring consists of measuring non-viable airborne particles, using a certified electronic non-viable air particle counter. Viable airborne particles are measured using an active air sampler, in conjunction with agar settling plates placed into defined locations. Viable particle measurements on specified surfaces are carried out using agar touch plates. Specific sampling maps were created for each room in the facility, based on room size and room classification, as outlined in the United States Pharmacopeia (USP). Alert and action levels were set accordingly, as again specified in the USP.

For each lot of any product manufactured, additional settling plate monitoring takes place during aseptic manufacturing processes and final fill operations. Settling plates are placed into biosafety cabinets before the process begins and removed immediately afterwards. This monitoring is separate from and in addition to the routine monitoring. Personnel monitoring also takes place immediately after an aseptic manufacturing process or a final fill operation. Touch plates are taken on gloves for each operator after the process is completed.

4.1 Recruiting and training of personnel

The least talked about, but most important aspect for a GMP facility is recruiting and training of appropriate manufacturing, QC and QA personnel for cell and gene therapy manufacturing. At UC Davis, in conjunction with California State University Sacramento (CSUS), we were able to implement a training program in Stem Cell Biology with a hands on training course in GMP manufacturing. The program offers a master's degree after successful completion. This program has been funded by the California Institute for Regenerative Medicine (CIRM) and is in its 3rd funding round, since its inception in 2011. Over 100 students, so far, have been trained successfully and have been supplying the work force for the UC Davis GMP facility. Personnel we hired from this program 10 years ago are still with us, and have helped shape our facility, with constant improvements achieved in our Quality System and in our manufacturing efforts. Without our own training program we would not have been able to come to the track record and the success in cell and gene therapy product manufacturing, from Phase I to Phase III products that we can display now.

4.2 Conclusion

We hope that our GMP facility with some of the pioneering features we implemented, will help others in the field to also design and construct even better and much improved GMP manufacturing facilities and implement the manufacturing of cell and gene therapy products, targeting currently incurable diseases. The application of cell and gene therapy, particularly in neurological diseases, will only be increasing over time. It is estimated that such therapies will bring improvements and maybe even cures to patients suffering from currently incurable, and devastating neurological diseases, such as Huntington's disease and Alzheimer's disease within the next 5 to 10 years. The future pioneers in this new field, I would like to encourage to think outside the box and pursue new avenues; many new therapies are needed.

References

Abou-El-Enein, M., Elsallab, M., Feldman, S. A., Fesnak, A. D., Heslop, H. E., Marks, P., et al. (2021). Scalable manufacturing of CAR T cells for cancer immunotherapy. *Blood Cancer Discov, 2*(5), 408–422. https://doi.org/10.1158/2643-3230 (BCD-21-0084. Epub 2021 Aug 3.PMID: 34568831 PMCID: PMC8462122).

Bauer, G., Valdez, P., Kearns, K., Bahner, I., Wen, S. F., Zaia, J. A., et al. (1997). Inhibition of human immunodeficiency virus-1 (HIV-1) replication after transduction of granulocyte colony-stimulating factor-mobilized CD34+ cells from HIV-1-infected donors using retroviral vectors containing anti-HIV-1 genes. *Blood, 89*(7), 2259–2267 (PMID: 9116267).

CFR. (March 29, 2022). *Code of Federal Regulations Title 21*. https://www.accessdata.fda.gov/scripts/cdrh/cfdocs/cfcfr/CFRsearch.cfm?CFRPart=210.

EudraLex. (February 02, 2011). *Volume 4—Good Manufacturing Practice (GMP) guidelines*. https://ec.europa.eu/health/medicinal-products/eudralex/eudralex-volume-4_en.

European Medicines Agency. (July 2018). *Guidance on good manufacturing practice and good distribution practice: Questions and answers*. https://www.ema.europa.eu/en/human-regulatory/research-development/compliance/good-manufacturing-practice/guidance-good-manufacturing-practice-good-distribution-practice-questions-answers.

Facts About the Current Good Manufacturing Practices (CGMPs). (June 1, 2021). https://www.fda.gov/drugs/pharmaceutical-quality-resources/facts-about-current-good-manufacturing-practices-cgmps.

Gabriel, R., Schmidt, M., & von Kalle, C. (2012). Integration of retroviral vectors. *Current Opinion in Immunology, 24*(5), 592–597. https://doi.org/10.1016/j.coi.2012.08.006 (Epub 2012 Sep 14. PMID: 22981243).

Guenechea, G., Gan, O. I., Inamitsu, T., Dorrell, C., Pereira, D. S., Kelly, M., et al. (2000). Transduction of human CD34+ CD38- bone marrow and cord blood-derived SCID-repopulating cells with third-generation lentiviral vectors. *Molecular Therapy, 1*(6), 566–573. https://doi.org/10.1006/mthe.2000.0077 (PMID: 10933981).

Guidance for Industry CGMP for Phase 1 Investigational Drugs. (July 2008). https://www.fda.gov/media/70975/download.

Hacein-Bey-Abina, S., Garrigue, A., Wang, G. P., Soulier, J., Lim, A., Morillon, E., et al. (2008). Insertional oncogenesis in 4 patients after retrovirus-mediated gene therapy of SCID-X1. *The Journal of Clinical Investigation, 118*(9), 3132–3142. https://doi.org/10.1172/JCI35700 (PMID: 18688285 PMCID: PMC2496963).

Hawkins, T. E., Marley, S. B., O'Brien, S. G., Gordon, M. Y., & Goldman, J. M. (1997). CD34+ cell selection in chronic phase chronic myeloid leukaemia: A comparison of laboratory grade columns. *Bone Marrow Transplantation, 20*(5), 409–413. https://doi.org/10.1038/sj.bmt.1700908 (PMID: 9339758).

Henig, I., & Zuckerman, T. (2014). Hematopoietic stem cell transplantation—50 years of evolution and future perspectives. *Rambam Maimonides Medical Journal, 5*(4), e0028. https://doi.org/10.5041/RMMJ.10162 (eCollection 2014 Oct. PMID: 25386344 PMCID: PMC4222417).

Holley, S. M., Kamdjou, T., Reidling, J. C., Fury, B., Coleal-Bergum, D., Bauer, G., et al. (2018). Therapeutic effects of stem cells in rodent models of Huntington's disease:

Review and electrophysiological findings. *CNS Neuroscience & Therapeutics*, *24*(4), 329–342. https://doi.org/10.1111/cns.12839 (Published online 2018 Mar 6. PMID: 2951229 PMCID: PMC6489814).

Morgan, R. A., Gray, D., Lomova, A., & Kohn, D. B. (2017). Hematopoietic stem cell gene therapy: Progress and lessons learned. *Cell Stem Cell*, *21*(5), 574–590. https://doi.org/10.1016/j.stem.2017.10.010 (PMID: 29100011 PMCID: PMC6039108).

Reidling, J. C., Relaño-Ginés, A., Holley, S. M., Ochaba, J., Moore, C., Fury, B., et al. (2018). Human neural stem cell transplantation rescues functional deficits in R6/2 and Q140 Huntington's disease mice. *Stem Cell Reports*, *10*(1), 58–72. https://doi.org/10.1016/j.stemcr.2017.11.005 (Epub 2017 Dec 7. PMID: 29233555 PMCID: PMC5768890).

Ross, C. A., Aylward, E. H., Wild, E. J., Langbehn, D. R., Long, J. D., Warner, J. H., et al. (2014). Huntington disease: Natural history, biomarkers and prospects for therapeutics. *Nature Reviews. Neurology*, *10*(4), 204–216. https://doi.org/10.1038/nrneurol.2014.24 (Epub 2014 Mar 11. PMID: 24614516).

Wilson, J. M. (2009). Lessons learned from the gene therapy trial for ornithine transcarbamylase deficiency. *Molecular Genetics and Metabolism*, *96*(4), 151–157. https://doi.org/10.1016/j.ymgme.2008.12.016 (Epub 2009 Feb 10).

CHAPTER SEVEN

Considerations for clinical trial design and conduct in the evaluation of novel advanced therapeutics in neurodegenerative disease

Cheney J.G. Drew[a,b,*] **and Monica Busse**[a,b]
[a]Centre For Trials Research, Cardiff University, Cardiff, United Kingdom
[b]Brain Repair and Intracranial Neurotherapeutics Unit (BRAIN), College of Biomedical and Life Sciences, Cardiff University, Cardiff, United Kingdom
*Corresponding author: e-mail address: drewc5@cardiff.ac.uk

Contents

International Review of Neurobiology, Volume 166
ISSN 0074-7742
https://doi.org/10.1016/bs.irn.2022.09.006

Abstract

The recent advances in the development of potentially disease modifying cell and gene therapies for neurodegenerative disease has resulted in the production of a number of promising novel therapies which are now moving forward to clinical evaluation. The robust evaluation of these therapies pose a significant number of challenges when compared to more traditional evaluations of pharmacotherapy, which is the current mainstay of neurodegenerative disease symptom management. Indeed, there is an inherent complexity in the design and conduct of these trials at multiple levels. Here we discuss specific aspects requiring consideration in the context of investigating novel cell and gene therapies for neurodegenerative disease. This extends to overarching trial designs that could be employed and the factors that underpin design choices such outcome assessments, participant selection and methods for delivery of cell and gene therapies. We explore methods of data collection that may improve efficiency in trials of cell and gene therapy to maximize data sharing and collaboration. Lastly, we explore some of the additional context beyond efficacy evaluations that should be considered to ensure implementation across relevant healthcare settings.

1. Introduction

Neurodegenerative disease has a large collective burden, not only in terms of the impact of disease experienced by affected families, but also in socio-economic terms (El-Hayek et al., 2019; Jones et al., 2016; Yang et al., 2020). In comparison to other chronic medical disorders, there have been relatively few advances in the development of new therapeutic agents that can ameliorate disease symptoms or significantly slow disease progression to a beneficial degree, with particular reference to Huntington's disease (HD), Parkinson's Disease (PD) and Alzheimer's Disease (AD). The traditional approach of exploiting fundamental understanding of disease mechanisms to identify novel drug targets and subsequently develop pharmaceutical agents to exert a beneficial biological effect at that target have so far failed to yield any disease modifying agents. However, in recent years, there have been several advances in the development of novel, advanced therapeutics for the treatment of neurodegenerative disease, specifically in the HD and PD arenas. We are now at a point where several cell and gene therapies, which have shown promise for HD and PD in pre-clinical testing, are moving forward to in-person evaluation.

In this chapter we will explore aspects of clinical trial design and conduct which should be considered in the evaluation of advanced therapeutics in neurodegenerative disease. In light of the recent focus on Advanced Therapeutic Medicinal Products (ATMPs) for the treatment of HD and

PD, the challenges associated with ATMP evaluation in these disorders (Barker, Cutting, & Daft, 2021; Rosser et al., 2022) will be the focus of the discussion. Challenges associated with the development of cell and gene therapies for HD and PD have been covered in other chapters in this volume but we include a brief summary potential ATMP strategies for these diseases here.

1.1 The potential use of cell and gene therapies in HD

The gene mutation that causes HD is a polyglutamine repeat in exon one of the *huntingtin* gene. Repeats of more than 39 are pathogenic with symptoms first occurring in mid-life with progression of motor, cognitive and behaviors across 15–20 years (McColgan & Tabrizi, 2018). This is underpinned by the selective loss of medium spiny interneurons in the striatum, leading to an overall deterioration of the thalamic neuronal networks essential for normal brain function and global brain atrophy in the latter stages of disease.

One potential way of combatting the effects of this specific neuronal death is to replace the cells that are lost with cells that are free from the underlying genetic mutation. The theory is that a successful graft will allow transplanted stem cells to grow and differentiate into cells of the required identify and thus restore the synaptic connections required for effective network functioning. There have been a few small studies examining transplantation of fetal cells, successfully demonstrating safety, but with limited functional benefit, likely due to a range of delivery issues (Rosser & Bachoud-Lévi, 2012). Currently donor cell source is limited to fetal tissue, which introduces a number of constraints with obtaining sufficient tissue and limited window of time from tissue collection to target delivery in which the tissue must be processed and assessed for quality (Bachoud-Lévi et al., 2020; Bachoud-Levi, Massart, & Rosser, 2020). Consequently, there are several lines of investigation into the generation of stem cell technology to allow the production of pluripotent and progenitor stem cells for provision of a more reliable donor cell source.

In contrast to the premise of cell replacement therapy, gene therapies may also be implemented to combat disease progression in HD. It is thought that if the production of the mutant huntingtin (htt) protein, with its toxic gain of function, can be limited, then the cellular effects of the mutant htt may be mitigated, thus preventing neuronal dysfunction and cell death. One method of doing this is by blocking the translation of *huntingtin* mRNA into protein using RNA interference technologies. One such example of this is the use of Anti-sense oligonucleotides (ASOs) are short (16-22 base), single-strand RNA or DNA sequences complementary to the target mRNA sequence and are readily taken up into neurons, with dose dependent and reversible effects (Leavitt & Tabrizi, 2020). Their ability to interfere with mRNA transcription can be achieved via multiple mechanisms depending on the nature of the specific target (Quemener et al., 2020). Other strategies include the introduction of microRNAs, short RNA sequences that bind the non-coding regions of target genes, into the cell to prevent protein translation (Quinlan, Kenny, Medina, Engel, & Jimenez-Mateos, 2018). There has already been success in the use of ASOs is degenerative disorders with the success of the ASO Nusinersen in the treatment of spinal muscular atrophy (Finkel et al., 2017). Given that HD, like spinal muscular atrophy, is a single gene disorder, the use of RNA interference technology to prevent the production of the pathogenic protein is an attractive strategy for achieving disease modification, with initial phase 1 trials showing significant promise (Tabrizi et al., 2019).

1.2 The potential use of cell and gene therapies for PD

The motor symptoms primarily associated with PD (tremor, dystonia, rigidity of movement) are caused by the loss of dopaminergic cells in the substantia nigra. Unlike the single genetic cause underlying the pathogenesis of HD, the etiology of PD is believed to be a combination of genetic and environmental risk factors (Jenner et al., 2013). These motor symptoms can be treated with various dopaminergic drugs which serve to restore some of the function lost as a result of the neuronal loss seen in PD, but the effects of these medications diminish with time and can be accompanied by significant unwanted effects (Borovac, 2016). Similarly to HD, it is thought the specific regional cell loss seen in PD can potentially be combatted through the introduction of donor cells unaffected by the disease to restore normal brain function. Trials of fetal cell transplantation have demonstrated safety but with variable effects on function (Freed et al., 1992, 2001; Lindvall et al., 1990; Olanow et al., 2003). Review of the fetal cell transplant trials suggests a series of design differences between these trials (Barker, Barrett, Mason, & Björklund, 2013) may account for the relative differences in function that were reported. These design aspects will be discussed in more detail in later sections of this chapter.

As with the use of fetal cells for transplantation in HD, the utility of fetal cells for transplantation in PD is limited by the availability of donated fetal tissue. The development of stem cell lines to provide a more reliable cell source for transplantation is more advanced in the PD field than in HD, to the point where the first trials of human embryonic stem cell derived cell products (NCT04802733) (Clinical Trials, 2022) and also induced pluripotent stem cell products (Takahashi, 2020; Takahashi & Price-Evans, 2019) have now entered clinical trials with further trials expected to start in 2022 (Kirkeby, Parmar, & Barker, 2017).

The lack of defined genetic cause and multiple etiologies associated with PD pathogenesis means that RNA interference strategies do not currently have a role to play in combatting disease progression. There are however other ways in which gene therapies can be used to treat PD. This includes both the delivery of biochemical elements to boost dopamine action, such as enzymes involved in dopamine synthesis or other factors involved in dopamine pathways and know to be down regulated in PD and the delivery of neurotrophic factors to stimulate the growth and repair of dopaminergic neurons in the substantia nigra (reviewed in (Axelsen & Woldbye, 2018; Blits & Petry, 2017)).

2. Principles of clinical trial design

The randomized controlled clinical trial (RCT) is widely regarded as the gold standard for the robust clinical evaluation of any novel therapeutic entity, where one seeks to establish a causal relationship between a treatment and the intended outcome (Hariton & Locascio, 2018).

The principal premise of the RCT is to eliminate the effects of bias and other confounding factors as far as possible through random assignation of a treatment or comparator, ideally a placebo, to the population of interest. The outcome of interest is assessed before the intervention and after a pre-determined follow-up period to allow evaluation of any potential effectiveness of that treatment, with the expectation that any between group differences are not observed due to chance alone.

The scale, or required group size, of any given RCT will be defined by the variability in the primary outcome of interest. Once the primary outcome has been selected, information on the inter- or intra-person variability of that outcome and the degree of expected change to represent a clinically meaningful difference are used in statistical calculation to determine how many participants are needed to ensure that any change measured in that primary outcome is due to the intervention being studied and not due to chance alone (An, Duong, Le-Rademacher, & Mandrekar, 2020;

Nottage & Siu, 2002; Stanley, 2007). The term "power" is often used in the description of the sample size calculation. In its simplest terms, power refers to the probability that an observed result has occurred due to an intervention than due to chance, with higher power reflecting a higher probability that the null hypothesis used to form the research question can be rejected (Sedgwick, 2015).

2.1 Methods for reducing bias in clinical trials

2.1.1 Randomization

The random assignation of participants exhibiting the necessary criteria for inclusion in the trial is used to eliminate *selection bias*, as far as is feasible, by ensuring that the participants in the treatment and control groups are as similar as possible (Odgaard-Jensen et al., 2011).

2.1.2 Allocation concealment

The implementation of participant randomization, coupled with suitable allocation concealment devices, is used to attenuate *allocation bias*. Allocation bias describes the unconscious potential for an altered approach or selection of participants if the group assignation can be determined prior to allocation, and therefore skew the participant characteristics in that group (Berger & Do, 2010). This can be attenuated through the implementation of participant randomization, coupled with suitable allocation concealment devices. Allocation concealment refers to not only the shielding of allocation for successively recruited participants to the researcher mediating that interaction so that allocation cannot be predicted from knowledge of previous assignations, but also the method of generating of the randomization sequence is also protected from direct observation (Berger & Do, 2010).

2.1.3 Blinding

Once a participant has been allocated to a specific treatment group, there is the potential for the assessor of any given outcome, when aware of the treatment allocation, to influence the empirical evaluation of that outcome (Hróbjartsson et al., 2012), which is known as *observer bias*. Similarly, where a participant is aware that they have been allocated to receive a specific treatment, there is the potential for them to alter their behavior or effort when it comes to outcome assessments, defined as *performance bias*. This is particularly pertinent in evaluations where measurement of the outcome of interest is subjective rather than objective. When applied to situations where the participant is not in receipt of any intervention but is aware of ongoing observation with subsequent behavior modification, this is known as the Hawthorn effect (McCarney et al., 2007). To mitigate against the influence these potential biases, the assignation of participants can be concealed from observers and participants, commonly known as blinding. The terms single and double blind refer to occasions where only the researcher or observer or both the researcher and observer as well as the participant are unaware of the treatment allocation. Best practice is to maintain a double blind with all personnel involved in the conduct of the trial from the point of randomization until such time as all data has been collated and relevant analyses performed. Only once the final analyses have been performed should the randomization code be broken to reveal the identity of treatment groups.

2.2 Specific considerations for clinical trial design in the context of cell and gene therapies in neurodegeneration

There are specific constraints in the development of clinical trials of ATMPS in neurodegenerative disease that need to be factored in when considering appropriate trial designs. First, and most obviously, the neurodegeneration occurs in the brain, although (and currently poorly understood) biological effects of neurodegenerative disease may be seen outside of the central nervous system. This means that any AMTP, be it cell, gene or other protein-based therapy needs to be able to reach the affected areas of the brain in order to restore lost function. In order to circumvent the blood brain barrier, which prevents any large molecule entering the brain (Kadry, Noorani, & Cucullo, 2020), therapies require direct intra-cranial or, in some cases, intra-thecal, administration, thus automatically increasing the complexity of delivery for the outset. Second, when considering cell and gene therapies it is important to remember that, unlike traditional pharmacological treatments, these may become a permanent biological feature- once cells or genes incorporated into viral vectors have been delivered, they cannot be removed or attenuated. Third, the natural trajectory of decline of neurodegenerative diseases such as HD and PD is over many years. With these slowly progressive disorders, any slowing or reversal of disease progression is likely to take years before the change is measurable, thus requiring lengthy and expensive follow-up periods in the clinical evaluation of new therapies. Fourthly, even though neurodegenerative diseases are relatively common as a whole, taken separately, particularly in the case of HD, which is considered a rare disease, patient numbers are relatively small when compared to other conditions such as stroke and cancer where ATMPs have been trialed previously. Lastly, the choice of trial design for any ATMP investigation in neurodegenerative disease will be significantly influenced by the primary objective of that trial, amongst other factors. For example, early phase trials aimed at exploring safety, tolerability and/or feasibility would be expected to include smaller numbers of participants and may only require a single arm, whereas a trial aimed at assessing efficacy would include numbers dependent on the features of the outcome assessment and should include additional measures to minimize bias.

2.2.1 Addressing the problem of small numbers using within patient trial designs

The most simple RCT design that introduces the least bias and is perhaps the most often used design is the randomized, parallel arm, placebo controlled, double-blind trial. For the reasons outlined previously, this design serves to reduce selection, allocation, observer and performance bias. If required, additional parallel arms can be added to allow for comparisons of more than one treatment or intervention if multiple options with clinical equipoise exist. However, the RCT using parallel groups more often than not requires a large sample size to be adequately depowered. This is a significant issue for trials involving people with HD given its rare disease status. This issue become even more acute in ATMP trials in HD.

One way to circumvent the requirement for large numbers is to implement a trial design where individual participants effectively act as their own act as their own controls, referred to as a "within patient" design. The cross over design is just one example of a within patient trial design which involves initial randomization of participants to intervention or comparator groups. This is followed by a washout period once the intervention has been delivered as intended, after which the allocation of the groups is switched, so that those participants originally in the comparator group go on to receive the intervention (Mills et al., 2009). Other within patient trial designs include n-of-1 (Lillie et al., 2011) and Latin square (Dean, 2001) where selection of any given repeated measures design for clinical evaluation of an intervention will be dependent on other factors such as participant population and the nature of the intervention.

These trial designs negate the influence of cofounding variates that may be present between groups as all participants serve as their own control. This allows for the trial to reach sufficient power with fewer numbers of participants. However, the utility of this design is limited by the characteristics of the intervention being studied. In pharmacological studies, the use of a cross over design would be dependent on the pharmacokinetic and pharmacodynamic properties of the medicinal product, where the investigators would need sufficient detail of these to implement a sufficiently long washout period between phases to ensure that any carry-over between treatment groups is minimized. This requirement for a washout period in between treatment phases means that this design is not suitable for cell and gene therapies. While some advanced therapies (such as ASOs) are not permanent interventions, they are still likely to have significant washout periods. Given that the long-term biological effects of repeated ASO administration are not yet fully elucidated, there would be a risk of ongoing contamination between groups.

2.2.2 Minimizing bias in ATMP trials in neurodegenerative disease

Trials of cell and gene therapies requiring direct intracranial administration pose a difficult ethical question- namely should sham surgeries be conducted in the evaluation of novel therapies? Sham surgery describes the scenario where a participant undergoes a surgical process up to the point of intervention delivery but stops short of any therapy being delivered (Beard et al., 2020). In the case of therapies delivered intra-cranially this includes anesthesia for the same time as would be taken to perform the surgical procedure itself and the completion of the steps required to deliver substances intra-cranially such as head shaving and the drilling of burr holes. Sham surgeries serve to maintain the blinding of group allocation to participants and assessors, reducing performance and observer bias but questions still remain over the ethics of the use of sham surgery in a tussle between the highest levels of scientific rigor and the highest level of ethical practice (Rogers, Hutchison, Skea, & Campbell, 2014). Some of the early trials of fetal cell transplantation in PD adopted sham procedures to minimize bias on the basis that surgical innovations require a greater degree of rigor in their evaluation owing to premature adoption of novel surgical practices later found to be lacking (Freeman et al., 1999). Latterly, with the introduction of gene therapies requiring intracranial administration, PD researchers were reported to be overwhelmingly in favor of the inclusion of sham procedures to minimize the false positive effects and limit the potential for adopting weakly effective of ineffective therapeutic interventions (Kim et al., 2005). This is in line with the stance of the Food and Drug Administration (FDA) who recommend the inclusion of sham procedures or placebos for the investigation of gene therapies in neurodegenerative disease (FDA; *Human Gene Therapy for Neurodegenerative Diseases; Draft Guidance for Industry*, 2022a), as long as the risk-benefit profile is reasonable and strategies are in place to minimize risk to participants (*Code of Federal Regulations Title 21*, 2022). Although there is a clear desire within the scientific community to conduct trials with the highest levels of rigor, there are still conversations to be had about the ethics of including a sham control, with relevant factors concerning the trial objectives, endpoints and study population influencing the decision on the inclusion of a sham control (Galpern et al., 2012). Indeed, despite the FDA recommendations, an analysis of cell therapy trials in PD that included sham procedures against open label trial designs found no evidence that the implementation of a sham surgery as integral aspect of trial design did not confer any methodological benefit for addressing the issue of placebo effect over other methods for controlling for the surgical intervention (Polgar & Mohamed, 2019). With the controversy surrounding sham controls for the evaluation of AMTPs in neurodegenerative disease in mind, the merits of trial designs that avoid the use of such a contentious design could be considered.

2.2.3 Alternative trial designs that negate the requirement for sham controls

Exploration of trial designs that use alternative methods to the use of sham controls to minimize bias is an important consideration in the design of ATMP trials in neurodegenerative disease and has been the subject of some discussion (Barker et al., 2019; Drew et al., 2021; Rosser et al., 2022). When considering alternative designs, particularly where small numbers are a concern, lessons learned both from clinical evaluations in rare diseases (Gagne, Thompson, O'Keefe, & Kesselheim, 2014; Hilgers, König, Molenberghs, & Senn, 2016) and from the designs implemented in previous clinical evaluations of ATMPS are important. Interestingly, a review of ATMP trial designs supporting marketing authorizations revealed a trend toward the implementation of small, open label trial designs even if the condition of interest was not traditionally associated with the problem of small numbers (Iglesias-Lopez, Agustí, Vallano, & Obach, 2021). This review also highlighted the relative popularity of conducting trials involving the use of historical control data (Iglesias-Lopez et al., 2021).

2.2.3.1 Use of historical control data

The use of historical control data as a method for benchmarking potential treatment effects in rare disease where the inclusion of more traditional control data is challenging, although not a new idea (Pocock, 1976), has gained popularity in recent years to the degree that the FDA have now produced guidance on its use to support clinical evaluations of new treatments (FDA; Rare Diseases: Natural History Studies for Drug Development Guidance for Industry, 2022b). The premise of historical control data is that previously collated outcome assessments data can be used to make comparisons to be made with similar populations who go on to receive potentially disease modifying interventions. There several methods for obtaining historical control data; exploitation of disease registries where routine longitudinal outcome assessment is used to build a comprehensive picture of disease progression and phenotypic variances within a population; prospective collection of natural disease history data; using *control* data already collected in previous clinical trials or by obtaining data through medical records (Ghadessi et al., 2020). Additional statistical methods should however be employed in the use of historical control data to address issues of missing data and to control for variability and the influence of confounders between subjects owing to the lack of randomization. It has been proposed that the latter can be addressed through the use of propensity score matching to achieve balance between analysis groups (Dahabreh et al., 2012; Lonjon et al., 2014; Markoulidakis et al., 2020, 2022).

The utility of historical controls is a tangible possibility in HD and PD. The HD field has long benefitted from the global, longitudinal observational study, Enroll-HD (www.enroll-hd.org) where participants attend annual assessments where a number of outcomes are assessed, including the most commonly used measures of disease progression (Enroll-HD, 2018). This study now has upwards of 20,000 participants worldwide with a wealth of historical data that is made accessible to researchers through a rigorous peer review process. The "natural history" approach to generating control data has been employed in recent trials of ASO therapies in HD by Hoffman La Roche and Wave Life Sciences. Whereas Hoffman La Roche designed a study for the prospective collection of observational, alongside additional biomarker collection, to augment control data for their ongoing investigations into their ASO therapies (Hooper et al., 2019), Wave Life Sciences utilized historical data from Enroll-HD and its precursors to enable comparison of their treatment cohort against expected disease trajectories (Viglietta, & Development, V. C., 2022).

Similar to the Enroll-HD platform, recent efforts by the Michael J Fox foundation have seen the establishment of the Parkinson's Progression Markers Initiative (www.ppmi-info.org) to track disease progression in people with PD. Both registries also benefit from adjunct studies that support the collection of biological samples to support pathological understanding and biomarker development, the advantages of which will be discussed in later sections (Parkinson's Progression Markers Initiative, 2022).

2.2.3.2 The randomized start design

The premise of randomized, or "delayed start" design is to allow for evaluation of a novel treatment in respect to demonstration of lasting effects on disease progression (D'Agostino, 2009). This design sees participants randomized to receive either the active treatment or comparator, which could be a placebo or standard care where placebos are not ethically or logistically valid. After a given length of time, the participants in the comparator group then receive the active treatment. In simple terms, the differences in the trajectory of disease course are used to determine treatment effects (from the first phase) and whether or not those effects are maintained (from the second phase). This design is ethically beneficial as it enables all participants to receive the active treatment and would have utility in disorders where there are no biomarkers to demonstrate disease modification, as is the case in many neurodegenerative disorders. However, there are several caveats to the use of the design, not least the requirement to have a good understanding of the disease process and selection of appropriate analytical techniques (D'Agostino, 2009). Analytical techniques to account for non-linear progression have been proposed (Wang, Robieson, Zhao, Wiener, & Koch, 2019) specifically for the use of the delayed start design and it has been employed in a PD drug trial (Olanow et al., 2009) and a fetal cell transplant trial in HD (Bachoud-Levi et al., 2020; Bachoud-Lévi et al., 2020).

2.2.3.3 Trials within a cohort

A similar premise to the use of historical control data particularly in relation to the use of disease registries for use in situations where blinding of the intervention is not possible, is the use of the pragmatic Trial Within a Cohort (TWiC) design (Relton, Torgerson, O'Cathain, & Nicholl, 2010). The underlying principle of the TWiC design is that a cohort of the population of interest is established with the assessment of outcomes of interest at a pre-determined frequency and participants eligible for any given intervention are randomized to receive the intervention under investigation. This feature of randomization serves to mitigate against any allocation bias that may introduced through participant's knowledge of assignation to a control group through the collection of control data from cohort participants who are unaware of their particular allocation. However, a fundamentally important aspect in the conduct of trials using this design is the way in which participants consent to be included in the research. It is an accepted international standard that before agreeing to participate in any clinical research, the participant must receive adequate information in relation to what will happen to them in the course of the research (ICH, 2016) and what will happen to their personal data. To meet the standard of informed consent in the TWiC design, when participants are recruited to the cohort they are made aware of the aspects of participating in the cohort and why it is being done, along with the possibility of being randomized to receive a therapeutic intervention at a later date, but at this stage they are not necessarily provided with information about that intervention (Kim, Flory, & Relton, 2018). At the point at which there is a desire to evaluate an intervention, eligible participants within the cohort are randomized to either the control or intervention arms, and those allocated to the intervention receive the additional information about the intervention to satisfy the standards of obtaining appropriate informed consent.

While the TWiC design has a number of pragmatic advantages, including the concurrent evaluation of more than one intervention and the ability to maintain long-term follow-up, its use is not without caveats (Weijer, Goldstein, & Taljaard, 2018). The presentation of group allocation to only those randomized to the intervention group may incur higher levels drop out following randomization than for traditional randomized designs, which has implications for the sample size calculation and analysis of trial outcomes (Reeves et al., 2018). Further, for studies that require long-term follow-up, the burden of repeated longitudinal assessments may place unfavorable burden on participants leading to issues with retention. It may be possible to modify this design to mitigate against potential confounders, particularly for use in early phase trials in neurodegenerative disease. For participants recruited into the main cohort there would be an expected and inevitable decline of function over time, and so participants would need to undergo screening for inclusion into an interventional study to ensure they meet the defined inclusion criteria, particularly if the routine data collection of the cohort is planned to occur with significant time intervals. Additionally, given the rigors of undergoing a neurosurgical procedure for the intra-cranial delivery of therapeutic agents, it is essential to ensure that participants are medically fit, again, requiring additional clinical assessments to be performed. For small investigator-led TWiCs, it would not be cost effective to perform clinical screening tests in all cohort participants in which case a staged approach to confirm eligibility has been suggested for use in cell therapy trials involving people with HD (Drew et al., 2021), while noting that this cohort design has been employed successfully in a fetal cell transplant trial in people with PD (Barker et al., 2019).

3. Outcome measures in cell and gene therapy trials in neurodegenerative disease

The decision as to what should be measured and when to fulfil the objectives of any given clinical investigation is a fundamental factor underpinning the overall choice of study design. The definition of the primary outcome measure is pivotal in determining sample size in efficacy trials, and to some degree the primary and secondary endpoints, which in turn influence the total length of the trial. The chosen outcome measure will be the critical determinant of a positive or negative outcome and so it is essential that the most applicable, valid measure with the capability to adequately detect the postulated change is selected for robust evaluation of the intervention being studied (Coster, 2013). Further, the choices of outcomes selected must retain relevance to all stakeholders, particularly those people the intervention being investigated is aimed at treating and the clinicians making treatment decisions (Treweek et al., 2015).

As the development of cell and gene therapies of neurodegenerative disease is still relatively nascent, the majority of trials that have been conducted to date or an in progress are small, early phase investigations. In these instances, the objectives of the research will always center around safety, tolerability and feasibility to ensure that the investigational product is safe and applicable in a small number of participants before moving forward to dose finding and efficacy studies (Deore, Dhumane, Wagh, & Sonawane, 2019). However, when the course of clinical investigation moves forward to determining the degree of efficacy a novel product may have in the modification of disease, selection of an appropriate primary outcome measure is key. While researchers (both academic and industry based) will have their own ideas and preference as to which outcome measure should form the primary endpoint, it is common that the final selection is influenced by medicines and healthcare product regulators such as the FDA and European Medicines Agency (EMA). However, best practice on outcome measure selection should also include the needs and preferences of participants (Andrejack & Mathur, 2020). The need for harmonization of selected outcomes measures across different studies will be explored in more depth in Section 3.2.

Outcomes can be grouped into five main domains; survival, clinical response, patient reported, events of interest and resource utilization with each domain featuring a number of individual measures that can be applicable to a clinical evaluation (Dodd et al., 2018). Not all domains are applicable to all clinical studies- for example, survival outcomes will be common in clinical evaluations of new chemotherapy agents for treating oncological diseases, but for chronic disorders, which would include neurodegenerative diseases, clinical response and patient reported outcomes would be more relevant.

Factors affecting the choice of outcome include the degree of bias which that measure exerts and its applicability to the main objectives of the clinical study. Ideally, outcome measures should be objectively assessed to mitigate against reporter bias and be sensitive to change (Coster, 2013), particularly in disorders such neurodegenerative disease where changes may be slowly progressive and testing is limited to small number of participants.

3.1 Outcomes applicable to cell and gene therapies in neurodegenerative disease

A number of generic and disease specific outcome measures exist to assess function and disease status in neurodegenerative disease. However, outcome measures currently available and routinely used to detect functional change in neurodegenerative disease may not adequately address all areas of living with a chronic condition. Measures such as the Unified Huntington's Disease Rating Scale (UHDRS) (Huntington Study Group, 1996) and Unified Parkinson's Disease Rating scale (UPDRS) (Goetz, 2003; Goetz et al., 2008) have been developed specifically to act as the gold standard for assessing disease progression in their respective disorders. However, it is accepted that even with these expert-developed measures, not all facets of living with HD or PD are captured in these functional motor-based measures, nor are they as sensitive as may be required to detect small changes in function (Quinn & Busse, 2017; Tosin et al., 2021). In addition to functional measures, one should also consider the inclusion of outcomes of importance to the participant population which more accurately reflect the effects of a novel treatment on their broader ability to live well (Andrejack & Mathur, 2020; FDA, 2016a, 2016b) as these may differ from clinically selected outcomes (Morel et al., 2022). For example, a common way of assessing motor function in movement disorders is measure finger tapping and while this can provide an accurate measure of motor ability, it does not particularly translate to the everyday activities of the participant. Instead, the participant may feel that their ability to perform simple motor tasks, such as the ease of buttoning a shirt or not, is a better reflection of restoration of function.

Core aspects of the evaluation of cell and gene therapies in neurodegeneration will affect the selection and range of outcome measures included in clinical evaluations. For instance, while effects on overall function are always going to be a cornerstone of investigations, the inherent nature of cell therapies, where restoration of neuronal networks is required to effect functional change, it is necessary to assess the degree to which a cell transplant has integrated within the host environment, and thus visualization of the transplanted brain region will likely be required. Additionally, as cell and gene therapies involve the introduction of foreign biological molecule to the body, one would reasonably expect activation of immune system modulators despite the brain being subject to a degree of immunoprivileged, therefore, it would seem pertinent to measure markers of immunoreactivity and inflammation in blood and cerebrospinal fluid (CSF) (Rosser et al., 2022).

Lastly, if cell and gene therapies prove to be efficacious and capable of disease modification, the market costs of these therapies are likely to be extremely high. For example, the cost of the ASO for spinal muscular atrophy, Nusinersen, marketed under the name Spinraza, was set at $125,000 dollars (Staton, 2017). This would be a total of $375,000 dollars for one years' worth of treatment and while this is not out with comparison to drugs for other orphan disorders, translation of this level of cost to a more common disorder such as PD could have significant economic implications. So, despite cell and gene therapies for neurodegenerative disorders still being in their developmental infancy, it would seem pertinent, even at these early stages, to include measures which can be used to evaluate the relative cost effectiveness of these potentially efficacious treatments. This would involve a full cost analysis of treatment delivery in comparison to the wider socioeconomic benefits for patients and their families in reducing disease burden, frequently assessed through the measurement of quality adjusted life years (Vergel & Sculpher, 2008).

3.1.1 Outcomes measures for use in cell and gene therapy trials with reference to HD and PD

The Unified Huntington's Disease Rating Scale (UHDRS) is the most commonly used measure of function in HD and provides and overall measure of disease burden, made up from the sub categories of motor, cognitive and behavioral assessment and functional capacity (Huntington Study Group, 1996). It is the only measure to be recommended by the Movement

Disorders Society for the assessment of motor function in HD (Mestre et al., 2018). The UHDRS, or specific sub-scales of the measure are used as both an endpoint for clinical research and as a screening tool for stratifying and selecting participants for inclusion in clinical trials. However, this clinician rated scale requires specialist training and is subject to potential inter-rate variability and changes in UHDRS score do not necessarily correlate with the degree of neurodegeneration in an individual. Efforts have been made to produce a more sensitive measure of disease progression that may be more representative of the degree of brain atrophy in individuals (Schobel et al., 2017). The composite score described uses a combination of clinical, motor, function and behavioral outcome measurements to produce a single metric that is more representative of disease burden in people with HD and may prove to be of greater utility in the assessment of novel therapies in HD. The concept of using such a composite outcome in HD has gained further support through it refinement to produce a composite measures with greater signal to noise ratio (Sewell, Penney, Jay, Zhang, & Paulsen, 2021) for use in clinical trials.

The revised Unified Parkinson's Disease Rating Scale (UDPRS) (Goetz et al., 2008) shares similarities with UHDRS as the most commonly used measure for assessing disease status in PD, covering non-motor and motor experiences of daily living, motor function and motor complications and subject to inter-rater variability. Other potential issues with UPDRS as a global measure of a disorder that exhibits significant heterogeneity between individuals, is that scores do not reflect the granularity of disease severity (for example a similar score can be achieved by a person suffering from significant tremor over many years and a person with a shorter disease course and where rigidity is their main motor symptom). Further, given the variable nature of PD symptoms within any given individual which are dependent upon medication sate, nutrition and other factors, the episodic "snap-shot in time" nature of UDPRS may not accurately represent the disease status of an individual. For both HD and PD, research into developing objective measures of motor impairment have been ongoing, with the development of assessments such as Q-Motor for HD (Reilmann & Schubert, 2017) and the "At-Home testing device" for PD (Goetz et al., 2009), with the use of Q-Motor now featuring in trials for HD treatments (Reilmann et al., 2019).

As we alluded to previously, it may not be sufficient to rely on the assessment of functional change alone in order to provide a more holistic picture of potential benefits. There are several patient reported outcomes that can be used to address wider aspects of the impact of disease and may provide insights to non-functional improvements. One might consider the inclusion of a generic quality of life measure, such as the SF-12 (Ware, Kosinski, & Keller, 1996), which can be used across a range of disorders. These are widely used measures, validated in a variety of languages, thus particularly useful for global studies. However, as it is a generic measure, it may lack sensitivity to measure aspects of quality of life that are specific to particular conditions. For both HD and PD, specific health-related quality of life measures have been developed and validated; the HD-Pro TRIAD (Carlozzi et al., 2014) and PDQ-39 (Peto, Jenkinson, & Fitzpatrick, 1998) to measure quality of life amongst other disease specific dimensions. However, these are not necessarily the only disease-specific measures of quality of life, leading to variability in the outcome measures employed across various studies. For trials involving people with PD, non-motor symptoms that affect activities of daily living have significant impacts on quality of life (Soh et al., 2013), so assessment of these should be strongly considered for inclusion in evaluations of novel therapies.

Another consideration for what should be included in terms of participant assessments in trials of cell and gene therapies in neurodegenerative disease are measures that record changes in behavior. In HD, behavioral changes including apathy, impulsivity, irritability and an increase in frontal behaviors are often some of the first symptoms to appear and are often reported as being the most problematic for patients and their family members (FDA, 2016b; Glidden et al., 2020). Despite the differing etiology between HD and PD, similar behavioral changes been reported (Trojano & Papagno, 2018). General, standardized measures exist to record cognitive function, depression and anxiety, apathy, irritability amongst others. These types of outcomes including measures such as the modified Frontal Behavioral Systems Scale (Duff et al., 2010), Apathy Evaluation Scale (Marin, Biedrzycki, & Firinciogullari, 1991), Irritability scale (Snaith, Constantopoulos, Jardine, & McGuffin, 1978) and Hospital Depression and Anxiety Scale (Bjelland, Dahl, Haug, & Neckelmann, 2002), which are all subjective patient reported outcomes. However, there are few disease-specific measures and even fewer that are subjective in nature in these areas. The recent European Consortium study REPAIR-HD (www.repair-hd.eu) (Repair-HD, 2020) worked toward the development of new objective measures to assess HD, including those for behavioral symptoms (McLauchlan, Lancaster, Craufurd, Linden, & Rosser, 2019).

Until now, discussions have been focused on how we assess the potential benefits of a novel treatment in the participants and patients receiving that product. However, given the significant burden experienced by care givers in neurodegenerative disease (Martinez-Martin et al., 2015), it would be pertinent to also consider recording how a novel therapy may affect the wider support network of participants. Care partners and family members may often provide insights into recent behaviors that the participant themselves, particularly in HD, may lack and so can provide a more accurate reflection of behavioral changes. Outcome measures exist to specifically assess carer burden (Epstein-Lubow, Gaudiano, Hinckley, Salloway, & Miller, 2010), which go some way to act as a surrogate measure for improvement in people with neurodegenerative disease following experimental treatment.

Given that trials of cell therapies are likely to be long-lasting on order to realistically detect functional benefits, there is likely to be an increased burden of participation on those volunteering their time to take part in these studies than for those participating in more acute investigations. Higher levels of burden may end up being reflected in higher levels of attrition, thus impacting on the overall data integrity of the investigation. Therefore, methods to reduce burden on participants should be contemplated. In recent years, the notion of developing methods for remote monitoring or at-home testing has gained popularity. The ability to conduct assessments without the requirement to travel to a specialist center would be beneficial in reducing overall burden. Such strategies were employed in the Hoffman La Roche trial of Tominersen (NCT03761849) where mobile digital devices were given to participants for functional monitoring and have been trialed in PD using specially developed at home devices (Goetz et al., 2009). However, for mainstream adoption into clinical trials, remote monitoring and assessment methods require further validation and standardization.

Important outcomes for intracranially delivered cell and gene therapies include brain imaging modalities, which for may include Magnetic Resonance Imaging (MRI) or Positron Emission Tomography (PET). Brain imaging markers may be used for a variety of reasons in trials of cell and gene therapies for neurodegeneration including in the selection of eligible participants (Drew et al., 2021), long-term safety assessments (Bachoud-Lévi et al., 2006; Hauser et al., 2002) and in monitoring graft placement and growth for cell transplants (Barker et al., 2019). While graft deposits can be detected using structural MRI, it is only possible to determine the functional activity of transplanted cells through the use of imaging methods such as PET, which involves the use of specific radio-labeled ligands to image cellular functions. The advantage of this is the ability to select ligands to identify the cellular functions of interest (Brooks, 2004), however, this is caveated by the limited availability and expense of custom radio-labeled ligands (MRC, 2017).

3.2 The development and utilization of core outcome sets

As outlined above, the number of outcome measures falling within the clinical response and patient reported outcomes is expansive and despite the existence of disease specific functional measures such as UHDRS and UPDRS, for holistic intervention evaluation, other outcome measures must be included to assess the breadth of potential impact of the treatment under investigation. The large scope of measures that are available for selection can result in disparity between outcomes measured across different studies. Not only can this lead to trial inefficiency, through the collection of numerous secondary outcomes which fail to be reported (O'Leary, Seow, Julian, Levine, & Pond, 2013) but it can also generate barriers to meta analytical studies which are critical when working in rarer disease domains (Chan et al., 2014).

It is highly likely given the complexity, cost and early-stage nature of cell and gene therapy trials, particularly in HD, that advancements will be made through a series of smaller, iterative trials rather than through large scale Phase III effectiveness studies. Therefore, it is of paramount importance that the outcomes included in these trials allow for comparison and collation of data across studies. One way of doing this is to include a core outcome set (COS) as a minimum data collection standard amongst the outcomes selected for any given clinical investigation. COS are designed to reduce outcome reporting bias and this issues of heterogeneity affecting systematic reviews, while ensuring that outcomes relevant to all key stakeholders for a given clinical condition are included (Williamson et al., 2012). The development of COS is supported by the Core Outcome Measures in Effectiveness Trials (COMET) initiative which provides a database for all COS under development and guidance for how to generate a COS (Williamson et al., 2017). A fundamental aspect of COS development is the consultation of a range of stakeholders, including patients and caregivers, to join in the process for ranking and prioritizing outcomes on interest for inclusion. This serves to ensure that decision makers are provided with the evidence necessary to make those decisions and that clinical trials maintain relevance for the participant population, driving efficiency through aiding recruitment and retention. This importance of this consensus process is highlighted by the differing perspectives of health care professionals involved in treating PD and those living with PD on the relative importance of incorporating specific outcomes in trials (Sjödahl, Nilsson, Idvall, Rosas, & Hagell, 2014). The adoption of specific core outcomes may increase the possibility for the calculation of composite scores which have shown potential value as markers for disease progression in HD (Schobel et al., 2017; Sewell et al., 2021), subsequently enabling the design of smaller trials with larger power.

There is no COS registered with and developed under COMET guidance for the evaluation of cell and/or gene therapies in HD or PD although the COMET database contains entries for the development of COS for both HD (www.COMET-Initiative.org) (COMET Initiative, 2022) and PD (De Roos et al., 2017). The former is focused on delivery of HD care and the latter generalized to the evaluation of all treatments for PD. Suggested assessment batteries for investigating cell therapies for both disorders do however exist. The Core Assessment Program for Intracerebal Transplants (CAPIT) was published 30 years ago (Langston et al., 1992) in the wake of the early cell transplant trials in PD. The assessment battery incorporated key motor and imaging assessments amongst stipulations for participant selection and scheduling of assessments. It is of note that this battery did not included any outcomes relating to non-motor symptoms or quality of life. The guidance was subsequently updated for application to the evaluation of other neurosurgical interventions such as deep brain stimulation with the generation of the Core Assessment Program for Surgical Interventional Therapies in Parkinson's disease (CAPSIT-PD) (Defer, Widner, Marié, Rémy, & Levivier, 1999) with the updated guidance recommending the inclusion of cognitive and quality of life assessments. Similar efforts have been made in standardizing the assessment of cell therapies in HD with the development of CAPIT-HD (Quinn et al., 1996) and latterly the CAPIT-HD2 (Bachoud-Lévi, 2017) batteries.

As discussed above, the use of COS in the evaluation of cell and gene therapies are beneficial for improving reporting biases and meta-analyses between studies, and can be relatively straightforward to embed into the data collection modules of disease registries. There are a number of methodological advantages in doing this, not least reducing research waste through the efficient collection of only the most meaningful outcomes and providing a wealth of historical data to negate the inclusion of large numbers of participants to provide control data in effectiveness trials. Adoption of COS in disease registries or longitudinal data platforms could further increase the validity of historical data through the rigorous application of data standards to maintain data integrity. This is type of initiative is already underway for rare diseases mediated by the Critical Path Institute (www.c-path.org), a collaborative mission to accelerate medical product development in response to the FDA's Critical Path Initiative program. This important collaboration has developed the Rare Disease Cure Accelerator—Data and Analytics platform (RDAC-DAP) which provides an accessible repository of patient level data and instils standards for data collection. The standardization of metrics then provides analytical power to enable mathematical modeling of disease progression. Similar efforts to standardize data collection within registries have been initiated by the development of the outcomes measures framework, driven by the Agency for Healthcare and Quality Research (Gliklich et al., 2014). This concept of harmonious data collection is highly attractive for maximizing the impact of data collected across trials of cell and gene therapies where trials are likely to remain small and iterative in their nature. This is particularly pertinent at this point in time in the development of cell and gene therapies for neurodegenerative disease where collaborative data sharing is key in order to maximize lessons learned through the progression of therapeutic development.

3.3 Considerations for the use of biomarkers and surrogate endpoints

In addition to assessments of clinical, functional and patient reported outcomes, it is also possible to include the use of surrogate endpoints to measure the effect of a therapeutic intervention. The term "surrogate endpoint" refers to the measurements of a biological feature that is expected to respond to a given treatment that may correspond to an intended clinical benefit of that treatment with necessarily being causally related. For instance, in cancer trials, tumor size or shrinkage can be used as a surrogate end-point to measure the effectiveness of a new chemotherapeutic agent, which would be expected to correlate to the clinical end-point of survival. The benefit of using surrogate endpoints is that a change in the surrogate end-point (i.e. tumor size) may be detected much earlier than the clinical end-point (i.e. overall survival). This is particularly relevant to the study of cell and gene therapies in neurodegenerative disease where one might expect any functional benefits of cell and gene therapies to manifest years following the administration of the intervention. Similarly, biomarkers are physiological aspects that are altered in a relatively predictable fashion by both disease and the treatment of that disease. Therefore, there is a potential for biomarkers to be used as surrogate end-points in clinical trials. Across neurodegenerative disease there has been significant interest in the identification and validation of biomarkers for use in clinical evaluations as described above. The ultimate goal would be to find a marker that can be easily obtained (i.e via simple blood draw) or measured in a simple and cost-effective fashion and in Alzheimer's disease, a number of recommendations for the use of biomarkers in clinical research have been suggested (Hansson et al., 2022).

For HD, a potential biomarker candidate for htt lowering therapies such as ASOs, would be the mutant htt protein itself, which is believed to enter the CSF from damaged neurons. While the measurement of mutant htt is possible and correlates with disease severity (Southwell et al., 2015; Wild et al., 2015), it can only be accessed via lumbar puncture and is only present in CSF in small concentrations. Further, the measurement of mutant htt in the CSF remains an indirect measurement of mutant htt in the central nervous system. More reliable would be the direct visualization of mutant htt in the brain and until recently this has not been possible, but the recent report of mutant htt sensitive PET ligand raises the exciting possibility of direct brain measurements and its use as an endpoint in clinical trials of htt lowering therapies.

A less invasive possibility for a reliable biomarker in HD is the measurement of neurofilament light protein (NfL). NfL is a neuronal cytoskeletal protein that is released from damaged neurons and has previously shown to be present in the CSF of people with other neurodegenerative disorders such as Alzheimer's disease, frontotemporal dementia and HD (Wild et al., 2015). NfL can be reliably detected in the CSF of people with HD, which excitingly correlate with blood plasma concentrations of NfL and function (Byrne et al., 2017). However, the use of NfL as a reliable marker of disease progression requires further validation (Parkin, Corey-Bloom, Snell, Castleton, & Thomas, 2021).

Large global observational studies such as TRACK-HD (Tabrizi et al., 2009) and PREDICT-HD (Paulsen et al., 2008) have gathered a wealth of longitudinal observational and biochemical data for use in the search for biomarkers and surrogate end-points. Meta-analyses of these and other longitudinal data sets have been used to identify potential biomarkers, amongst which it was determined that imaging markers could potentially be used as surrogate endpoints (Wijeratne et al., 2020). The continued longitudinal monitoring of clinical and functional data through studies such Enroll-HD will serve to support the search for reliable biomarkers in conjunction with the bioresource collection study HDClarity and soon to be recruiting imaging study ImageClarity (Rodrigues et al., 2022).

The Parkinson's Progression Markers Initiative (www.ppmi-info.org) and other longitudinal studies such as the Personalised Parkinson's Project (Bloem et al., 2019) developed with the aim of identifying biomarkers for use in PD research, have similarly supported the longitudinal collection of clinical and functional data alongside biosamples. The identification of reliable biomarkers in PD is particularly problematic given the variation seen between people with PD in key aspects such as disease progression and response to medication. A number of potential candidate biomarkers have been identified ranging from the use of imaging modalities across MRI, PET (Saeed et al., 2017) and Single Photon Emission Computer Tomography (SPECT) (Brooks, 2016), to biofluid biomarkers ranging across microRNAs (dos Santos et al., 2018), α-synuclein (Fayyad et al., 2019) and inflammatory markers (Li & Le, 2020). Yet, despite the identification of these potential biomarkers, as with HD, there are no accepted biomarkers standards in PD to predict disease onset and monitor disease progression.

Beyond the use of imaging and biochemical evaluations in biomarker development, much work is being undertaken to develop digital biomarkers in PD (Stephenson, Badawy, Mathur, Tome, & Rochester, 2021). The prospect of developing digital biomarkers for monitoring neurodegenerative disease is exciting for; their ability to reduce burden on participants in the context of long-term follow-up and present a more holistic assessment of disease status (as opposed to the single moment in time evaluations conducted at study visits) (Dorsey, Papapetropoulos, Xiong, & Kieburtz, 2017; Gold et al., 2018). The principles underpinning the development of digital biomarkers for PD can be translated to similar initiatives in other neurodegenerative disorders.

4. Technical and practical factors for consideration in operationalizing the intracranial delivery of cell and gene therapies

The effectiveness of cell and gene therapies is neurodegenerative disease is heavily reliant on the accurate delivery of said investigational products to the requisite target area of the brain. Potential methodological differences may have, hitherto unknown, effects on functional outcomes mediated through unsuccessful product transfer, off-target delivery and innate physiological reactions. Any of these factors have the potential, synergistically and in isolation to affect the success of an intra-cranially delivered treatments. Each factor should be considered in the context of others when designing trials of cell and gene therapy to maximize the likelihood of success.

4.1 Devices for delivering cell and gene therapies to the brain

In neurodegenerative diseases such as HD and PD, while there are global effects on brain function, the majority of pathology is restricted to discrete areas, which is why cell and gene therapies are an attractive therapeutic option. However, restoration of function within these areas will be limited by the accurate delivery of the cell or gene therapy. Reaching the target location requires the use of some sort of penetrative delivery device to navigate through the brain, often with multiple entries to ensure coverage of the whole area of interest and whereas the development of cell and gene therapies has progressed apace, the advancement of technology to deliver these advanced therapies has lagged (Potts, Silvestrini, & Lim, 2013).

Historically, methods for the intra-cranial delivery of substances to the brain have relied on simple, straight cannulas inserted directly to the target using stereotactic means (Freed et al., 2001; Hauser et al., 2002; Kopyov, Jacques, Lieberman, Duma, & Rogers, 1997; Olanow et al., 2003), coupled with a simple syringe driver or plunger. For both cell and gene therapies multiple tracts are required to achieve sufficient coverage of the target area, which inherently increases the risk of surgical complications such as hemorrhage (Potts et al., 2013). While the need for multiple tracts may be avoided in the delivery of gene therapies through large volume infusions, this is not possible for cell therapies, where it has been shown that a large injection bolus of cells leads to sub-optimal graft survival (Skuk, Paradis, Goulet, & Tremblay, 2007) and contribute to increased damage of surrounding tissues (Perry, Andersson, & Gordon, 1993).

The simple catheter design is highly susceptible to the significant reflux of products being delivered to the brain. This describes the process by where the fluid introduced at the injection site is constrained by the biomechanical properties of the tissue it is injected into and travels back up the tract formed by the catheter as a path of least resistance, leading to sub-optimal and off target delivery (Potts et al., 2013). Newer device designs aim to combat this issue by using cannulas that confer convection enhanced delivery (CED), whereby the maintenance of a pressure gradient in the cannula during infusion enhances distribution of injected macromolecules compared to simple diffusion alone (Hunt Bobo et al., 1994). In-dwelling delivery devices employing CED, that use a an implanted catheter system with an external port for drug administration have been used the investigation of growth factors in PD (Whone et al., 2019) (NCT03775538), however, the success of this system remains reliant on the accurate placement of the drug delivery catheters. CED devices have also been used in the single administration of gene therapies (Christine et al., 2009; Kaplitt et al., 2007). While the use of CED confers enhanced, more accurate delivery of substances to the brain, one still needs to be confident that if trials fail to meet the primary outcome (as in (Whone et al., 2019; Whone, Luz, et al., 2019), that this result is related to the lack of efficacy of the investigational product and not from a failure in product delivery. To circumvent this issue, it is possible using MRI guidance for product delivery (Richardson et al., 2011). This involves the use of specially developed, MRI-compatible stereotactic equipment and the labeling of the product with gadolinium to allow real-time visualization of infusions (Richardson et al., 2020). This enables monitoring of infusion progress and placement, enabling adjustments to be made if required thus providing a degree of certainty that the gene product has been deposited where it needs to be. This technology has been employed with some success in gene therapy trials in PD (Heiss et al., 2019) and its use has been expanded to gene therapy trials in HD (NCT04120493 (Rodrigues & Wild, 2020).

Special consideration needs to be given to the device selection for delivering cells to brain. Even with the use of CED to prevent reflux of cell deposits at the target site, a number of design aspects of the selected canula and its use can affect graft survival. These include sedimentation of cells within the cannula (Torres, Trigano, & Dunnett, 2015), the size of the cannula (Steiner et al., 2008) and the rate at which cells are delivered (Wahlberg, Ghuman, Liu, & Modo, 2018). Innovations in cannula design have led to the development of devices that use alternatives to the distal end aperture of traditional straight cannulas (Kondziolka, Gobbel, Fellows-Mayle, Chang, & Uram, 2011) or enable three dimensional delivery (Mendez, Hong, Smith, Dagher, & Desrosiers, 2000). However, none have yet received regulatory approval to allow their widespread use in clinical research and so, the choice of delivery device for cell therapy trials remains problematic.

4.2 Considerations for increasing the scale of cell and gene therapy trials in neurodegeneration

4.2.1 Defining the surgical procedure

Clinical trials of ATMPs in neurodegenerative disease feature a number of critical aspects that meaning that the delivery of cell and gene therapies meets the definition of a complex intervention as set out in the Medical Research Council of the guidance their evaluation (Skivington et al., 2021). This guidance further suggests that the evaluation of complex surgical interventions, such as cell replacement therapy, are likely to require a number of small, iterative studies as opposed to the traditional advancement through the phases of clinical trial associated with more traditional pharmacological evaluations. While this evaluation structure would allow for those initial studies to remain in single, specialist centers, it is accepted that ultimately, effectiveness trials will necessitate the involvement of multiple research centers globally. Further, as we have already outlined, the success of cell and gene therapies will rely on not only the treatment, but also how that therapy is delivered, and with what success, to the intended target. Therefore, the surgical process is something that needs to be standardized across differing centers and operators and so we must undertake evaluation of the surgical process in these early phase trials in order to define the vital components. A robust and systematic way of doing this would be to adopt the principles of the Idea, Development, Exploration, Assessment, Long-term follow-up (IDEAL) framework (McCulloch et al., 2009). This framework outlines the stages of surgical innovation and how one should approach each stage to generate evidence of effectiveness. The framework has already been adapted for innovations in medical devices (Sedrakyan et al., 2016) and physiotherapy interventions (Beard et al., 2018) and could be further adapted to include interventions such as cell and gene therapies. The application of the framework would enable the fidelity monitoring of the surgical process required for ATMP delivery in neurodegenerative disease and subsequently engender confidence in the results observed.

4.2.2 Defining peri-operative requirements for trial delivery and patient care

Accessory to defining core aspects of the surgical process, it is also important to understand what the needs of both participants and trial delivery staff before and after the surgery takes place. It is usual for all staff involved in the delivery of a treatment or complex intervention to receive study specific training prior to initiation of a study at that site as a regulatory requirement. However, the complexity of intra-cranial administration procedures is such that numerous personnel over a number of different specialities will inevitably come into contact with a participant during the course of the intervention delivery. In the design of trials of this nature it is important to define what should be regarded as clinical management (such as immediate post-operative care) versus what is a specific requirement for trial follow-up. Training should be provided to all those involved in the conduct of protocol defined tasks, but we would also advocate wider engagement of staff that may have peripheral contact with trial participants during in-patient hospital stays or within the community during the follow-up period.

The need for long-term follow up of participants to ensure safety in investigations of these novel therapies has wider implications for care of participant outside of the trial. Currently, it is accepted that participants undergoing cell transplants should receive some degree of immunosuppression to maximize the chances of graft survival and differentiation (Barker et al., 2019; Drew et al., 2021), but how long this should be for and what regimen is most appropriate is still to be determined (Rosser et al., 2022). Nevertheless, if ongoing immunotherapy is mandated, this will need to be managed between researchers and primary health care providers, once again, necessitating additional engagement and communication of trial procedures to those outside the core delivery team.

Another area for consideration is the need for rehabilitation of participants who have received cell and gene therapies. Evidence suggests that specific motor rehabilitation and additional environmental stimulants can improve the survival and integration of cell grafts (Berlet, Galang Cabantan, Gonzales-Portillo, & Borlongan, 2022; Clinch, Busse, Döbrössy, & Dunnett, 2017) and that adoption of rehabilitation strategies to maximize the success of grafts should be contemplated for the move to effectiveness trials once safety and feasibility of current investigational products has been demonstrated. The inclusion of such strategies should remain evidence based and theoretically driven which will help to determine the specific drivers of success in these highly complex interventions (Skivington et al., 2021).

5. Considerations for the inclusion of participants in trials of cell and gene therapies

The success of any given clinical trial is dependent on achieving recruitment and ongoing retention of the required number of participants for confident evaluation of the primary outcome. However, not all people are able or willing to either take part or remain involved in clinical trials for a variety of reasons. Here we contemplate how best to identify and engage with trial participants to maximize both the efficiency of trials and equity of access to all.

5.1 Participant selection

All clinical trial protocols explicitly specify the determinants for inclusion in that particular study. A set of inclusion and exclusion criteria is not only used to define the population that one wishes to study, but can also be used as a means of mitigating against potential safety risks to participants. For early phase studies, these criteria can be used to stratify the population of interest to those where researchers hypothesize either the maximum effect of benefit may be observed, or the widest safety margins may lie. Latterly, with larger effectiveness trials, the selection of participants should be geared toward recruiting a broadly representative population to ensure the generalizability of result. At current stages of development, the specific definition of a trial population poses both scientific and ethical dilemmas for the evaluation of cell and gene therapies in neurodegenerative disease.

Given the degree of risk associated with unproven therapies requiring neurosurgical delivery, there is a balance, in terms of participant selection, between including those who have reached a certain stage in their disease course where a serious complication arising from the research may have less impact than someone at an earlier disease stage and selecting participants where potential benefit may be observed. There is a school of thought that in HD, by the time at which patients are overtly symptomatic (and commonly included in clinical trials), the degree of damage in the brain may be too advanced for restoration of function to occur (Rubinsztein & Orr, 2016) and advocates for the inclusion of participants at an early disease stage. This reasoning is compounded by other factors associated with the delivery of cell and gene therapies, including the size of the striatum and degree of peri-vascular space. Significantly atrophied basal ganglia provide a small target in which to insert a cannula for the delivery of substances, compounded by increased peri-vascular space resulting in off target "leakage" of infused product.

In HD, the selection of participants for clinical trials has historically relied on measures of disease progression such as total functional capacity (TFC), total motor score or diagnostic confidence level for motor manifest patients. While these are widely used selection criteria, it could be argued that TFC is more a measure of resilience in participants that a direct measure of brain function. Further, the selection of prodromal participants presents difficulties due to the lack of overt symptoms to characterize disease stage for definition of the study population. However, meta-analyses of longitudinal observational data across TRACK-HD, PREDICT-HD and Enroll-HD has led to the development and validation of the prognostic index, or PIN, for HD (Long et al., 2017) enabling participant selection for trials and has further been demonstrated to have sensitivity for the stratification of prodromal participants (Langbehn & Hersch, 2020). Further, more detailed interrogation of longitudinal data sets, plus the employment of machine learning methods, have led to the identification of additional disease states (Mohan et al., 2022) and novel disease stage classification system (Tabrizi et al., 2022) which could be used to stratify participants for inclusion in clinical trials.

The same issues for PD also exist, but the picture is further complicated by the availability of medicines and interventions such as deep brain stimulation, that are relatively successful in treating PD symptoms- something that is lacking for HD. People with PD are frequently selected for trials using disease stage, primarily assessed by measures such as the Hoehn and Yahr scale (Goetz et al., 2004), but beyond that there are few guidelines for stratification in what is a more heterogenous disorder.

Early studies of cell therapy in PD suggest that younger people with less advanced disease show greater benefit from cell transplants (Freed et al., 2001; Olanow et al., 2003) and that the ability for cell grafts to survive may be increased by enhanced levels of trophic factors in participants displaying significant neurodegeneration (Kirik, Winkler, & Bjö, 2001; Piccini, Pavese, Hagell, Reimer, & Brain, 2005). The inclusion of people with earlier stage PD may not be without significant risk through exposure to neurosurgical procedures and all that encompasses combined with the highly experimental nature of cell and gene therapies. This level of risk is further compounded by the irreversibility of cell and gene therapies which risks the exclusion of participants from future trials of other potentially disease modifying therapies. In the context of people with relatively early disease, this poses an interesting and complicated ethical question as to whether or not they should be included in early phase trials of cell and gene therapies. There is an argument that this is not a decision that researchers should be making for people with PD, but should be a joint discussion between clinician and patient, based on likely treatment preference and risk-benefit profiles to enable that person to reach a truly informed decision.

5.2 Communicating with participants in cell and gene therapy investigations

Clinical trials are not possible without willing volunteers from the specified population who are prepared to give freely of their time and are willing to subject themselves to highly experimental interventions. Given the lack of long-term safety data for cell and gene therapies delivered to brain, it is vital that people approached to take part in such experimental studies are fully and frankly informed of all possible outcomes. This should be achieved though the consent process which involves information sharing to facilitate an understanding and acknowledgment of the risks. However, this can be significantly affected by differing levels of education and scientific literacy within the participant population. Globally recognized good clinical practice standards dictate that informed consent should be an ongoing discussion through the course of any research and not a singular exercise conducted at the beginning of the study (ICH, 2016). In practice this is usually achieved through the verbal confirmation of consent prior to any kind of assessment or procedure at follow-up visits combined with the assumption that a lack of a decision to withdraw consent implies continuing consent from the participant. For cell and gene therapy trials, where there is a requirement for invasive, risky procedures and the need for long-term follow up, it may be suggested that the consent process should be viewed as a dynamic rather than static process, with due care and attention paid the ensuring that the initial provision of consent is done in a fully informed fashion. A key aspect of this process is the careful management of expectations, particularly for neurodegenerative disorders where up until now there has been a dearth of disease modifying therapeutics, and now call and gene therapies are providing tangible hope of advancing treatment. It is during the informed consent process that participants should be properly prepared for potentially negative outcomes, and there is a need to do further work, ideally with past participants, to elicit the best strategies for doing this effectively. Recommendations, in reference to gene therapies for hemophilia but still pertinent to ATMP trials in neurodegeneration, have been made to improve the consent process (Woollard, Gorman, & Rosenfelt, 2021). Focusing on the detail of the consent process during the evaluation of cell and gene therapies will uncover key determinants of decision-making processes, that will have relevance to involving people in making choices about their own healthcare should these technologies become adopted.

Following recruitment of participants, thought should also be given to how trial progress and results are communicated to participants during and at the end of a trial. There is little evidence supporting distinct recruitment strategies outside monetary incentives (Brueton et al., 2013), but the evaluation of real-world experiences suggests the use of a variety of communication strategies are commonly used in the ongoing engagement of participants in long lasting clinical trials (Abshire et al., 2017; Brueton et al., 2014). Additionally, even though it is strongly recommended, communicating trial results to participants is generally poorly done (Bruhn et al., 2021). Owing to the need for long-term follow-up, but where little is required from participants beyond annual or bi-annual assessments, maintaining engagement of participants is vital to avoid losing important longitudinal data necessary for advancing the development of cell and gene therapies. Further, channels of communication between researchers and participants for the dissemination of new information or while there is scant evidence for best practice in ongoing communication with participants, we would advocate consultation with those with lived experience of the disease being investigated, ideally with additional experience of participating in long-term invasive trials to ensure the acceptability of strategies employed. Similar efforts should be made to ensure that participant preferences for how and when trial results are received are included as integral part of trial conduct.

5.3 Promoting diversity and inclusion across trial participants

One aspect of clinical trial recruitment that has been somewhat overlooked in the past but is now the subject of a number of initiatives, is the diversity and inclusivity of trial populations (NIHR, 2020). Recruitment into clinical trials is an ongoing challenge, with reports of several barriers to inclusion across a range of diseases and populations (Bodicoat et al., 2021; Rodríguez-Torres, González-Pérez, & Díaz-Pérez, 2021) and it appears that recruitment into trials of ATMPs in neurodegenerative disease is no different (Moore, Guzman, Mason, Williams-Gray, & Barker, 2014). Efforts should be made to extend recruitment to more diverse and underserved populations to ensure the trial population is more broadly representative of a disease population and trial designs should be sympathetic to structural barriers that may prevent diverse inclusion. This may require specific additional training for those involved in recruitment (Niranjan et al., 2019).

Recruitment into clinical trials can be limited by geography and a person's proximity to the recruiting center, or the involvement of their clinician in recruiting to a trial, which has significant implications for the inclusion of diverse and representative population. This is particularly acute in cell and gene therapy trials where the required combination of highly specialized medical and surgical facilities and expertise automatically restricts the locations of research sites to a handful of specialist centers. When planning future trials, consideration to how participation can be widened from the immediate geographical locale should be given and will likely necessitate funding relatively high levels of travel and accommodation expenses to enable people from further afield to participate. Consequently, decisions around after care and follow-up will need to account for the needs of participants who are resident a distance from the treating center, particularly for these highly specialized trials.

6. Summary

The evaluation of cell and gene therapies poses a range of novel and complex challenges. In the context of progressive neurodegenerative disorders, specific consideration should be given to how trials are designed and conducted to ensure the efficient collation of data for maximum impact. The potential for cell and gene therapies to modify the progression of degenerative disease and have a real and sustained impact on function is immense. However, this potential will only be realized through the robust and efficient evaluation of newly developed medicinal products. A carefully considered approach to each complexity, either in isolation of combination will advance the field in a meaningful manner and determination of the fundamental requirements for progressing to scale will improve the efficiency of subsequent trials. It is important that lessons from individual studies are shared within research communities to maximize learning in an effort to reduce wastage and improve efficiencies, thereby accelerating the progress of potential therapies to market where they are so desperately needed. And so, despite the infancy of cell and gene therapies for neurodegeneration, the careful consideration of innovative methods for holistic long-term evaluation will pave the way for the smooth implementation of efficacious therapies into global healthcare systems.

References

Abshire, M., Dinglas, V. D., Cajita, M. I. A., Eakin, M. N., Needham, D. M., & Himmelfarb, C. D. (2017). Participant retention practices in longitudinal clinical research studies with high retention rates. *BMC Medical Research Methodology*, *17*(1), 30. https://doi.org/10.1186/s12874-017-0310-z.

An, M. W., Duong, Q., Le-Rademacher, J., & Mandrekar, S. J. (2020). Principles of good clinical trial design. *Journal of Thoracic Oncology*, *15*(8), 1277–1280. https://doi.org/10.1016/j.jtho.2020.05.005.

Andrejack, J., & Mathur, S. (2020). What people with Parkinson's disease want. *Journal of Parkinson's Disease*, *10*(1), S5–S10. IOS Press BV. https://doi.org/10.3233/JPD-202107.

Axelsen, T. M., & Woldbye, D. P. D. (2018). Gene therapy for Parkinson's disease, an update. *Journal of Parkinson's Disease, 8*(2), 195–215. IOS Press. https://doi.org/10.3233/JPD-181331.

Bachoud-Lévi, A.-C. (2017). From open to large-scale randomized cell transplantation trials in Huntington's disease. *Progress in Brain Research, 230*, 227–261. https://doi.org/10.1016/bs.pbr.2016.12.011.

Bachoud-Lévi, A.-C., Gaura, V., Brugières, P., Lefaucheur, J.-P., Boissé, M.-F., Maison, P., et al. (2006). Effect of fetal neural transplants in patients with Huntington's disease 6 years after surgery: A long-term follow-up study. *The Lancet Neurology, 5*(4), 303–309. https://doi.org/10.1016/S1474-4422(06)70381-7.

Bachoud-Lévi, A. C., Schramm, C., Remy, P., Aubin, G., Blond, S., Bocket, L., et al. (2020). Human fetal cell therapy in Huntington's disease: A randomized, multicenter, Phase II trial. *Movement Disorders, 35*(8), 1323–1335. https://doi.org/10.1002/mds.28201.

Bachoud-Levi, C., Massart, R., & Rosser, A. E. (2020). Cell therapy in Huntington's disease: Taking stock of past studies to move the field forward. *Stem Cells*. In revisio.

Barker, R. A., Farrell, K., Guzman, N. V., He, X., Lazic, S. E., Moore, S., et al. (2019). Designing stem-cell-based dopamine cell replacement trials for Parkinson's disease. *Nature Medicine, 25*(7), 1045–1053. https://doi.org/10.1038/s41591-019-0507-2.

Barker, R. A., Barrett, J., Mason, S. L., & Björklund, A. (2013). Fetal dopaminergic transplantation trials and the future of neural grafting in Parkinson's disease. *The Lancet Neurology, 12*(1), 84–91. Lancet Publishing Group. https://doi.org/10.1016/S1474-4422(12)70295-8.

Barker, R. A., Cutting, E. V., & Daft, D. M. (2021). Bringing advanced therapy medicinal products (ATMPs) for Parkinson's disease to the clinic: The Investigator's Perspective. *Journal of Parkinson's Disease, 11*(s2), S129–S134. IOS Press BV. https://doi.org/10.3233/jpd-212563.

Beard, D., Hamilton, D., Davies, L., Cook, J., Hirst, A., McCulloch, P., et al. (2018). Evidence-based evaluation of practice and innovation in physical therapy using the IDEAL-physio framework. *Physical Therapy, 98*(2), 108–121. https://doi.org/10.1093/ptj/pzx103.

Beard, D. J., Campbell, M. K., Blazeby, J. M., Carr, A. J., Weijer, C., Cuthbertson, B. H., et al. (2020). Considerations and methods for placebo controls in surgical trials (ASPIRE guidelines). *The Lancet, 395*(10226), 828–838. Lancet Publishing Group. https://doi.org/10.1016/S0140-6736(19)33137-X.

Berger, V. W., & Do, A. C. (2010). Allocation concealment continues to be misunderstood. *Journal of Clinical Epidemiology, 63*(4), 468–469. Elsevier. https://doi.org/10.1016/j.jclinepi.2009.09.004.

Berlet, R., Galang Cabantan, D. A., Gonzales-Portillo, D., & Borlongan, C. V. (2022). Enriched environment and exercise enhance stem cell therapy for stroke, Parkinson's disease, and Huntington's disease. *Frontiers in Cell & Developmental Biology, 10*, 322. Frontiers Media S.A. https://doi.org/10.3389/fcell.2022.798826.

Bjelland, I., Dahl, A. A., Haug, T. T., & Neckelmann, D. (2002). The validity of the hospital anxiety and depression scale. An updated literature review. *Journal of Psychosomatic Research, 52*(2), 69–77. http://www.ncbi.nlm.nih.gov/pubmed/11832252.

Blits, B., & Petry, H. (2017). Perspective on the road toward gene therapy for Parkinson's disease. *Frontiers in Neuroanatomy, 10*, 128. https://doi.org/10.3389/fnana.2016.00128.

Bloem, B. R., Marks, W. J., Silva De Lima, A. L., Kuijf, M. L., Van Laar, T., Jacobs, B. P. F., et al. (2019). The personalized Parkinson project: Examining disease progression through broad biomarkers in early Parkinson's disease. *BMC Neurology, 19*(1), 160. https://doi.org/10.1186/s12883-019-1394-3.

Bodicoat, D. H., Routen, A. C., Willis, A., Ekezie, W., Gillies, C., Lawson, C., et al. (2021). Promoting inclusion in clinical trials—A rapid review of the literature and recommendations for action. *Trials*, *22*(1), 880. https://doi.org/10.1186/s13063-021-05849-7.

Borovac, J. A. (2016). Side effects of a dopamine agonist therapy for Parkinson's disease: A mini-review of clinical pharmacology. *Yale Journal of Biology and Medicine*, *89*(1), 37–47. Yale Journal of Biology and Medicine Inc. /pmc/articles/PMC4797835/.

Brooks, D. J. (2004). Positron emission tomography imaging of transplant function. *NeuroRx*, *1*(4), 482–491. https://doi.org/10.1602/neurorx.1.4.482.

Brooks, D. J. (2016). Molecular imaging of dopamine transporters. *Ageing Research Reviews*, *30*, 114–121. Elsevier Ireland Ltd. https://doi.org/10.1016/j.arr.2015.12.009.

Brueton, V. C., Tierney, J., Stenning, S., Harding, S., Meredith, S., Nazareth, I., et al. (2013). Strategies to improve retention in randomised trials. *Cochrane Database of Systematic Reviews*, *2013*(12), MR000032. https://doi.org/10.1002/14651858.MR000032.pub2.

Brueton, V. C., Stevenson, F., Vale, C. L., Stenning, S. P., Tierney, J. F., Harding, S., et al. (2014). Use of strategies to improve retention in primary care randomised trials: A qualitative study with in-depth interviews. *BMJ Open*, *4*(1), 3835. https://doi.org/10.1136/bmjopen-2013-003835.

Bruhn, H., Cowan, E. J., Campbell, M. K., Constable, L., Cotton, S., Entwistle, V., et al. (2021). Providing trial results to participants in phase III pragmatic effectiveness RCTs: A scoping review. *Trials*, *22*(1), 361. https://doi.org/10.1186/s13063-021-05300-x.

Byrne, L. M., Rodrigues, F. B., Blennow, K., Durr, A., Leavitt, B. R., Roos, R. A. C., et al. (2017). Neurofilament light protein in blood as a potential biomarker of neurodegeneration in Huntington's disease: A retrospective cohort analysis. *The Lancet Neurology*, *16*(8), 601–609. https://doi.org/10.1016/S1474-4422(17)30124-2.

Carlozzi, N. E., Victorson, D., Sung, V., Beaumont, J. L., Cheng, W., Gorin, B., et al. (2014). HD-PRO-TRIAD™ validation: A patient-reported Instrument for the symptom triad of Huntington's disease. *Tremor & Other Hyperkinetic Movements (New York, N.Y.)*, *4*, 223. https://doi.org/10.7916/D8PN93NZ.

Chan, A. W., Song, F., Vickers, A., Jefferson, T., Dickersin, K., Gøtzsche, P. C., et al. (2014). Increasing value and reducing waste: Addressing inaccessible research. *The Lancet*, *383*(9913), 257–266. Elsevier B.V. https://doi.org/10.1016/S0140-6736(13)62296-5.

Christine, C. W., Starr, P. A., Larson, P. S., Eberling, J. L., Jagust, W. J., Hawkins, R. A., et al. (2009). Safety and tolerability of putaminal AADC gene therapy for Parkinson disease. *Neurology*, *73*(20), 1662–1669. https://doi.org/10.1212/WNL.0b013e3181c29356.

Clinch, S., Busse, M., Döbrössy, M. D., & Dunnett, S. B. (2017). Rehabilitation training in neural restitution. *Progress in Brain Research*, *230*, 305–329. Elsevier B.V. https://doi.org/10.1016/bs.pbr.2016.12.006.

Clinical Trials. (2022). *NCT04802733: Phase 1 Safety and Tolerability Study of MSK-DA01 Cell Therapy for Advanced Parkinson's Disease—Full Text View - ClinicalTrials.gov*. Retrieved July 31, 2022 from. https://clinicaltrials.gov/ct2/show/NCT04802733.

Code of Federal Regulations Title 21. (2022). Retrieved July 20, 2022 from. https://www.accessdata.fda.gov/scripts/cdrh/cfdocs/cfcfr/CFRSearch.cfm?fr=56.111.

COMET Initiative. (2022). *Towards improvement in care delivery in Huntington's Disease (HD): ImproveHD*. Retrieved July 24, 2022, from. https://www.comet-initiative.org/Studies/Details/820.

Coster, W. J. (2013). Making the best match: Selecting outcome measures for clinical trials and outcome studies. *American Journal of Occupational Therapy*, *67*(2), 162–170. https://doi.org/10.5014/ajot.2013.006015.

Dahabreh, I. J., Sheldrick, R. C., Paulus, J. K., Chung, M., Varvarigou, V., Jafri, H., et al. (2012). Do observational studies using propensity score methods agree with randomized trials? A systematic comparison of studies on acute coronary syndromes. *European Heart Journal*, *33*(15), 1893–1901. https://doi.org/10.1093/eurheartj/ehs114.

De Roos, P., Bloem, B. R., Kelley, T. A., Antonini, A., Dodel, R., Hagell, P., et al. (2017). A consensus set of outcomes for Parkinson's disease from the International Consortium for Health Outcomes Measurement. *Journal of Parkinson's Disease*, 7(3), 533–543. https://doi.org/10.3233/JPD-161055.

Dean, A. M. (2001). Experimental Design: Overview. *International Encyclopedia of the Social & Behavioral Sciences*, 5090–5096. Elsevier. https://doi.org/10.1016/b0-08-043076-7/00417-4.

Defer, G. L., Widner, H., Marié, R. M., Rémy, P., & Levivier, M. (1999). Core assessment program for surgical interventional therapies in Parkinson's disease (CAPSIT-PD). *Movement Disorders*, *14*(4), 572–584. https://doi.org/10.1002/1531-8257(199907)14:4<572::AID-MDS1005>3.0.CO;2-C.

Deore, A. B., Dhumane, J. R., Wagh, R., & Sonawane, R. (2019). The stages of drug discovery and development process. *Asian Journal of Pharmaceutical Research and Development*, 7(6), 62–67. https://doi.org/10.22270/ajprd.v7i6.616.

Dodd, S., Clarke, M., Becker, L., Mavergames, C., Fish, R., & Williamson, P. R. (2018). A taxonomy has been developed for outcomes in medical research to help improve knowledge discovery. *Journal of Clinical Epidemiology*, *96*, 84–92. https://doi.org/10.1016/j.jclinepi.2017.12.020.

Dorsey, E. R., Papapetropoulos, S., Xiong, M., & Kieburtz, K. (2017). The first frontier: Digital biomarkers for neurodegenerative disorders. *Digital Biomarkers*, *1*(1), 6–13. S. Karger AG. https://doi.org/10.1159/000477383.

dos Santos, M. C. T., Barreto-Sanz, M. A., Correia, B. R. S., Bell, R., Widnall, C., Perez, L. T., et al. (2018). miRNA-based signatures in cerebrospinal fluid as potential diagnostic tools for early stage Parkinson's disease. *Oncotarget*, *9*(25), 17455–17465. https://doi.org/10.18632/oncotarget.24736.

D'Agostino, R. B. (2009). The delayed-start study design. *New England Journal of Medicine*, *361*(13), 1304–1306. https://doi.org/10.1056/NEJMsm0904209.

Drew, C. J. G., Sharouf, F., Randell, E., Brookes-Howell, L., Smallman, K., Sewell, B., et al. (2021). Protocol for an open label: phase I trial within a cohort of foetal cell transplants in people with Huntington's disease. *Brain Communications*, *3*(1). https://doi.org/10.1093/braincomms/fcaa230.

Duff, K., Paulsen, J. S., Beglinger, L. J., Langbehn, D. R., Wang, C., Stout, J. C., et al. (2010). "Frontal" behaviors before the diagnosis of Huntington's disease and their relationship to markers of disease progression: Evidence of early lack of awareness. *Journal of Neuropsychiatry*, *22*(2), 196–207. https://doi.org/10.1176/appi.neuropsych.22.2.196.

El-Hayek, Y. H., Wiley, R. E., Khoury, C. P., Daya, R. P., Ballard, C., Evans, A. R., et al. (2019). Tip of the iceberg: Assessing the global socioeconomic costs of Alzheimer's disease and related dementias and strategic implications for stakeholders. *Journal of Alzheimer's Disease*, *70*(2), 321–339. IOS Press. https://doi.org/10.3233/JAD-190426.

Enroll-HD. (2018). *Clinical data and biosamples.* Retrieved July 25, 2018, from. https://www.enroll-hd.org/for-researchers/enroll-hd-available-clinical-data-and-biosamples/.

Epstein-Lubow, G., Gaudiano, B. A., Hinckley, M., Salloway, S., & Miller, I. W. (2010). Evidence for the validity of the American medical association's caregiver self-assessment questionnaire as a screening measure for depression. *Journal of the American Geriatrics Society*, *58*(2), 387–388. https://doi.org/10.1111/j.1532-5415.2009.02701.x.

Fayyad, M., Salim, S., Majbour, N., Erskine, D., Stoops, E., Mollenhauer, B., et al. (2019). Parkinson's disease biomarkers based on α-synuclein. *Journal of Neurochemistry*, *150*(5), 626–636. Blackwell Publishing Ltd. https://doi.org/10.1111/jnc.14809.

FDA. (2016a). *The Voice of the Patient—Parkinson's Disease*. Retrieved July 24, 2022, from. https://fda.report/media/97680/The-Voice-of-the-Patient--Parkinson%27s-Disease-Public-Meeting-September-22- -2015.pdf.

FDA. (2016b). *The Voice of the Patient*. Retrieved July 24, 2022, from. https://www.fda.gov/media/124393/download.

FDA. (2022a). *Human Gene Therapy for Neurodegenerative Diseases; Draft Guidance for Industry*. Retrieved July 20, 2022, from. https://www.fda.gov/regulatory-information/search-fda-guidance-documents/human-gene-therapy-neurodegenerative-diseases.

FDA. (2022b). *Rare diseases: Natural history studies for drug development guidance for industry DRAFT GUIDANCE*. Retrieved July 20, 2022 from. https://www.fda.gov/regulatory-information/search-fda-guidance-documents/rare-diseases-natural-history-studies-drug-development.

Finkel, R. S., Mercuri, E., Darras, B. T., Connolly, A. M., Kuntz, N. L., Kirschner, J., et al. (2017). Nusinersen versus Sham control in infantile-onset spinal muscular atrophy. *New England Journal of Medicine*, *377*(18), 1723–1732. https://doi.org/10.1056/nejmoa1702752.

Freed, C. R., Breeze, R. E., Rosenberg, N. L., Schneck, S. A., Kriek, E., Qi, J., et al. (1992). Survival of implanted fetal dopamine cells and neurologic improvement 12 to 46 months after transplantation for Parkinson's disease. *New England Journal of Medicine*, *327*(22), 1549–1555. https://doi.org/10.1056/nejm199211263272202.

Freed, C. R., Greene, P. E., Breeze, R. E., Tsai, W.-Y., DuMouchel, W., Kao, R., et al. (2001). Transplantation of embryonic dopamine neurons for severe Parkinson's disease. *New England Journal of Medicine*, *344*(10), 710–719. https://doi.org/10.1056/nejm200103083441002.

Freeman, T. B., Vawter, D. E., Leaverton, P. E., Godbold, J. H., Hauser, R. A., Goetz, C. G., et al. (1999). Use of placebo surgery in controlled trials of a cellular-based therapy for Parkinson's disease. *New England Journal of Medicine*, *341*(13), 988–992. https://doi.org/10.1056/nejm199909233411311.

Gagne, J. J., Thompson, L., O'Keefe, K., & Kesselheim, A. S. (2014). Innovative research methods for studying treatments for rare diseases: Methodological review. *BMJ Clinical Research*, *349*. https://doi.org/10.1136/bmj.g6802.

Galpern, W. R., Ce, O., Corrigan-Curay, J., Lang, A. E., Rey Kahn, J., Tagle, D., et al. (2012). Personal view Sham neurosurgical procedures in clinical trials for neurodegenerative diseases: Scientific and ethical considerations. *Lancet Neurology*, *11*. https://doi.org/10.1016/S1474-4422(12)70064-9.

Ghadessi, M., Tang, R., Zhou, J., Liu, R., Wang, C., Toyoizumi, K., et al. (2020). A roadmap to using historical controls in clinical trials—By Drug Information Association Adaptive Design Scientific Working Group (DIA-ADSWG). *Orphanet Journal of Rare Diseases*, *15*(1), 69. BioMed Central Ltd. https://doi.org/10.1186/s13023-020-1332-x.

Glidden, A. M., Luebbe, E. A., Elson, M. J., Goldenthal, S. B., Snyder, C. W., Zizzi, C. E., et al. (2020). Patient-reported impact of symptoms in Huntington disease: PRISM-HD. *Neurology*, *94*(19), e2045–e2053. https://doi.org/10.1212/WNL.0000000000008906.

Gliklich, R. E., Leavy, M. B., Karl, J., Campion, D. M., Levy, D., & Berliner, E. (2014). A framework for creating standardized outcome measures for patient registries. *Journal of Comparative Effectiveness Research*, *3*(5), 473–480. https://doi.org/10.2217/cer.14.38.

Goetz, C. C. (2003). The Unified Parkinson's disease rating scale (UPDRS): Status and recommendations. *Movement Disorders*, *18*(7), 738–750. Mov Disord. https://doi.org/10.1002/mds.10473.

Goetz, C. G., Poewe, W., Rascol, O., Sampaio, C., Stebbins, G. T., Counsell, C., et al. (2004). Movement Disorder Society Task Force report on the Hoehn and Yahr staging scale: Status and recommendations The *Movement* Disorder Society Task Force on rating scales for Parkinson's disease. *Movement Disorders*, *19*(9), 1020–1028. https://doi.org/10.1002/mds.20213.

Goetz, C. G., Tilley, B. C., Shaftman, S. R., Stebbins, G. T., Fahn, S., Martinez-Martin, P., et al. (2008). Movement Disorder Society-Sponsored Revision of the Unified Parkinson's Disease Rating Scale (MDS-UPDRS): Scale presentation and clinimetric testing results. *Movement Disorders*, *23*(15), 2129–2170. https://doi.org/10.1002/mds.22340.

Goetz, C. G., Stebbins, G. T., Wolff, D., DeLeeuw, W., Bronte-Stewart, H., Elble, R., et al. (2009). Testing objective measures of motor impairment in early Parkinson's disease: Feasibility study of an at-home testing device. *Movement Disorders*, *24*(4), 551–556. https://doi.org/10.1002/mds.22379.

Gold, M., Amatniek, J., Carrillo, M. C., Cedarbaum, J. M., Hendrix, J. A., Miller, B. B., et al. (2018). Digital technologies as biomarkers, clinical outcomes assessment, and recruitment tools in Alzheimer's disease clinical trials. *Alzheimer's and Dementia: Translational Research and Clinical Interventions*, *4*, 234–242. Elsevier Inc. https://doi.org/10.1016/j.trci.2018.04.003.

Hansson, O., Edelmayer, R. M., Boxer, A. L., Carrillo, M. C., Mielke, M. M., Rabinovici, G. D., et al. (2022). The Alzheimer's Association appropriate use recommendations for blood biomarkers in Alzheimer's disease. *Alzheimer's & Dementia*. https://doi.org/10.1002/alz.12756.

Hariton, E., & Locascio, J. J. (2018). Randomised controlled trials—The gold standard for effectiveness research: Study design: Randomised controlled trials. *BJOG: An International Journal of Obstetrics and Gynaecology*, *125*(13), 1716. Blackwell Publishing Ltd. https://doi.org/10.1111/1471-0528.15199.

Hauser, R. A., Furtado, S., Cimino, C. R., Delgado, H., Eichler, S., Schwartz, S., et al. (2002). Bilateral human fetal striatal transplantation in Huntington's disease. *Neurology*, *58*(5), 687–695. https://doi.org/10.1212/WNL.58.5.687.

Heiss, J. D., Lungu, C., Hammoud, D. A., Herscovitch, P., Ehrlich, D. J., Argersinger, D. P., et al. (2019). Trial of magnetic resonance–guided putaminal gene therapy for advanced Parkinson's disease. *Movement Disorders*, *34*(7), 1073–1078. https://doi.org/10.1002/mds.27724.

Hilgers, R.-D., König, F., Molenberghs, G., & Senn, S. (2016). Design and analysis of clinical trials for small rare disease populations. *Journal of Diseases Research & Treatment*, *1*(3). www.rarediseasesjournal.com.

Hooper, G., Palermo, G., Hlavac, F., Finnegan, C., Frick, E., Boak, L., et al. (2019). The Roche HD natural history study—An external comparator by design. *Journal of the Neurological Sciences*, *405*, 312. https://doi.org/10.1016/j.jns.2019.10.1409.

Hróbjartsson, A., Thomsen, A. S. S., Emanuelsson, F., Tendal, B., Hilden, J., Boutron, I., et al. (2012). Observer bias in randomised clinical trials with binary outcomes: Systematic review of trials with both blinded and non-blinded outcome assessors. *BMJ (Online)*, *344*(7848). https://doi.org/10.1136/bmj.e1119.

Hunt Bobo, R., Laske, D. W., Akbasak, A., Morrison, P. F., Dedrick, R. L., & Oldfield, E. H. (1994). Convection-enhanced delivery of macromolecules in the brain. *Proceedings of the National Academy of Sciences of the United States of America*, *91*(6), 2076–2080. https://doi.org/10.1073/pnas.91.6.2076.

Huntington Study Group. (1996). Unified Huntington's disease rating scale: Reliability-and-Consistency. *Movement Disorders*, *11*(2), 136–142. https://doi.org/10.1002/mds.870110204.

ICH. (2016). *International Council for Harmonisation of technical requirements for pharmaceuticals for human use (ICH) ICH harmonised guideline integrated addendum to ICH e6(r1): guideline for Good Clinical Practice E6(R2).*

Iglesias-Lopez, C., Agustí, A., Vallano, A., & Obach, M. (2021). Methodological characteristics of clinical trials supporting the marketing authorisation of advanced therapies in the European Union. *Frontiers in Pharmacology*, *12*, 3228. https://doi.org/10.3389/fphar.2021.773712.

Jenner, P., Morris, H. R., Robbins, T. W., Goedert, M., Hardy, J., Ben-Shlomo, Y., et al. (2013). Parkinson's disease-the debate on the clinical phenomenology, aetiology, pathology and pathogenesis. *Journal of Parkinson's Disease*, *3*(1), 1–11. IOS Press. https://doi.org/10.3233/JPD-130175.

Jones, C., Busse, M., Quinn, L., Dawes, H., Drew, C., Kelson, M., et al. (2016). The societal cost of Huntington's disease: Are we underestimating the burden? *European Journal of Neurology*, *23*(10), 1588–1590. https://doi.org/10.1111/ene.13107.

Kadry, H., Noorani, B., & Cucullo, L. (2020). A blood–brain barrier overview on structure, function, impairment, and biomarkers of integrity. *Fluids and Barriers of the CNS*, *17*(1), 69. BioMed Central Ltd. https://doi.org/10.1186/s12987-020-00230-3.

Kaplitt, M. G., Feigin, A., Tang, C., Fitzsimons, H. L., Mattis, P., Lawlor, P. A., et al. (2007). Safety and tolerability of gene therapy with an adeno-associated virus (AAV) borne GAD gene for Parkinson's disease: An open label, phase I trial. *Lancet*, *369*(9579), 2097–2105. https://doi.org/10.1016/S0140-6736(07)60982-9.

Kim, S. Y. H., Frank, S., Holloway, R., Zimmerman, C., Wilson, R., & Kieburtz, K. (2005). Science and ethics of sham surgery: A survey of Parkinson disease clinical researchers. *Archives of Neurology*, *62*(9), 1357–1360. Arch Neurol. https://doi.org/10.1001/archneur.62.9.1357.

Kim, S. Y. H., Flory, J., & Relton, C. (2018). Ethics and practice of trials within cohorts: An emerging pragmatic trial design. *Clinical Trials*, *15*(1), 9–16. https://doi.org/10.1177/1740774517746620.

Kirik, D., Winkler, C., & Bjö, A. (2001). Growth and functional efficacy of intrastriatal nigral transplants depend on the extent of nigrostriatal degeneration. *The Journal of Neuroscience*, *21*(8). https://www.jneurosci.org/content/21/8/2889.short.

Kirkeby, A., Parmar, M., & Barker, R. A. (2017). Strategies for bringing stem cell-derived dopamine neurons to the clinic: A European approach (STEM-PD). In *Vol. 230. Progress in brain research* (pp. 165–190). Elsevier B.V. https://doi.org/10.1016/bs.pbr.2016.11.011.

Kondziolka, D., Gobbel, G. T., Fellows-Mayle, W., Chang, Y. F., & Uram, M. (2011). Injection parameters affect cell viability and implant volumes in automated cell delivery for the brain. *Cell Transplantation*, *20*(11–12), 1901–1906. https://doi.org/10.3727/096368911X566190.

Kopyov, O. V., Jacques, D. S., Lieberman, A., Duma, C. M., & Rogers, R. L. (1997). Outcome following intrastriatal fetal mesencephalic grafts for Parkinson's patients is directly related to the volume of grafted tissue. *Experimental Neurology*, *146*(2), 536–545. https://doi.org/10.1006/exnr.1997.6577.

Langbehn, D. R., & Hersch, S. (2020). Clinical outcomes and selection criteria for prodromal Huntington's disease trials. *Movement Disorders*, *35*(12), 2193–2200. https://doi.org/10.1002/mds.28222.

Langston, J. W., Widner, H., Goetz, C. G., Brooks, D., Fahn, S., Freeman, T., et al. (1992). Core assessment program for intracerebral transplantations (CAPIT). *Movement Disorders*, 7(1), 2–13. https://doi.org/10.1002/mds.870070103.

Leavitt, B. R., & Tabrizi, S. J. (2020). Antisense oligonucleotides for neurodegeneration. *Science*, *367*(6485), 1428–1429. American Association for the Advancement of Science. https://doi.org/10.1126/science.aba4624.

Li, T., & Le, W. (2020). Biomarkers for Parkinson's disease: How good are they? *Neuroscience Bulletin*, *36*(2), 183–194. Springer. https://doi.org/10.1007/s12264-019-00433-1.

Lillie, E. O., Patay, B., Diamant, J., Issell, B., Topol, E. J., & Schork, N. J. (2011). The n-of-1 clinical trial: The ultimate strategy for individualizing medicine? *Personalized Medicine*, *8*(2), 161–173. https://doi.org/10.2217/pme.11.7.

Lindvall, O., Brundin, P., Widner, H., Rehncrona, S., Gustavii, B., Frackowiak, R., et al. (1990). Grafts of fetal dopamine neurons survive and improve motor function in Parkinson's disease. *Science*, *247*(4942), 574–577. https://doi.org/10.1126/science.2105529.

Long, J. D., Langbehn, D. R., Tabrizi, S. J., Landwehrmeyer, B. G., Paulsen, J. S., Warner, J., et al. (2017). Validation of a prognostic index for Huntington's disease. *Movement Disorders, 32*(2), 256–263. https://doi.org/10.1002/mds.26838.

Lonjon, G., Boutron, I., Trinquart, L., Ahmad, N., Aim, F., Nizard, R., et al. (2014). Comparison of treatment effect estimates from prospective nonrandomized studies with propensity score analysis and randomized controlled trials of surgical procedures. *Annals of Surgery, 259*(1), 18–25. Ann Surg. https://doi.org/10.1097/SLA.0000000000000256.

Marin, R. S., Biedrzycki, R. C., & Firinciogullari, S. (1991). Reliability and validity of the apathy evaluation scale. *Psychiatry Research, 38*(2), 143–162. https://doi.org/10.1016/0165-1781(91)90040-V.

Markoulidakis, A., Taiyari, K., Holmans, P., Pallmann, P., Busse, M., Godley, M. D., et al. (2020). *A tutorial comparing different covariate balancing methods with an application evaluating the causal effects of substance use treatment programs for adolescents*. http://arxiv.org/abs/2010.09563.

Markoulidakis, A., Holmans, P., Pallmann, P., Griin, B. A., Griffin, B. A., & Uk, M. C. (2022). *Understanding how balance and sample size impact bias in the estimation of causal treatment effects: A simulation study*. https://doi.org/10.21203/rs.3.rs-1742290/v1.

Martinez-Martin, P., Rodriguez-Blazquez, C., Forjaz, M. J., Frades-Payo, B., Agüera-Ortiz, L., Weintraub, D., et al. (2015). Neuropsychiatric symptoms and caregiver's burden in Parkinson's disease. *Parkinsonism and Related Disorders, 21*(6), 629–634. https://doi.org/10.1016/j.parkreldis.2015.03.024.

McCarney, R., Warner, J., Iliffe, S., Van Haselen, R., Griffin, M., & Fisher, P. (2007). The Hawthorne effect: A randomised, controlled trial. *BMC Medical Research Methodology*, 7, 30. https://doi.org/10.1186/1471-2288-7-30.

McColgan, P., & Tabrizi, S. J. (2018). Huntington's disease: A clinical review. *European Journal of Neurology, 25*(1), 24–34. Blackwell Publishing Ltd. https://doi.org/10.1111/ene.13413.

McCulloch, P., Altman, D. G., Campbell, W. B., Flum, D. R., Glasziou, P., Marshall, J. C., et al. (2009). No surgical innovation without evaluation: The IDEAL recommendations. *The Lancet, 374*(9695), 1105–1112. Elsevier. https://doi.org/10.1016/S0140-6736(09)61116-8.

McLauchlan, D. J., Lancaster, T., Craufurd, D., Linden, D. E. J., & Rosser, A. E. (2019). Insensitivity to loss predicts apathy in Huntington's disease. *Movement Disorders, 34*(9), 1381–1391. https://doi.org/10.1002/mds.27787.

Mendez, I., Hong, M., Smith, S., Dagher, A., & Desrosiers, J. (2000). Neural transplantation cannula and microinjector system: Experimental and clinical experience—Technical note. *Journal of Neurosurgery, 92*(3), 493–499. https://doi.org/10.3171/jns.2000.92.3.0493.

Mestre, T. A., Forjaz, M. J., Mahlknecht, P., Cardoso, F., Ferreira, J. J., Reilmann, R., et al. (2018). Rating scales for motor symptoms and signs in Huntington's disease: Critique and recommendations. *Movement Disorders Clinical Practice, 5*(2), 111–117. https://doi.org/10.1002/mdc3.12571.

Mills, E. J., Chan, A. W., Wu, P., Vail, A., Guyatt, G. H., & Altman, D. G. (2009). Design, analysis, and presentation of crossover trials. *Trials, 10*(1), 1–6. https://doi.org/10.1186/1745-6215-10-27.

Mohan, A., Sun, Z., Ghosh, S., Li, Y., Sathe, S., Hu, J., et al. (2022). A machine-learning derived Huntington's disease progression model: Insights for clinical trial design. *Movement Disorders, 37*(3), 553–562. https://doi.org/10.1002/mds.28866.

Moore, S. F., Guzman, N. V., Mason, S. L., Williams-Gray, C. H., & Barker, R. A. (2014). Which patients with Parkinson's disease participate in clinical trials? One centre's experiences with a new cell based therapy trial (TRANSEURO). *Journal of Parkinson's Disease, 4*(4), 671–676. https://doi.org/10.3233/JPD-140432.

Morel, T., Cleanthous, S., Andrejack, J., Barker, R. A., Blavat, G., Brooks, W., et al. (2022). Patient experience in early-stage Parkinson's disease: Using a mixed methods analysis to identify which concepts are cardinal for clinical trial outcome assessment. *Neurology and Therapy*. https://doi.org/10.1007/s40120-022-00375-3.

MRC. (2017). *Review of Positron Emission Tomography (PET) within The Medical Imaging Research Landscape*.

NIHR. (2020). *Improving inclusion of under-served groups in clinical research: Guidance*. INCLUDE Project. Retrieved July 31, 2022 from. https://www.nihr.ac.uk/documents/improving-inclusion-of-under-served-groups-in-clinical-research-guidance-from-include-project/25435.

Niranjan, S. J., Durant, R. W., Wenzel, J. A., Cook, E. D., Fouad, M. N., Vickers, S. M., et al. (2019). Training needs of clinical and research professionals to optimize minority recruitment and retention in cancer clinical trials. *Journal of Cancer Education, 34*(1), 26–34. https://doi.org/10.1007/s13187-017-1261-0.

Nottage, M., & Siu, L. L. (2002). https://doi.org/10.1200/JCO.2002.07.537.

Odgaard-Jensen, J., Vist, G. E., Timmer, A., Kunz, R., Akl, E. A., Schünemann, H., et al. (2011). Randomisation to protect against selection bias in healthcare trials. *Cochrane Database of Systematic Reviews, 2011*(4). John Wiley and Sons Ltd. https://doi.org/10.1002/14651858.MR000012.pub3.

Olanow, C. W., Goetz, C. G., Kordower, J. H., Stoessl, A. J., Sossi, V., Brin, M. F., et al. (2003). A double-blind controlled trial of bilateral fetal nigral transplantation in Parkinson's disease. *Annals of Neurology, 54*(3), 403–414. https://doi.org/10.1002/ana.10720.

Olanow, C. W., Rascol, O., Hauser, R., Feigin, P. D., Jankovic, J., Lang, A., et al. (2009). A double-blind, delayed-start trial of Rasagiline in Parkinson's disease. *New England Journal of Medicine, 361*(13), 1268–1278. https://doi.org/10.1056/nejmoa0809335.

O'Leary, E., Seow, H., Julian, J., Levine, M., & Pond, G. R. (2013). Data collection in cancer clinical trials: Too much of a good thing? *Clinical Trials, 10*(4), 624–632. https://doi.org/10.1177/1740774513491337.

Parkin, G. M., Corey-Bloom, J., Snell, C., Castleton, J., & Thomas, E. A. (2021). Plasma neurofilament light in Huntington's disease: A marker for disease onset, but not symptom progression. *Parkinsonism and Related Disorders, 87*, 32–38. https://doi.org/10.1016/j.parkreldis.2021.04.017.

Parkinson's Progression Markers Initiative. (2022). *The study that could change everything*. Retrieved July 20, 2022, from. https://www.ppmi-info.org/.

Paulsen, J. S., Langbehn, D. R., Stout, J. C., Aylward, E., Ross, C. A., Nance, M., et al. (2008). Detection of Huntington's disease decades before diagnosis: The Predict-HD study. *Journal of Neurology, Neurosurgery and Psychiatry, 79*(8), 874–880. https://doi.org/10.1136/jnnp.2007.128728.

Perry, V. H., Andersson, P. B., & Gordon, S. (1993). Macrophages and inflammation in the central nervous system. *Trends in Neurosciences, 16*(7), 268–273. Trends Neurosci. https://doi.org/10.1016/0166-2236(93)90180-T.

Peto, V., Jenkinson, C., & Fitzpatrick, R. (1998). PDQ-39: A review of the development, validation and application of a Parkinson's disease quality of life questionnaire and its associated measures. *Journal of Neurology, S10-4*.

Piccini, P., Pavese, N., Hagell, P., Reimer, J., & Brain, A. B. (2005). Factors affecting the clinical outcome after neural transplantation in Parkinson's disease. *Brain*. Retrieved July 30, 2022, from. https://academic.oup.com/brain/article-abstract/128/12/2977/420488.

Pocock, S. J. (1976). The combination of randomized and historical controls in clinical trials. *Journal of Chronic Diseases, 29*(3), 175–188. https://doi.org/10.1016/0021-9681(76)90044-8.

Polgar, S., & Mohamed, S. (2019). Evidence-based evaluation of the ethics of sham surgery for Parkinson's disease. *Journal of Parkinson's Disease, 9*(3), 565–574. IOS Press. https://doi.org/10.3233/JPD-191577.

Potts, M. B., Silvestrini, M. T., & Lim, D. A. (2013). Devices for cell transplantation into the central nervous system: Design considerations and emerging technologies. *Surgical Neurology International, 4*(Suppl 1), S22. https://doi.org/10.4103/2152-7806.109190.

Quemener, A. M., Bachelot, L., Forestier, A., Donnou-Fournet, E., Gilot, D., & Galibert, M. D. (2020). The powerful world of antisense oligonucleotides: From bench to bedside. *Wiley Interdisciplinary Reviews: RNA, 11*(5), e1594. Blackwell Publishing Ltd. https://doi.org/10.1002/wrna.1594.

Quinlan, S., Kenny, A., Medina, M., Engel, T., & Jimenez-Mateos, E. M. (2018). MicroRNAs in neurodegenerative diseases. In *Vol. 334. International Review of Cell and Molecular Biology* (pp. 309–343). Elsevier Inc. https://doi.org/10.1016/bs.ircmb.2017.04.002.

Quinn, L., & Busse, M. (2017). The role of rehabilitation therapy in Huntington disease. *Handbook of Clinical Neurology, 144*, 151–165. https://doi.org/10.1016/B978-0-12-801893-4.00013-4.

Quinn, N., Brown, R., Craufurd, D., Goldman, S., Hodges, J., Kieburtz, K., et al. (1996). Core assessment program for intracerebral transplantation in Huntington's disease (CAPIT-HD). *Movement Disorders, 11*(2), 143–150. https://doi.org/10.1002/mds.870110205.

Reeves, D., Howells, K., Sidaway, M., Blakemore, A., Hann, M., Panagioti, M., et al. (2018). The cohort multiple randomized controlled trial design was found to be highly susceptible to low statistical power and internal validity biases. *Journal of Clinical Epidemiology, 95*, 111–119. https://doi.org/10.1016/j.jclinepi.2017.12.008.

Reilmann, R., & Schubert, R. (2017). Motor outcome measures in Huntington disease clinical trials. *Handbook of Clinical Neurology, 144*, 209–225. https://doi.org/10.1016/B978-0-12-801893-4.00018-3.

Reilmann, R., McGarry, A., Grachev, I. D., Savola, J. M., Borowsky, B., Eyal, E., et al. (2019). Safety and efficacy of pridopidine in patients with Huntington's disease (PRIDE-HD): A phase 2, randomised, placebo-controlled, multicentre, dose-ranging study. *The Lancet Neurology, 18*(2), 165–176. https://doi.org/10.1016/S1474-4422(18)30391-0.

Relton, C., Torgerson, D., O'Cathain, A., & Nicholl, J. (2010). Rethinking pragmatic randomised controlled trials: Introducing the "cohort multiple randomised controlled trial" design. *BMJ, 340*, c1066. http://www.ncbi.nlm.nih.gov/pubmed/20304934.

Repair-HD. (2020). Retrieved April 5, 2020, from. http://www.repair-hd.eu/.

Richardson, R. M., Kells, A. P., Martin, A. J., Larson, P. S., Starr, P. A., Piferi, P. G., et al. (2011). Novel platform for MRI-guided convection-enhanced delivery of therapeutics: Preclinical validation in nonhuman primate brain. *Stereotactic and Functional Neurosurgery, 89*(3), 141–151. https://doi.org/10.1159/000323544.

Richardson, R. M., Bankiewicz, K. S., Christine, C. W., Van Laar, A. D., Gross, R. E., Lonser, R., et al. (2020). Data-driven evolution of neurosurgical gene therapy delivery in Parkinson's disease. *Journal of Neurology, Neurosurgery and Psychiatry, 91*(11), 1210–1218. BMJ Publishing Group. https://doi.org/10.1136/jnnp-2020-322904.

Rodrigues, F. B., & Wild, E. J. (2020). Huntington's Disease Clinical Trials Corner: April 2020. *Journal of Huntington's Disease, 9*(2), 185–197. IOS Press. https://doi.org/10.3233/JHD-200002.

Rodrigues, F. B., Owen, G., Sathe, S., Pak, E., Kaur, D., Ehrhardt, A. G., et al. (2022). Safety and feasibility of research lumbar puncture in Huntington's disease: The HDClarity cohort and bioresource. *Journal of Huntington's Disease, 11*(1), 59–69. https://doi.org/10.3233/JHD-210508.

Rodríguez-Torres, E., González-Pérez, M. M., & Díaz-Pérez, C. (2021). Barriers and facilitators to the participation of subjects in clinical trials: An overview of reviews. *Contemporary Clinical Trials Communications*, *23*, 100829. https://doi.org/10.1016/j.conctc.2021.100829.

Rogers, W., Hutchison, K., Skea, Z. C., & Campbell, M. K. (2014). Strengthening the ethical assessment of placebo-controlled surgical trials: Three proposals. *BMC Medical Ethics*, *15*(1). BioMed Central Ltd. https://doi.org/10.1186/1472-6939-15-78.

Rosser, A. E., & Bachoud-Lévi, A.-C. (2012). Clinical trials of neural transplantation in Huntington's disease. *Progress in Brain Research*, *200*, 345–371. https://doi.org/10.1016/B978-0-444-59575-1.00016-8.

Rosser, A. E., Busse, M. E., Gray, W. P., Badin, R. A., Perrier, A. L., Wheelock, V., et al. (2022). Translating cell therapies for neurodegenerative diseases: Huntington's disease as a model disorder. *Brain*, *145*(5), 1584–1597. https://doi.org/10.1093/brain/awac086.

Rubinsztein, D. C., & Orr, H. T. (2016). Diminishing return for mechanistic therapeutics with neurodegenerative disease duration?: There may be a point in the course of a neurodegenerative condition where therapeutics targeting disease-causing mechanisms are futile. *BioEssays*, *38*(10), 977–980. https://doi.org/10.1002/bies.201600048.

Saeed, U., Compagnone, J., Aviv, R. I., Strafella, A. P., Black, S. E., Lang, A. E., et al. (2017). Imaging biomarkers in Parkinson's disease and Parkinsonian syndromes: Current and emerging concepts. *Translational Neurodegeneration*, *6*(1). BioMed Central Ltd. https://doi.org/10.1186/s40035-017-0076-6.

Schobel, S. A., Palermo, G., Auinger, P., Long, J. D., Ma, S., Khwaja, O. S., et al. (2017). Motor, cognitive, and functional declines contribute to a single progressive factor in early HD. *Neurology*, *89*(24), 2495–2502. https://doi.org/10.1212/WNL.0000000000004743.

Sedgwick, P. (2015). Randomised controlled trials: Understanding power. *BMJ*, *350*. https://doi.org/10.1136/bmj.h3229. BMJ Publishing Group.

Sedrakyan, A., Campbell, B., Merino, J. G., Kuntz, R., Hirst, A., & McCulloch, P. (2016). IDEAL-D: A rational framework for evaluating and regulating the use of medical devices. *BMJ (Clinical Research Edition)*, *353*, i2372. https://doi.org/10.1136/bmj.i2372.

Sewell, D. K., Penney, J., Jay, M., Zhang, Y., & Paulsen, J. S. (2021). Predicting an optimal composite outcome variable for Huntington's disease clinical trials. *Journal of Applied Statistics*, *48*(7), 1339–1348. https://doi.org/10.1080/02664763.2020.1759034.

Sjödahl, C., Nilsson, M. H., Idvall, M., Rosas, S. R., & Hagell, P. (2014). Conceptualizing and prioritizing clinical trial outcomes from the perspectives of people with Parkinson's disease versus health care professionals: A concept mapping study. Quality of Life Research, *23*(6), 1687–1700. https://doi.org/10.1007/s11136-013-0614-3.

Skivington, K., Matthews, L., Simpson, S. A., Craig, P., Baird, J., Blazeby, J. M., et al. (2021). A new framework for developing and evaluating complex interventions: Update of Medical Research Council guidance. *The BMJ*, *374*. https://doi.org/10.1136/bmj.n2061.

Skuk, D., Paradis, M., Goulet, M., & Tremblay, J. P. (2007). Ischemic central necrosis in pockets of transplanted myoblasts in nonhuman primates: Implications for cell-transplantation strategies. *Transplantation*, *84*(10), 1307–1315. https://doi.org/10.1097/01.tp.0000288322.94252.22.

Snaith, R. P., Constantopoulos, A. A., Jardine, M. Y., & McGuffin, P. (1978). A clinical scale for the self-assessment of irritability. *British Journal of Psychiatry*, *132*(2), 164–171. https://doi.org/10.1192/bjp.132.2.164.

Soh, S. E., McGinley, J. L., Watts, J. J., Iansek, R., Murphy, A. T., Menz, H. B., et al. (2013). Determinants of health-related quality of life in people with Parkinson's disease: A path analysis. *Quality of Life Research*, *22*(7), 1543–1553. https://doi.org/10.1007/s11136-012-0289-1.

Southwell, A. L., Smith, S. E. P., Davis, T. R., Caron, N. S., Villanueva, E. B., Xie, Y., et al. (2015). Ultrasensitive measurement of huntingtin protein in cerebrospinal fluid demonstrates increase with Huntington disease stage and decrease following brain huntingtin suppression. *Scientific Reports*, *5*. https://doi.org/10.1038/srep12166.

Stanley, K. (2007). Design of randomized controlled trials. *Circulation*, *115*(9), 1164–1169. https://doi.org/10.1161/CIRCULATIONAHA.105.594945.

Staton. (2017). *Biogen's $375K Spinraza price puts a Sovaldi-style spotlight on rare disease meds*. Fierce Pharma. Retrieved July 31, 2022, from. https://www.fiercepharma.com/pharma/biogen-s-375k-spinraza-price-puts-a-sovaldi-style-spotlight-rare-disease-meds.

Steiner, B., Winter, C., Blumensath, S., Paul, G., Harnack, D., Nikkhah, G., et al. (2008). Survival and functional recovery of transplanted human dopaminergic neurons into hemiparkinsonian rats depend on the cannula size of the implantation instrument. *Journal of Neuroscience Methods*, *169*(1), 128–134. https://doi.org/10.1016/j.jneumeth.2007.11.032.

Stephenson, D., Badawy, R., Mathur, S., Tome, M., & Rochester, L. (2021). Digital progression biomarkers as novel endpoints in clinical trials: A multistakeholder perspective. *Journal of Parkinson's Disease*, *11*(s1), S103–S109. IOS Press BV. https://doi.org/10.3233/JPD-202428.

Tabrizi, S. J., Langbehn, D. R., Leavitt, B. R., Roos, R. A., Durr, A., Craufurd, D., et al. (2009). Biological and clinical manifestations of Huntington's disease in the longitudinal TRACK-HD study: Cross-sectional analysis of baseline data. *The Lancet Neurology*, *8*(9), 791–801. https://doi.org/10.1016/S1474-4422(09)70170-X.

Tabrizi, S. J., Leavitt, B. R., Landwehrmeyer, G. B., Wild, E. J., Saft, C., Barker, R. A., et al. (2019). Targeting Huntingtin expression in patients with Huntington's disease. *New England Journal of Medicine*, *380*(24), 2307–2316. https://doi.org/10.1056/NEJMoa1900907.

Tabrizi S.J., Schobel S., Gantman E.C., Mansbach A., Borowsky B., Konstantinova P., Mestre T.A., et al., (2022). A biological classification of Huntington's disease: The Integrated Staging System. *Lancet Neurology* 21 (7), 632-644. doi: 10.1016/S1474-4422(22)00120-X. PMID: 35716693.

Takahashi, J. (2020). iPS cell-based therapy for Parkinson's disease: A Kyoto trial. *Regenerative Therapy*, *13*, 18–22. https://doi.org/10.1016/j.reth.2020.06.002.

Takahashi, J., & Price-Evans, A. (2019). Preparing for first human trial of induced pluripotent stem cell-derived cells for Parkinson's disease: An interview with Jun Takahashi. *Regenerative Medicine*, *14*(2), 93–95. Future Medicine Ltd. https://doi.org/10.2217/rme-2018-0158.

Torres, E. M., Trigano, M., & Dunnett, T. B. (2015). Translation of cell therapies to the clinic: Characteristics of cell suspensions in large-diameter injection cannulae. *Cell Transplantation*, *24*(4), 737–749. https://doi.org/10.3727/096368914X685429.

Tosin, M. H. S., Stebbins, G. T., Comella, C., Patterson, C. G., Hall, D. A., Schenkman, M. L., et al. (2021). Does MDS-UPDRS provide greater sensitivity to mild disease than UPDRS in de novo Parkinson's disease? *Movement Disorders Clinical Practice*, *8*(7), 1092–1099. https://doi.org/10.1002/mdc3.13329.

Treweek, S., Altman, D. G., Bower, P., Campbell, M., Chalmers, I., Cotton, S., et al. (2015). Making randomised trials more efficient: Report of the first meeting to discuss the Trial Forge platform. *Trials*, *16*, 261. https://doi.org/10.1186/s13063-015-0776-0.

Trojano, L., & Papagno, C. (2018). Cognitive and behavioral disorders in Parkinson's disease: An update. II: behavioral disorders. *Neurological Sciences*, *39*(1), 53–61. Springer-Verlag Italia s.r.l. https://doi.org/10.1007/s10072-017-3155-7.

Vergel, Y. B., & Sculpher, M. (2008). Quality-adjusted life years. *Practical Neurology*, *8*(3), 175–182. https://doi.org/10.1136/pn.2007.140186.

Viglietta, V., & Development, V. C. (2022). *A Ph1b/2a study of WVE-003, an investigational allele-selective, mHTT-lowering oligonucleotide for the treatment of early manifest Huntington's disease, and review of PRECISION-HD results.*

Wahlberg, B., Ghuman, H., Liu, J. R., & Modo, M. (2018). Ex vivo biomechanical characterization of syringe-needle ejections for intracerebral cell delivery. *Scientific Reports*, *8*(1). https://doi.org/10.1038/s41598-018-27568-x.

Wang, D., Robieson, W., Zhao, J., Wiener, C., & Koch, G. (2019). Statistical considerations in a delayed-start design to demonstrate disease modification effect in neurodegenerative disorders. *Pharmaceutical Statistics*, *18*(4), 407–419. https://doi.org/10.1002/pst.1931.

Ware, J. E., Kosinski, M., & Keller, S. D. (1996). A 12-item short-form health survey: Construction of scales and preliminary tests of reliability and validity. *Medical Care*, *34*(3), 220–233. https://doi.org/10.1097/00005650-199603000-00003.

Weijer, C., Goldstein, C. E., & Taljaard, M. (2018). TwiC or treat? Are trials within cohorts ethically defensible? *Clinical Trials (London, England)*, *15*(1), 21–24. https://doi.org/10.1177/1740774517746622.

Whone, A., Luz, M., Boca, M., Woolley, M., Mooney, L., Dharia, S., et al. (2019). Randomized trial of intermittent intraputamenal glial cell line-derived neurotrophic factor in Parkinson's disease. *Brain*, *142*(3), 512–525. https://doi.org/10.1093/brain/awz023.

Whone, A. L., Boca, M., Luz, M., Woolley, M., Mooney, L., Dharia, S., et al. (2019). Extended treatment with glial cell line-derived neurotrophic factor in Parkinson's disease. *Journal of Parkinson's Disease*, *9*(2), 301–313. https://doi.org/10.3233/JPD-191576.

Wijeratne, P. A., Johnson, E. B., Eshaghi, A., Aksman, L., Gregory, S., Johnson, H. J., et al. (2020). Robust markers and sample sizes for multicenter trials of Huntington disease. *Annals of Neurology*, *87*(5), 751–762. https://doi.org/10.1002/ana.25709.

Wild, E. J., Boggio, R., Langbehn, D., Robertson, N., Haider, S., Miller, J. R. C., et al. (2015). Quantification of mutant huntingtin protein in cerebrospinal fluid from Huntington's disease patients. *Journal of Clinical Investigation*, *125*(5), 1979–1986. https://doi.org/10.1172/JCI80743.

Williamson, P. R., Altman, D. G., Blazeby, J. M., Clarke, M., Devane, D., Gargon, E., et al. (2012). Developing core outcome sets for clinical trials: Issues to consider. *Trials*, *13*. https://doi.org/10.1186/1745-6215-13-132.

Williamson, P. R., Altman, D. G., Bagley, H., Barnes, K. L., Blazeby, J. M., Brookes, S. T., et al. (2017). The COMET handbook: Version 1.0. *Trials*, *18*, 1–50. BioMed Central Ltd. https://doi.org/10.1186/s13063-017-1978-4.

Woollard, L., Gorman, R., & Rosenfelt, D. J. (2021). Improving patient informed consent for haemophilia gene therapy: The case for change. *Therapeutic Advances in Rare Disease*, *2*. https://doi.org/10.1177/26330040211047244. 263300402110472.

Yang, W., Hamilton, J. L., Kopil, C., Beck, J. C., Tanner, C. M., Albin, R. L., et al. (2020). Current and projected future economic burden of Parkinson's disease in the U.S. *NPJ Parkinson's Disease*, *6*(1), 1–9. https://doi.org/10.1038/s41531-020-0117-1.

CHAPTER EIGHT

More than a participant in trials of cell and gene therapy: Hearing the voices of people living with neurodegenerative diseases

Emma L. Lane[a,*], Lyndsey Isaacs[b], and Soania Mathur[c]
[a]School of Pharmacy and Pharmaceutical Sciences, Cardiff University, Cardiff, United Kingdom
[b]Cure Parkinson's, London, United Kingdom
[c]Unshakeable MD, Toronto, ON, Canada
*Corresponding author: e-mail address: laneel@cardiff.ac.uk

Contents

Abstract

With the advent of novel advanced therapy medicinal products (ATMPs) for neurodegenerative diseases, their pathway to clinical trials and the therapeutic landscape has highlighted some new challenges, many of which are outlined in other chapters of this volume. The practical considerations of all these aspects from basic research and animal

International Review of Neurobiology, Volume 166
ISSN 0074-7742
https://doi.org/10.1016/bs.irn.2022.09.007

models through to clinical trials and eventual clinical implementation are significant. By and large, the major voices surrounding these challenges are the scientists and clinical teams who both develop the interventions and design and deliver the clinical trials to test these novel ATMPs. Their expertise is of course essential, but there is a key voice that can add considerable benefit to the pipeline, that of the lived experience of the disease being treated and the new intervention being considered. While still in their relative infancy in neurodegenerative disease, some ATMPs are already in clinical application in other disease areas, mainly cancer and inherited disorders. This more advanced status has raised some interesting questions about the role of the patient voice across all aspects of the therapeutic research and clinical delivery pipeline. This chapter highlights what has been learnt from the patient voice in their understanding and perspectives of ATMPs and in their experiences of clinical trials in neurodegenerative diseases to date. We discuss when, and how, including people living with neurodegenerative disease is of value in the development and implementation of ATMPs and the questions this collaborative effort can allow us to answer.

Abbreviations

ASO antisense oligonucleotide
ATMP advanced therapy medicinal product
DBS deep brain stimulation
GDNF glial cell-derived neurotrophic factor
HD Huntington's disease
MS multiple sclerosis
PD Parkinson's disease

1. Introduction

Advanced therapeutic medicinal products (ATMPs) are a new class of therapies based on genes, cells or tissues, that seek to stop the progress of a disease or indeed correct the dysfunction associated with a disorder through repair and regeneration strategies. While interventions are largely specific to each disease, there are several common approaches which are being employed. Three main classes of ATMP are described by the European regulatory authority, the European Medicines Agency (EMA); gene therapy, somatic cell therapy and tissue engineered medicines (collectively referred to here as ATMPs or cell and gene therapy). Gene therapy uses tools to insert genes into cells that are intended to manage or treat the disease, while the cell therapy approach uses cells themselves to either replace, repair or regenerate human tissues. While some ATMPs have reached the point of clinical implementation in other fields (for example, oncology and spinal muscular atrophy), ATMP approaches for neurodegenerative diseases are still in their relative infancy, being either in late-stage preclinical development or phase 2 safety and initial efficacy studies.

Active development of cell and gene therapies is occurring across a spectrum of neurodegenerative disease including Parkinson's (PD), Alzheimer's disease and multiple sclerosis (MS) as well as rarer disorders such as Huntington's disease (HD) (see reviews, Jensen, Gøtzsche, & Woldbye, 2021; Merola et al., 2021; Prasad & Hung, 2021). Each of these disorders, responsible for progressive motor and/or cognitive dysfunctions, involves a pathological cell death process leading to the loss of specific neuronal subtypes and in many cases the additional involvement of at least one toxic protein species. Cell therapies, delivered directly to the brain, are proposed generally as mechanisms to replace the lost neuronal connectivity and/or neurotransmitters. Gene therapy approaches offer a range of functions, either replacing missing neurotransmission or proposing to slow or stall the pathological processes by either reducing production of the toxic proteins or providing neurotrophic support to the remaining cells. In the case of stroke and spinal cord injury where the insult is more acute, reparative and restorative therapies are also being developed using the same strategies.

Clinical trials of ATMPs for neurodegenerative diseases are highly involved, complex trials requiring brain surgery or lumbar punctures with extensive pre and post intervention outcomes assessment for evaluation of safety and efficacy. The design and conduct of clinical trials in this area is varied but often, from a participant's perspective, differs significantly to pharmacological trials, being akin more to neurosurgical trials for elective tissue ablations or deep brain stimulation (DBS). While in some instances in ATMP trials there are repeat administrations of the therapeutic agent, many are delivered as a single "one-hit" procedure. As such, these trials require a significant commitment from a participant to undergo a procedure at considerable personal risk. The cells or viral vectors are irreversibly delivered into the brain and while tested to the best of abilities in the laboratory, their safety is at this point unknown. Yet for participants significant hope may be engendered; hope that they may see a long-lasting impact not just on their symptoms but for the first time altering the course of their disease (Cho, Miller, & Kim, 2020; de Bot, 2019; Keränen, Pasternack, & Halkoaho, 2017). The reality, however, is that the success rate of trials in neurodegenerative diseases is remarkably poor overall, especially when neuroprotective, disease modifying or regenerative approaches have been trialed

(Feustel, MacPherson, Fergusson, Kieburtz, & Kimmelman, 2020). This "hope" is perhaps therefore better described therefore as therapeutic optimism, the belief that they will experience direct medical benefit from early-stage interventions. It is against this unique backdrop that here we consider how those living with the disease are heard, with a specific look at how the participant voice is considered in trials. While it is not in doubt that there is a significant demand from the patient community for treatments to better manage their disease (Andrejack & Mathur, 2020), we explore what is known, how the public perceive these advanced interventions and consider this alongside the challenges they may face in bravely undertaking clinical trials.

2. Public and patient perceptions of cell and gene therapy

A number of studies have explored the extent of public knowledge and understanding, beliefs and perceptions towards both cell and gene therapy across the globe. In a recent systematic review the vast majority (82%) of papers on this topic were published in the last 10 years (Aiyegbusi et al., 2020), highlighting how contemporary this work is. However, information gathered to date largely consists of the views of the general public, with only a few studies touching on specific disease affected populations. With advances in oncology and pulmonary disorders preceding the neurological applications of ATMPs, the vast majority of data does not relate to neurological conditions.

The first trials of cell therapy in neurodegenerative diseases involved either autologous transplantation (adrenal glands used for PD) or the use of fetal brain tissue collected from elective terminations of pregnancy, each study showing varying degrees of success. With the majority of studies being conducted in PD, the autologous approach evidenced limited efficacy, while various small trials of fetal cell therapy demonstrated efficacy in some patients but provoked an understandably charged ethical debate (Doerflinger, 1999; Fletcher, 1992; Gillon, 1988; Hoffer & Olson, 1991; Lowy, 1989). Despite this ardent academic opining, there is limited evidence to suggest that there were many direct efforts to systematically obtain the views either of the donors of the fetal tissue or the recipients of the transplants. Fetal tissues from elective terminations of pregnancy were initially viewed as a waste material, with little information given to donors about the ultimate use for the tissues (Vawter & Gervais, 1995). Additional practical and legal obstacles alongside the ethical concerns, slowed progress but fetal cell trials have remained the proof of principle cell therapy product, prior to the advent of stem cell-based therapies, and thus small trials have recently been re-initiated in a bid to optimize elements of methods for evaluating this approach. TransEUro (Barker & Consortium, 2019) and TRIDENT (Drew et al., 2021) are two of the most recent fetal cell therapy trials, seeking to reinvigorate a path for cell transplantation in PD and HD respectively. As research has moved towards other sources of cells the discourse has changed, in particular with the use of human embryonic stem cells as a source. While the early public debate on stem cells focused largely on the donor views of the contribution of the unwanted fetuses or fertilized blastocysts used to isolate stem cells to be developed into cell lines and was largely positive (Hug, 2008; Peddie et al., 2009; Pfeffer, 2008), more recent studies have explored the public and potential recipients views (de Melo-Martín, Holtzman, & Hacker, 2020; Grauman & Drevin, 2022).

While cell replacement therapy focuses on restoring missing neuronal circuitry in some form, gene therapy trials have been more varied in their aims. In PD, gene therapy approaches have been used to provide neuroprotective growth factors such as glial derived neurotrophic factor (GDNF) and Nurturin, and also used to restore dopamine producing machinery by transfecting cells in the caudate putamen with the necessary enzymes (amino acid decarboxylase with or without tyrosine hydroxylase). In other disorders such as HD, the goal is to reduce or prevent production of the mutated huntingtin protein in an effort to slow disease progression which can be achieved by delivering antisense oligonucleotides (ASO) or siRNA, with similar approaches used in spinal muscular atrophy and amyotrophic lateral sclerosis (reviewed by Sun & Roy, 2021). As with cell therapy, while there has been some work to explore more general the publics' view on gene therapy, little work has been conducted in patient specific populations and in particular those with neurodegenerative diseases (Aiyegbusi et al., 2020; Delhove, Osenk, Prichard, & Donnelley, 2020).

2.1 Acceptance and the importance of information

In general views from the public and patient populations are largely positive towards the use of ATMPs (Aked et al., 2017; de Melo-Martín et al., 2020; Delhove et al., 2020; Unsworth, Mathias, Dorstyn, & Koblar, 2020b), for example the majority of stroke patients surveyed (63%), despite limited knowledge, were positive about stem cell therapy with a third expressing a willingness to take part in a trial and very small number expressing ethical concerns (Aked et al., 2017). Various characteristics have all been associated with different beliefs; negative views have been associated with older age, being female or religious beliefs, while knowledge, years of education and severity of disease are strongly linked to acceptance (Delhove et al., 2020; Kim et al., 2006). The only study of perceptions of gene therapy in Parkinson's disease illustrated that the appetite for risk was the biggest differentiator between would be participants in a hypothetical gene therapy trial, vs those who would not take part. Risk appetite was also recognized as a factor in potential participation in early phase stem cell trials in PD (de Melo-Martín et al., 2020). The motivation that actively doing something was better than waiting for a cure, along with the belief that their efforts would benefit society as a whole were also the positive reasons cited for taking part (Kim et al., 2006). Repeatedly across studies of both cell and gene therapies, knowledge and understanding are recognized as being key factors in acceptance, with relatively limited understanding in the general population and much of what is known obtained from less reliable sources (Aiyegbusi et al., 2020). There is a need for more reliable and accessible information from trusted sources (de Melo-Martín et al., 2020), which becomes all the more critical when acceptance of these therapies has been shown to increase with information provision and higher levels of education, especially scienced-based (Aiyegbusi et al., 2020; Aked et al., 2017; Delhove et al., 2020). Without this readily accessible information, patient groups also have a relatively limited insight into the inherent risks of these interventions and indeed a Japanese study suggested greater concern about practical issues such as cost and liability, than risks (Shineha, Inoue, Ikka, Kishimoto, & Yashiro, 2018). It is therefore incumbent upon the scientists and clinicians to ensure both adequate educational strategies are designed and utilized for the informed evaluation of risk, and that safeguards and frameworks are implemented to support decision making at the point of access to trials.

Information provision is also critical to preventing stem cell tourism (traveling overseas to obtain stem cell therapy), which remains a significant concern. Previously it had been relatively straightforward to determine that with no proven therapies, clinics promoting stem cell therapies were far from legitimate, medically assured ventures. However, as stem cell therapies become more common place, supporting patients and their families to discriminate between these and proven therapies will need to be considered. A study with stroke patients found that information did help reduce the number of patients considering unproven stem cell treatments but that, within a relatively short period of time the impact of this education intervention was lost (Unsworth, Mathias, Dorstyn, & Koblar, 2020a). Engaging in continued dialogues with the patient populations across demographics to ascertain their understanding and provide appropriate tools for them to make informed decisions therefore has to be an imperative (Lyons, Salgaonkar, & Flaherty, 2022).

2.2 The global context

There are significant cultural contexts that factor into the global ATMP market, from the legality of utilizing cells derived from specific sources (human embryonic stem cell derived products are still not legal in some countries) (Alahmad, Aljohani, & Najjar, 2020) to the nature of the healthcare system will also impact directly on the perceptions of the patients. For countries which lack a free at point of access healthcare system, cost and acceptance by health insurance companies is a significant consideration that participants are alert to (Benvenuti, Wang, & Borroni, 2021). The cost of ATMP development is such that the majority are unsustainable within a public/academic context, even at phase 2 clinical trial stage (only 1 of 5 stem cells trial in PD for example) trials increasingly have backing from larger pharmaceutical companies. Given an interesting perspective arising in an Australian study that differentiated between research in private and public contexts, that "public" scientists were more trusted than those in the private sector, the increasing early stage role of industry partners may do little to increase the trust of the public in therapies of this nature (Critchley, Bruce, & Farrugia, 2013). With greater involvement of the private sector also comes limited insights into participant experiences as data may well be collected but, as they pertain to their proprietary products and ultimately could support their own sales, the data does not reach the public domain. The natural consequence is that there is limited data sharing and information is not available to those clinical trial teams still operating in the academic environment.

3. The experiences of clinical trial participants with neurodegenerative diseases

Akin to being an explorer, or the first in space, participants in early phase ATMP trials are largely forging a previously untrodden path. Therefore, designing patient-centered trials needs firstly to build on our understanding of how previous trials of other interventions have been experienced in that disease, secondly to have good patient input in supporting the design of trials and lastly requires an iterative approach that captures the experiences of participants as these trials progress in conjunction with current perceptions. Collectively these can be used to inform future education, trial design and clinical implementation. With no formal definitions of patient experience in trials, this concept can be described in different ways; in understanding the research burden by metrics (time spent involved in the trial), evaluations which rate or describe the experience of elements of the participation, or participants preferences for how it should take place (Planner et al., 2019). There are tools that have been developed to support these types of data capture, but the depth of information obtained is often limited and highly dependent on the motivations of those seeking the feedback (Planner et al., 2019). Even while there is a substantial literature exploring the lived experience of the different neurodegenerative diseases themselves, there has been little work in chronicling of how these same populations experience clinical trials of any sort (Gorzynska, McGoohan, & Velayudhan, 2022) and no studies that detail the experience of individuals who have received cell or gene therapy across any medical area (Aiyegbusi et al., 2020). As with the data on perceptions, the majority of this limited literature focuses on PD and stroke, with little on HD or other neurodegenerative diseases.

Below we consider what is known about the experience across the whole pipeline of the participant journey on a ATMP clinical trial in neurodegenerative diseases (see Table 1). While challenges for the clinical trial team are covered elsewhere (see chapter "Considerations for clinical trial design and conduct in the evaluation of novel advanced therapeutics in neurodegenerative disease" by Drew and Busse, this volume), some potential issues for the participants can readily be imagined through the multiple points of the trial, some of which have been identified by existing studies but others which may be specific to ATMP studies. The focus of much of the literature to date has been on motivations to participate, the informed consent process and the issues of sham-controlled surgeries.

3.1 Motivations and barriers for trial participation

When we consider the participant at the start of their trial journey, they must first be motivated to take part, seek access to the trial or be approached to participate prior to undertaking an informed consent process. Access to trials from the outset can be significantly impacted by education, race and socioeconomic factors. Motivations to take part in trials are, not unexpectedly,

Table 1 A brief time line of events within a clinical trial.

Event	Challenge
Identification of a trial	How does a potential participant identify trials? Who has access to this information?
(Discussion with own neurologist/network/ online to get more information)	Does the neurologist know enough to advise or support? Are online materials reliable and accurate?
Contact trial team	Is there an easily identifiable way to contact someone knowledgeable about the trial?
Undertake screening assessment	What are the cost and logistics of travel to site and is reimbursement full and easy? What if you are considered ineligible?
Undergo informed consent procedure	Is there full understanding of the process in an informed way, including any placebo randomization.
Baseline motor/ cognitive/psychiatric function assessments	Is there perceived pressure to perform? What are the physical and mental impacts of being "off" for assessments on both participant and support network (in Parkinson's) How are fluctuating conditions accounted for?
Baseline procedures including brain scans (MRI/PET) or lumbar punctures	Do brain imaging teams understand the needs of neurological patients undergoing PET scanner (up to 90 min, potentially "off" medication/with chorea/ dystonia/dyskinesia) How laborious is the procedure on the patient?
Undergo brain surgery (implantation of hardware, cell or gene therapy products)	Do surgical teams/recovery teams understand the specific needs of patients in these surgeries?
Regular motor/ cognitive/psychiatric function assessments (frequency might be monthly to quarterly)	What are the effects and demands of these repeated assessments on the patient? Are they barriers to participation?
The end of the study	How does it feel if the study is unsuccessful? How is the sense of responsibility/involvement/part of a team acknowledged?
Returning to neurologist care	Does the regular care neurologist have enough knowledge to adequately support the participant? Can the participant adapt to less frequent interaction with the healthcare team?
Post-trial monitoring	How can engagement be maintained? What form does long term side effect monitoring take? Is there financial provision to support post-trial monitoring and participant needs?

largely centered around accessing better treatments for themselves primarily, although the advance of science for the benefit of others is also a major driver (Kehagia et al., 2022; Kim et al., 2012; Ravina et al., 2010; Valadas et al., 2011). In general, significant barriers include eligibility, travel, concerns about side effects or that they won't benefit personally (DasMahapatra, Raja, Gilbert, & Wicks, 2017; Kim, Wilson, et al., 2012). Secondary to these, are disease specific motivators and indeed barriers to participation. In MS there are good disease modifying oral and subcutaneous treatments, in PD a significant range of symptomatic therapies that work very effectively in the early stages of the disease exist, whereas in HD, motivation and/or risk appetite appears to be higher due to both the highly limited options available to manage the symptoms and, as a wholly genetic disorder, a reflection of patients experiencing the disease course in relatives who are also gene positive.

Patient advocates in Parkinson's disease have actively contributed to the conversations around different aspect of trials (Cohen et al., 2012; Mathur, DeWitte, Robledo, Isaacs, & Stamford, 2015a, 2015b; Mathur & Stamford, 2021; Tarrant et al., 2015). One of the more contentious topics is the inclusion of sham or placebo procedures. There is a transatlantic divide on this issue with Europe and the UK being less accepting of sham surgical procedures than US authorities. While patients conceptually understand the need for sham surgery controlled groups, recognizing the need to mitigate for the placebo effect (Kim et al., 2012), their view quite broadly is that these are not ethical, and maybe a significant contributing barrier to participation as this acceptance is not reflected in trial participation behavior (Swift, 2012). The debate is still rife in scientific as well as patient circles, some scientists considering this approach to be "a potentially dangerous and possible unethical." One study interviewed the participants of 3 sham surgery controlled trials (including 2 gene therapy trials) finding that participants were well informed about the nature of the placebo arm but admitted to a negative response when first hearing about it (Kim, De Vries, et al., 2012). Those who chose to decline the opportunity to take part were also interviewed although their explicit on the impact of the possibility of sham surgery on their decision is not mentioned (Kim, Wilson, et al., 2012). One mechanism to deal with the issues of receiving sham treatment is to commit to offering the intervention at the end of the trial, but this is perhaps not an adequate substitute given the potential that benefits may wane, and feasibility may be compromised, by the lengthy interval to the end of the trial. It is noteworthy though that sham treated participants have still sought to have the intervention even if trials report negative outcomes (Pollack, 2005). For more than a decade patients themselves have been advocating for alternatives to the double blind randomized design as a way to circumvent the need for a sham group (Cohen et al., 2012). Indeed alternatives such as Trial with Cohort Designs were used to build TransEUro (Barker & consortium, 2019) and is being applied in early phase studies in both PD and HD (Drew et al., 2021).

3.2 Diversity

It is well documented that under-served populations lack access to basic healthcare, and have poorer healthcare outcomes as a result (Rochester & Carroll, 2022). This persists into clinical trials with a startling lack of diversity across all trials, with a recent call to action in Parkinson's highlighting that reporting of ethnicity was highly variable (only 43% of studies reporting ethnicity data) and with very poor evidence of black minority representation (Lau et al., 2022). Studies suggest that the contributing factors are many, with one of the major challenges to trial participation being access (Adrissi & Fleisher, 2022). As ATMP trials are, by their nature, often led by centers of research excellence in PD, the recruited phase 2 participants may be largely from the lead clinical investigators own patient list (Moore, Guzman, Mason, Williams-Gray, & Barker, 2014) this may be exacerbated. Other contributing factors are likely systemic, a lack of culturally relevant resources, staffing that doesn't reflect the population seeking to be recruited and poor engagement (Adrissi & Fleisher, 2022; Massett et al., 2021). Some work within cancer in the US suggests that when offered, the uptake of clinical trials between black and white patients is not different (Riner et al., 2022), but that potentially unforeseen consequences of certain eligibility criteria and unconscious biases from the recruiting team could contribute to their exclusion (Snyder, 2022). Diversity across the trial team and public contributors would help address such issues if indeed they contribute to the bias seen in neurological trials.

Considering diversity in both participants and patient voices is critical both for trial representation of the wider PD community and in the preparation of the community for clinical implementation if the therapy is successful. As greater availability of different cell or gene products become available and an element of choice is introduced, education will be become ever more important and acknowledging the concerns of different communities vital. Trust between patients and scientists/healthcare professionals is essential and there may be some barriers relating to the cultural histories and interactions with healthcare in different communities (Allison, Patel, & Kaur, 2022). In the US, the story of Henrietta Lacks, the original unwitting donor of the HeLa cells so widely used in research, still reverberates and could do so even more loudly with the use of cells as therapies (Allison et al., 2022; Kemet, 2019). In the US, the Henrietta Lacks Enhancing Cancer Research Act in the US was passed which was designed to provoke the examination of access to cancer research by traditionally overlooked groups, work which may also have impacts which reach beyond oncology (Boffa, Churchwell, & Maduka, 2021).

3.3 Does disease stage matter?

With the exception of stroke and spinal cord injury, the majority of conditions considered in this chapter progress over several years if not decades. With this feature in common, it is striking that the therapeutic landscape differs dramatically both between diseases and through their lifetime. As such the stage of disease can impact on considerations for ATMPs for different reasons. From a functional perspective, sufficient preservation of the target brain region for successful delivery of the ATMP is essential for the product to perform its function, whether that be promoting survival of remaining neurites, supporting regrowth, or replacing lost transmitter systems. As diseases progress, the anatomy of the target structures may change. In HD and multiple system atrophy there may be significant atrophy of the striatum, the target brain region, at relatively early disease stages creating neurosurgical challenges to deliver the ATMP. While in PD, anatomically the integrity of the target structure, the caudate putamen, is relatively preserved but neuroprotective and neurorestorative agents may require delivery at a time critical window during which there are enough remaining nigrostriatal innervation to effect preservation. For this reason, and that the interventions are largely intended to be either neuroprotective or to extend the life time of existing treatments, delivery relatively early in the disease course would be optimal.

In the context of this chapter, while clinically earlier may be preferable, is there a conflict with regulators concerns over safety in early phase trials or with the acceptability to the patient group? As might be anticipated, the patient view is heavily influenced by the specific disease they are living with, affecting their motivations to participate in trials and consideration of more invasive therapies. PD and HD could be considered as being at opposite ends of the spectrum in this regard. HD patients have limited symptomatic relief and management opportunities at any stage of the disease, but critically while the genetic nature of the disease offers the positive feature of the opportunity for early detection of gene positive status and risk of developing the disease, it also sadly brings a greater awareness of their future. This lack of other options and observation of family members living with later stages of the disease leads to a greater willingness to engage in higher risk studies at an earlier stage in the disease. A survey of perceptions of a hypothetical clinical trial in HD found that patients (those symptomatic more than those asymptomatically gene positive) also harbored greater risks of therapeutic optimism and misconception (Cotter, Siskind, Sha, & Hanson-Kahn, 2019; de Bot, 2019). In contrast, idiopathic PD largely lacks any genetic marker but many symptomatic treatments are available in early stage disease. Here cell therapy has been compared to the well-established neurosurgical intervention Deep Brain Stimulation (DBS) in an attempt to define the target patient population (Barker, Bjorklund, Frucht, & Svendsen, 2021; Rehncrona, Lund, & Neurotransplantation, 2006). DBS is largely considered as an intervention to be explored when patients start to experience issues with oral medication and motor complications at latter disease stages. In some instances, cell and gene trial eligibility criteria may require participants to be eligible for, and possibly to have been offered DBS and turned it down, recognizing that good DBS candidates will also meet the criteria for ATMP trials. However, despite all being neurosurgically delivered, the "immediate" relief offered by DBS fundamentally differs to ATMPs; depending on the gene being transfected, timelines could be 6–12 months for gene therapy, while cell transplantation could take in excess of 2 years to produce significant optimal motor recovery (as demonstrated in previous fetal cell trials (Kefalopoulou et al., 2014). This suggests that including participants from earlier diseases stages might be preferential to get optimal benefit, but this introduces an important dichotomy. A trial of DBS delivered earlier in the disease, EARLYSTIM, found it to be highly effective but patient uptake remains very limited, with patients often preferring to remain with oral symptomatic therapies until medication side effects and disease progression leave them with few other options (Cabrera et al., 2021; Das et al., 2019; Stoehr et al., 2022). Although a small sample size, a look at patient preferences for treatment modalities placed neurosurgery as the most negative, with oral intake being the most positive (Weernink, van Til, Groothuis-Oudshoorn, & Ijzerman, 2017). Furthermore, where safety is the primary outcome, and in recognition of the invasiveness and irreversibility of

ATMPs in early trials, caution (and at times regulators) dictates that where practicable patients in latter stages of the disease are considered for trials. This can lead to its own issues of interpretation, as the participant demographic may differ significantly from the optimal eventual target population of early disease stage. Understanding these current experiences with DBS may therefore provide insights for future challenges in clinical implementation of ATMPs.

3.4 Informed consent—What does that mean for ATMP trials?

Advanced therapies hold promise for disease modification, changing the actual trajectory of neurodegenerative diseases either by slowing decline or by improving current disability. This potential remains the highest aspiration for patients. As such, if there are potential benefits reflected by improved quality of life and diminishing disability, people living with PD are often willing to take risks—that hope is a powerful motivator. Therefore, there is a question as to whether consent to participate in the trial is genuinely "informed" or compromised by clouded judgment. There is a concern that decisions may be made, not based on educated information, but predicated on the very real desire to halt their inevitable decline into disability. That is why it is imperative that the potential risks be accurately and realistically communicated to patients, allowing them to make an educated decision.

Although there are clearly concerns about the effectiveness of advanced therapies, their safety and risk profiles, advanced treatments that may potentially be disease modifying are urgently needed. These cannot be developed without patient participation in robust clinical trials. In order to engage patients and have ethical buy-in from the patient community, the process of recruiting for clinical trials must involve clear communication of all potential risks to the proposed intervention, potential benefits and possible limitations of the treatment being studied in order to make an educated decision. The process of informed consent for ATMP trials must be even more robust given the level of invasiveness that these treatments incur.

A cornerstone in effective clinical trial recruitment and informed consent, is communication. Successful communication requires written materials as well as verbal descriptions to be easily understood by the lay person. Communications should be pitched to the understanding of a 12 year old, but commonly documents are written to a significantly higher level of comprehension (Hellmers, Obeng-Aduasare, de Melo-Martín, &

Henchcliffe, 2016; Nishimura et al., 2013). These efforts must not only take level of the material into account but also consider a number of other influences that are part of a potential participant's decision-making process. These include age, stage or duration of disease, disability and quality of life. The process may also be impacted by the amount of family support, work commitments, family responsibilities, finances, comorbidities, cultural nuances and pressure as well as personal ethics. Each person must tailor their choice depending on these considerations, ultimately making a decision whether or not to participate in a trial involving an advanced therapy compared to a standard therapeutic that may incur less risk. The information given to potential participants must be fully complete, written clearly and understandable to those who may not have any prior scientific background. Informed consent must be truly informed. This challenges the clinical team and research ethics committees to ensure everything is explained, while the challenge from a participant's perspective is the pages and pages of documentation, often in technical, jargonistic or "legalese" language. Conflict between the ethical and legal requirements and the accessibility of the documents, can lead to participants making their decision without in-depth reading of the literature they are provided. Participants often show good understanding of some elements of the trial process such as voluntary participation and withdrawal (good is considered over 50%) but this drops to worryingly low levels on safety issues, risks and side effects (Pietrzykowski & Smilowska, 2021). There has been extensive work to understand informed consent processes and how these can be improved in more complex trials which can involve repeated consenting sessions and different methods of presenting the information (without overloading the participant) and assessing understanding (de Melo-Martín, Hellmers, & Henchcliffe, 2015; Halpern, Paolo, & Huang, 2019; Hellmers et al., 2016). However, it is noteworthy that truly informed consent is complex to both measure and achieve, notwithstanding that the participants' perspective on that consent can potentially change based on their experiences through the trial and their end of trial experiences and release of overall and personal trial outcomes (Tarrant et al., 2015).

Expectation management of the potential benefits poses another challenge. ATMPs may not necessarily target the underlying pathology of PD or HD, for example, meaning that the disease will continue to progress in some form after treatment has been delivered. Expectations of the potential of therapies currently used in late-stage PD are high across multiple domains of symptoms (Reddy et al., 2014). With the discreet delivery of these new entities directly into brain areas, very few therapies will be able to target all symptoms of the disease, most are targeted at motor symptoms while it is the non-motor symptoms prevalent in many neurodegenerative diseases, which for many patients have the greatest impact on quality of life (Santos Garcia et al., 2019). Not only that, but any adverse effects or complications resulting from the use of advanced therapies, would be difficult to mitigate due to the general irreversibility of these procedures. Additional concerns exist in relation to possible future treatment, as there may be genuine worries of whether an invasive procedure now may preclude involvement in future trials or whether it may limit future treatment options (de Melo-Martín et al., 2020). In addition to the complexities of the trial itself all of these issues must be fully understood by participants within the informed consent process.

3.5 Being on the trial

Although limited, especially when we consider pharmacological and surgical interventions, the reports of clinical trials that do collate participant experience are largely positive (Gorzynska et al., 2022). The benefits of trial participation are reportedly far wider than the intervention itself, having a more enriched life through the various interactions created by the trial and intellectual challenge of understanding the intervention. Participants feel a sense of achievement which increases morale and self-confidence, and mental and physical well-being (Kunkel et al., 2017). Confidence in the clinical team or the person delivering the intervention was critical to the positive experience, as trust was built and their expertise valued along with the regular monitoring and support provided (Kehagia et al., 2022; Sturkenboom et al., 2013).

A recent report of pharmacological clinical trial in PD identified some specific burdens, some of which have been previously overlooked and should be factored into trial designs. In a series of interviews and focus groups with the participants through the trial, which explored a statin as a putative neuroprotective agent, practical considerations were raised. This included accessing and administering the medication (swallowing can be compromised in PD), and completing diary entries for positive and adverse events which influenced their feelings of self-worth and value that they were contributing to the trial (Kehagia et al., 2022). Importantly, assessments conducted in "clinically defined off" (12h after the last dose of anti-Parkinson's medication) states were identified as being a particular challenge to participation. The impact of these assessments was in several domains, physical (as participants were motorically impaired by the lack of dopamine) and psychological from both the physiological loss of dopamine and the realization of how impaired they were, impacting not only on the participants (to the extent that many cited this as a reason for withdrawal) but also on their support partners. Importantly however, there was a recognition that if these were more effectively addressed, participants would take part in trials in the future.

Unlike other chronic illnesses like diabetes or hypertension, there are no objective biomarkers to track the progress of most neurodegenerative diseases or an individual's response to proposed interventions, correlated with improved outcomes for patients. Decisions by the medical team in terms of clinical care or whether or not a research trial is "successful" are based on inadequate clinometric scales that are often biased on clinical judgment or patient recall. These outcome measures are often not truly reflective of the patient's lived experience in their own environment. Indeed, they are not always a true measure of impact on patients' perceived quality of life which is what is most important and best determined by patients themselves. Outcome measures that are most scientifically relevant may not be what is most patient relevant and therefore, true patient input is central to developing treatments that will truly impact the PD community.

3.6 The end of the trial

The end of a trial often involves the cessation of the intervention; exercise plans stop, pharmacotherapy is halted and importantly contact with the trial team ceases. Yet, as described above, ATMP trials are more aligned with surgical trials in that the interventions are often experienced acutely, but they differ in that longer term monitoring is required to ascertain the full impact, for better or worse. Understanding the events that occur at the end of a trial is particularly prescient here due to the permanence of the interventions, but also the high failure rate of the trials. Recent trials of disease modifying drugs or ATMPs exemplify the issues that can arise; they have either failed to meet primary endpoints (Whone et al., 2019) or been halted prematurely because of concerns that benefits no longer outweighed the risks (Largent & Karlawish, 2020). A PD study explored the neuroprotective ability of a growth factor, glial-derived neurotrophic factor (GDNF), directly infused into the brain monthly via a surgically implanted striatal cannula with subcutaneous port. They showed a trend towards improvement in motor function following 40 weeks of administration in the double-blind phase but unfortunately there was an improvement of similar magnitude in the placebo arm (Whone et al., 2019). Participants on the Tominersen trial, an anti-huntingtin ASO for HD, received monthly lumbar punctures for treatment but administration of the ASO was halted prematurely following review by the data monitoring committee. The trial itself continued to follow the progression of participants (Tabrizi et al., 2022). Similarly, although not an ATMP, various phase 3 trials for the proposed disease modifying anti-amyloid antibodies and BACE inhibitors for mild cognitive impairment and dementia caused by Alzheimer's disease have also stopped early, following concerns that benefits no longer outweighed the risks (Largent & Karlawish, 2020). Recognizing the frequency of these eventualities should prompt consideration for what constitutes the end of the study for the participant, how prepared they are for this eventuality and their experience as the trial ends.

Trials have a defined end point, specified within the ethically approved protocol, and influenced to a degree by the funding envelope, a point which will have been transparent to the participant at the time of recruitment. The "end," practically, as far as the participant is concerned, may be associated with the last official assessments. A secondary endpoint for participants may come after analysis is completed, where the end of study results may be released to participants indicating whether the trial has met its primary outcomes and therefore been a success. This is a relatively recent practice and it is still far from universal that outcomes are made available to patients in an accessible way despite data showing that this is a priority for nearly all patients and would impact on their willingness to take part in future studies (Sood et al., 2009). For the participant in an ATMP trial, a clear concern may be the impact the outcome has on their continued access to treatment. In the GDNF study, for example, participants were given access to their individual data and many found this incredibly useful but also contributed to their frustration that they could no longer access the treatment (L. Isaacs, Personal communication). Breaking the blind but also providing the detail on who has truly benefitted as opposed to a placebo effect, confusion as to why they improved personally but the trial overall failed to meet its primary outcomes and a feeling of responsibility towards the study after a significant investment both practically and emotionally, are just some of the other feelings participants might experience (Tarrant et al., 2015). With frequent site visits, a rapport may have been built with the study team and potentially with other participants which contributes to the additional emotional impact once this ceases to exist or can evoke a feeling of betrayal if the results have not met expectations (Tarrant et al., 2015). The therapeutic misconception also contributes to the sense of loss, as participants return to a more normal (i.e. much less frequent) interaction with healthcare professionals, who may have little insight into the trial they took part in and their experience through it as a participant.

Hope has both direct and indirect positive effects on our subjective sense of wellbeing, and also on better physical health behaviors (Long et al., 2020). Hope in both participants and carers may be particularly affected by the situation in which a trial ends prematurely with no warning. If a situation arises in which, for safety reasons, a trial is halted or delivery of the investigational product ceases, commercially pharmaceutical companies cannot inform participants ahead of a formal announcement to shareholders. The first participants may hear about changes to the trial could therefore be through a press release, which has been rationalized as reassurance that at least they are not "the last ones to be notified." Despite personalized follow up, trials ending prematurely have been likened to experiencing a trauma for some participants, with a need to be "rescued" (Kwon, 2021; Largent & Karlawish, 2020). Although this is more likely to be a concern for studies in which the intervention has been ongoing, ATMPs may fall into this category of trials depending on the mode of delivery.

Critically, with ATMPs there will be a need for much longer follow up beyond the official end of the trial, for safety reasons. The trial experience and events at the end of the trial will likely have a significant impact on this follow-up, recognizing this early planning for how it can be mitigated will be vital to that retention. Possible mechanisms for this might be to establish sustainable patterns of communication with participants that can serve multiple purposes; improving retention within the course of the trial, continuing to inform and educate about the trial and the progress in the field as a whole, preparing participants for what they may experience at the end of the trial but also creating a sense of being part of a team that will keep them involved beyond the end of the trial. Both participants and chief investigators may benefit from establishing a Patient Registry for ATMP trials within specific disease areas, which would provide a means to maintain communication and a platform for data collection and sharing in longer term follow up.

4. How is the voice of those living with degenerative disease engaged with?

The main ways in which scientists or clinicians engage with people with lived experience of degenerative diseases are broadly captured as PIE (Participation, Involvement and Engagement see Fig. 1). These are vital

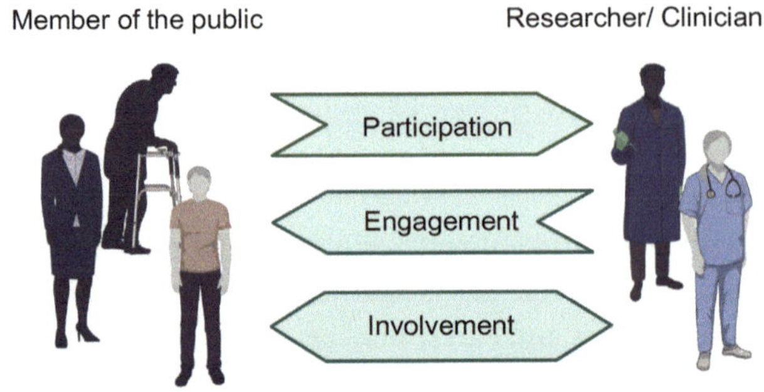

Fig. 1 An illustration of the directional nature of participation, engagement and involvement, the three core facets of PIE and the ways in which members of the public engage in trials. Involvement can be hugely varied in nature but at its essence is the two-way communication and equality between the public contributor and researcher/scientist. Figure created using www.biorender.com

tools to ensure good communication with, education of, and shared journeys with people with chronic conditions but the three are often conflated. Each has a very distinct definition, especially in relation to clinical trials and clinical roll-out. Participation is perhaps the most obvious, the focus of much of the above text, involving recruitment and consent to be a participant on a trial, undertaking any assessments to determine the efficacy of an intervention, receiving the experimental treatment or providing data to the trial team in some form. (https://www.invo.org.uk/public-involvementparticipationengagement-in-research). Engagement is the dissemination of information or knowledge, this may be as part of a trial or study, or to promote the research in a broader sense to either the lay public, those living with the condition in question or to a broader scientific community. In essence these involve unidirectional communication in opposing directions, between participants or members of the public, and scientific or clinical teams (see Fig. 1). Conversely, the third approach, involvement, should be much more of a bidirectional conversation, and is when members of the public are actively included in the planning or delivery of research projects or organizations. This can take different forms such as (but not limited to) being on a research or trial implementation advisory board, being a grant co-applicant, identifying research priorities and consulting on patient-facing documentation and study processes. Evidence shows that trials conducted in this way do not just benefit the participant, they benefit everyone. Data from Pharma Intelligence and Economist Intelligence Group (EIU) highlights that in neurology, patient centric trials were able to recruit in half the time as traditional trials, which provides a significant time and cost saving to the trial (The Economist Intelligence Unit, 2018). Drug products are also 20% more likely to reach launch suggesting that patient centric trials speed up delivery of the trial and its outcomes, potentially benefitting future recipients.

4.1 Not all involvement is equal—Who's voice should be heard?

Clinical trials have been occurring for decades in neurodegenerative diseases, but as described above there is little documented information to draw upon in previous literature that captures the lived participant experiences (Harding, Robinson, Wilson, Crutch, & Mummery, 2021; Planner et al., 2019). The desire in a clinical trial is always to balance the participant burden with the needs of the study, but understanding where that balance lies is challenging, and involvement of individuals with lived experience of the disease can greatly inform this. While Involvement, as defined by the NIHR, should not require specific expertise, when we engage with the public, the context of the public contributors is always important. This is also true when we consider the different voices we can hear from within each community. There are people living with neurodegenerative diseases who variously fall into different groups; those who have not engaged to date in research, those who are interested in research opportunities but have not taken part (through lack of awareness, ineligibility or historically excluded for other reasons), those who have participated in observational studies, non-pharmacological interventional studies, pharmacological studies and/or surgical interventions. A public contributor may fall into any one of these classes and added to this are the diverse lived experiences of the disorders themselves, the different stages of the disease course and experiences with existing therapies as well as different socioeconomic, ethnic and geographical backgrounds. If we consider genetic diseases such as HD, and increasingly as we know more about the genetics of other disorders, gene positive prodromal patients may also be contributing to these conversations. Together, this results in a broad set of unique experiences and knowledge of the events around a clinical trial.

When getting involved as a public/patient contributor advising on a project, all of the experiences above will be valid and their perspective incredibly valuable to the success of the project, but we must also be cognizant of the inherent bias a single voice or even multiple voices with similar experiences could bring. As described above, empirical data collection on trial experience collected alongside active trials, exploring metrics, surveys or qualitative interviews can provide a cross sectional perspective of the breadth of experiences providing better context. To achieve this, some studies have sought to embed a process evaluation within the clinical trial which captures the experiences in real time such as the fetal cell transplantation trial in HD, TRIDENT (Drew et al., 2021). Not only does this serve to inform the trial team of any issues arising and provide a sense of the universality of said issue, but may also serve to aid retention by giving the participants a voice.

While all views will be valid, meaningfully inputting to the design of complex ATMP trials arguably requires either a unique profile of trial experience or extensive education and support. Charities are proactively involved in supporting identification of appropriate public contributors to match to trial management boards but with the extent of underrepresentation of minority ethnic groups in all clinical trials (Lau et al., 2022) as well as ATMP trials (Moore et al., 2014) the lack of diversity of trial participants has a knock on effect that public contributors are from a similarly narrow range of ethnic and socioeconomic backgrounds, risking the self-perpetuation of any inherent barriers to participation. Diversity in this regard is challenging much in the way that diversity in recruitment is., with a similar under-representation in public contributors (Ekezie, Routen, Denegri, & Khunti, 2021).

4.2 Beyond the participant—The voice of the support network

While "involvement" has become a buzz word in research, much of the work appears focused on the "user." There have been minimal studies looking at the impact of being involved in a trial from the carer/family member experience (Ashcroft, Wykes, Taylor, Crowther, & Szmukler, 2016). This is despite evidence that care or support partners influence treatment decision-making in chronic illnesses (Lamore, Montalescot, & Untas, 2017), indeed in studies of shared decision making around surgical treatments it has been shown that involving the family member/care partner can help to reduce anxiety, enable advocacy, and help to convey important information to the participant (Hirpara, Cleghorn, Sockalingam, & Quereshy, 2016). For the participant in a clinical trial, the family member or care partner takes on a range of roles, offering support both emotionally and socially. In a trial which involves dementia, or a surgically invasive intervention that support is even more pertinent (Black, Taylor, Rabins, & Karlawish, 2016; Gorzynska et al., 2022).

For major interventional trials a tangible support network is highly desirable for the wellbeing of the patient. For some, the whole family will feel part of the trial, or others can be very isolated if they have no one to share it with. While trial recruiters may ascertain support structures informally, sometimes this may also be defined as part of the inclusion/exclusion criteria for a trial (more frequent where dementia is a symptom) (Drew et al., 2021; Largent, Karlawish, & Grill, 2018). This is not as simple as it may sound, as it can be considered paternalistic, especially if the definition of "support network" is too restrictive and that individuals with limited support structures could be excluded from taking part. Yet the physical and mental challenges that such trials can pose may make it almost a necessity that early phase studies should take available support for the participant into consideration. Many of the neurodegenerative diseases involve an element of cognitive decline. Even if, at the point of recruitment participants present with limited cognitive impairment, in lengthy trials it would expected that this could deteriorate. Involving proxies for consent at the earliest stage in the process would then be a necessity and for ATMP trials to openly consider what the participants advanced wishes may be to support the decision making of the proxies at a future date (de Melo-Martín et al., 2015).

Caregivers can also offer mutual support to other carers and provide recognition of their expertise. Indeed, involvement in a trial can be of benefit to that individual themselves—learning new skills, knowledge, and personal development and they can be frustrated when their value is overlooked (Kehagia et al., 2022). This support can, however, come at a personal cost; caregiver burden is recognized as an important factor in chronic diseases but how does this come into play in clinical trials? The inherent risks that these trials carry, plus the risks of failure are all too real, as we have discussed, but it is not only participants who experience this see-saw of risk versus hope of what stem cell or gene therapies can offer, which can cause tension in the patient-carer relationship. Carers experience "conflict burden," the desire to support their loved one in accessing a treatment that may help them or others, weighted against fear of adding to their existing care load during the trial "it was the first time in my marriage that I felt like a carer not a wife" (L. Isaacs, Personal communication) and feelings of regret or guilt if the trial fails to bring benefit or indeed causes harm (Cunningham, Abhyankar, Cowie, Galinsky, & Methven, 2018). In PD the "off" state assessments are not only a burden for the participants but have been described as being "emotionally distressing and practically challenging" by the support partners (Kehagia et al., 2022). The practical impacts described relate to the physical demands of bringing the participant to assessments with reduced mobility, but also included financial and time burdens depending on the location of the trial center and length of the trial amongst other factors. These complexities increase the urgency for ATMP trials to consider providing trial information framed around this context, taking into account carer-participant relationships, and considering how to raise awareness, in *both* parties, of the pressures the decision-making process can bring, and the consequences of enrolling in a trial.

5. Factors that contribute to a meaningful patient voice

Through this chapter we have emphasized the value of the patient voice but it is not an insignificant task to create trials that consider their views, not only trial design at its outset but that also take into account their perspective and are truly committed to a patient-centred approach both during their journey within the trial and beyond it. There are many resources that support the planning and implementation of PPI, many of which have been collated into a directory by EuroGCT https://www.eurogct.org/fr/themes/participation-de-patient. Early planning and clear investment from the Chief Investigator *and* wider clinical team in PPI allow it to be more meaningful and minimize tokenism (Dudley et al., 2015). While steering committees are often utilized to bring in a range of voices, positioning PPI members directly on to trial management teams and responsive advisory groups can have bigger impacts on perceptions of influence (Drew et al., 2021). Training for both researchers and public contributors is key to obtaining the mutual understanding and respect that is required for effective collaboration (Dudley et al., 2015). The need for identification of roles from the outset, perhaps in the form of role descriptors, supports this relationship by determining the boundaries and levels of input required (Meinders, Donnelly, Sheehan, & Bloem, 2022). Explicit demonstration of the impact of PPI to justify the investment of time and money is yet another administrative hurdle that seeks to diminish the contribution to numbers, yet qualitative studies are showing that not all PPI is equal. In an era where metrics are sought to benchmark everything, the explicit value of the guidance and impact PPI can have is harder to define but is nonetheless highly tangible for participants when trials have been designed and conducted with them in mind (Dudley, Gamble, Preston, et al., 2015). Drawing on evidence from well conducted studies evaluating the impact of PPI on informed consent materials and improved enrolment approaches for example, are powerful tools to evidence why this should be the default (Crocker et al., 2018).

6. The patient voice in clinical delivery

The need for improved patient education will persist beyond clinical trials and into clinical delivery. As we look ahead, considering how ATMPs might be positioned in the therapeutic landscape could provide important insights. Although we are several years, if not a decade or more away from clinical implementation of cell and gene therapy for PD, experiences in other disease areas are quite rightly leading patient advocates to look ahead to challenges that may need similar length of time to resolve in parallel with scientific development, adding their voice to clinical and academic fora (Benvenuti et al., 2021; Mathur & Stamford, 2021). With diverse healthcare and reimbursement mechanisms there are concerns of cost and access which risk the perpetuation and exaggeration of the health inequalities which are already in existence.

While posing challenges for trial recruitment and delivery, knowing when the therapy will be best implemented will also shape the ultimate target population for clinical implementation. For many interventions being developed, this raises the imperative to educate patients very early in the disease course about the possibility of these interventions, to increase access to trials. This is a singular advantage afforded to the HD patients by the genetic nature of the disease and thus possibility to identify (and maybe even treat) gene positive individuals at a prodromal stage. The challenges described above to achieve this in different disease areas will need to be carefully considered and can only be done with the patient voice. Assessing the experience of patients within healthcare settings is a much more common occurrence than within clinical trials and provides the impetus for quality improvement. In this same way, measuring patient experience can lead to improved patient-centered trials. Further work to understand patients' treatment preferences and what lies beneath them, is much warranted ahead of large scale ATMP delivery.

7. Conclusion

This chapter has drawn together what is known of current understandings and perspectives of the public and those living with degenerative diseases on ATMPs, with the documented experiences of clinical trials of the same populations. The focus in ATMP development to date has been in realizing the potential of this exciting avenue of therapeutics for degenerative diseases, but with its novelty brings new territory in designing and implementing trials. Listening to the patient voice in a meaningful way, as a collective with a range of experiences, and not as a single voice becomes even more of an imperative. There are different ways the patient and participant voice can input into these, each with its own values and limitations. What cannot be argued against, is a weight of evidence from multiple voices with lived experience of trials, yet that is rarely collected and/or reported. Survey and focus group data on perceptions, alongside advocate opinion pieces are hugely valuable in highlighting the areas that need to be considered in the development of ATMP products, in identifying potential areas of concern, or reassuringly, lack of concern in some domains. Data of this nature provides a base for healthcare providers to consider when exploring clinical implementation. Until we accurately collect and collate that information alongside understanding the beliefs and desires of our patient population, we cannot know how it may influence current and future trials or the clinical roll out of any successful intervention. Knowledge and education are much sought after by the patient communities but producing accessible and trusted sources of information requires partnership. Trust between clinicians and patients starts with the clinical trials but is vital for successful wider roll out in the future. These collaborative efforts between the public and researchers should allow greater consideration of the challenges faced by ATMP trial participants and how those living with these diseases and their support network can come on this exciting journey in an informed, involved and equitable manner.

Acknowledgments

EL is funded by Parkinson's UK G-002, Cure Parkinson's EL01 and is supported by Health and Care Research Wales Infrastructure funding BRAIN Unit.

The authors

Given the scope of this article it is important to acknowledge the experience of the authors. Lyndsey Isaacs is a trustee for Cure Parkinson's and partner of Tom Isaacs who founded the Cure Parkinson's Trust and lived with Parkinson's for 22 years. Soania Mathur is a General Practitioner who lives with Parkinson's disease and is a patient advocate and founding member of the PD Avengers. Both contributed to production of this review in concept and writing.

References

Adrissi, J., & Fleisher, J. (2022). Moving the dial toward equity in Parkinson's disease clinical research: A review of current literature and future directions in diversifying PD clinical trial participation. *Current Neurology and Neuroscience Reports*, *22*(8), 475–483. https://doi.org/10.1007/s11910-022-01212-8.

Aiyegbusi, O. L., Macpherson, K., Elston, L., Myles, S., Washington, J., Sungum, N., et al. (2020). Patient and public perspectives on cell and gene therapies: A systematic review. *Nature Communications, 11*(1), 6265. https://doi.org/10.1038/s41467-020-20096-1.

Aked, J., Delavaran, H., Lindvall, O., Norrving, B., Kokaia, Z., & Lindgren, A. (2017). Attitudes to stem cell therapy among ischemic stroke survivors in the Lund stroke recovery study. *Stem Cells and Development, 26*(8), 566–572. https://doi.org/10.1089/scd.2016.0343.

Alahmad, G., Aljohani, S., & Najjar, M. F. (2020). Ethical challenges regarding the use of stem cells: Interviews with researchers from Saudi Arabia. *BMC Medical Ethics, 21*(1), 35. https://doi.org/10.1186/s12910-020-00482-6.

Allison, K., Patel, D., & Kaur, R. (2022). Assessing multiple factors affecting minority participation in clinical trials: Development of the clinical trials participation barriers survey. *Cureus, 14*(4), e24424. https://doi.org/10.7759/cureus.24424.

Andrejack, J., & Mathur, S. (2020). What people with Parkinson's disease want. *Journal of Parkinson's Disease, 10*(s1), S5–s10. https://doi.org/10.3233/jpd-202107.

Ashcroft, J., Wykes, T., Taylor, J., Crowther, A., & Szmukler, G. (2016). Impact on the individual: What do patients and carers gain, lose and expect from being involved in research? *Journal of Mental Health, 25*(1), 28–35. https://doi.org/10.3109/09638237.2015.1101424.

Barker, R. A., Bjorklund, A., Frucht, S. J., & Svendsen, C. N. (2021). Stem cell-derived dopamine neurons: Will they replace DBS as the leading neurosurgical treatment for Parkinson's disease? *Journal of Parkinsons Disease, 11*(3), 909–917. https://doi.org/10.3233/jpd-219008.

Barker, R. A., & Consortium, T. (2019). Designing stem-cell-based dopamine cell replacement trials for Parkinson's disease. *Nature Medicine, 25*(7), 1045–1053. https://doi.org/10.1038/s41591-019-0507-2.

Benvenuti, S., Wang, C. M., & Borroni, S. (2021). Perspectives, expectations, and concerns of European patient advocates on advanced therapy medicinal products. *Frontiers in Medicine, 8*, Article 728529. https://doi.org/10.3389/fmed.2021.728529.

Black, B. S., Taylor, H. A., Rabins, P. V., & Karlawish, J. (2016). Study partners perform essential tasks in dementia research and can experience burdens and benefits in this role. *Dementia (London)*. https://doi.org/10.1177/1471301216648796.

Boffa, D. J., Churchwell, K. B., & Maduka, R. C. (2021). Diversity, equity, and representativeness: Coming to terms with the Henrietta lacks act. *Journal of the National Comprehensive Cancer Network, 19*(8), 993–996. https://doi.org/10.6004/jnccn.2021.7071.

Cabrera, L. Y., Mitchell, S. D., Bender, A., Tvedten, E., Sidiropoulos, C., & Sarva, H. (2021). Attitudes toward use and timing of deep brain stimulation: A patient's with DBS perspective. *Clinical Neurology and Neurosurgery, 203*. https://doi.org/10.1016/j.clineuro.2021.106553. Article 106553.

Cho, H. L., Miller, D. G., & Kim, S. Y. H. (2020). Understanding people's 'unrealistic optimism' about clinical research participation. *Journal of Medical Ethics, 46*(3), 172–177. https://doi.org/10.1136/medethics-2019-105377.

Cohen, P. D., Isaacs, T., Willocks, P., Herman, L., Stamford, J., Riggare, S., et al. (2012). Sham neurosurgical procedures: The patients' perspective. *Lancet Neurology, 11*(12), 1022. https://doi.org/10.1016/s1474-4422(12)70277-6.

Cotter, K., Siskind, C. E., Sha, S. J., & Hanson-Kahn, A. K. (2019). Positive attitudes and therapeutic misconception around hypothetical clinical trial participation in the Huntington's disease community. *J Huntingtons Dis, 8*(4), 421–430. https://doi.org/10.3233/jhd-190382.

Critchley, C. R., Bruce, G., & Farrugia, M. (2013). The impact of commercialisation on public perceptions of stem cell research: Exploring differences across the use of induced pluripotent cells, human and animal embryos. *Stem Cell Reviews and Reports, 9*(5), 541–554. https://doi.org/10.1007/s12015-013-9445-4.

Crocker, J. C., Ricci-Cabello, I., Parker, A., Hirst, J. A., Chant, A., Petit-Zeman, S., et al. (2018). Impact of patient and public involvement on enrolment and retention in clinical trials: Systematic review and meta-analysis. *BMJ*, *363*, k4738. https://doi.org/10.1136/bmj.k4738.

Cunningham, N. A., Abhyankar, P., Cowie, J., Galinsky, J., & Methven, K. (2018). Regenerative medicine: Stroke survivor and carer views and motivations towards a proposed stem cell clinical trial using placebo neurosurgery. *Health Expectations*, *21*(1), 367–378. https://doi.org/10.1111/hex.12632.

Das, S., Ramesh, S., Baldassari, M., Rasool, M., Ratliff, J., Kremens, D., et al. (2019). Evaluating perceptions of deep brain stimulation in DBS-naive patients with Parkinson's disease—a pilot study. *Movement Disorders*, *34*, S851–S852. <go to ISI>://WOS:000487785005047.

DasMahapatra, P., Raja, P., Gilbert, J., & Wicks, P. (2017). Clinical trials from the patient perspective: Survey in an online patient community. *BMC Health Services Research*, *17*(1), 166. https://doi.org/10.1186/s12913-017-2090-x.

de Bot, S. T. (2019). Raising awareness of therapeutic misconception and optimism around clinical trials in Huntington's disease. *J Huntingtons Dis*, *8*(4), 431–433. https://doi.org/10.3233/jhd-199006.

de Melo-Martín, I., Hellmers, N., & Henchcliffe, C. (2015). First-in-human cell transplant trials in Parkinson's disease: The need for an improved informed consent process. *Parkinsonism & Related Disorders*, *21*(8), 829–832. https://doi.org/10.1016/j.parkreldis.2015.05.011.

de Melo-Martín, I., Holtzman, M., & Hacker, K. S. (2020). "I want to do it, but I want to make sure that I do it right." Views of patients with Parkinson's disease regarding early stem cell clinical trial participation. *AJOB Empir Bioeth*, *11*(3), 160–171. https://doi.org/10.1080/23294515.2020.1775721.

Delhove, J., Osenk, I., Prichard, I., & Donnelley, M. (2020). Public acceptability of gene therapy and gene editing for human use: A systematic review. *Human Gene Therapy*, *31*(1–2), 20–46. https://doi.org/10.1089/hum.2019.197.

Doerflinger, R. M. (1999). The ethics of funding embryonic stem cell research: A Catholic viewpoint. *Kennedy Institute of Ethics Journal*, *9*(2), 137–150. https://doi.org/10.1353/ken.1999.0011.

Drew, C. J. G., Sharouf, F., Randell, E., Brookes-Howell, L., Smallman, K., Sewell, B., et al. (2021). Protocol for an open label: Phase I trial within a cohort of foetal cell transplants in people with Huntington's disease. *Brain Commun*, *3*(1), fcaa230. https://doi.org/10.1093/braincomms/fcaa230.

Dudley, L., Gamble, C., Allam, A., Bell, P., Buck, D., Goodare, H., et al. (2015). A little more conversation please? Qualitative study of researchers' and patients' interview accounts of training for patient and public involvement in clinical trials. *Trials*, *16*, 190. https://doi.org/10.1186/s13063-015-0667-4.

Dudley, L., Gamble, C., Preston, J., Buck, D., Hanley, B., Williamson, P., et al. (2015). What difference does patient and public involvement make and what are its pathways to impact? Qualitative study of patients and researchers from a cohort of randomised clinical trials. *PLoS One*, *10*(6), e0128817. https://doi.org/10.1371/journal.pone.0128817.

Ekezie, W., Routen, A., Denegri, S., & Khunti, K. (2021). Patient and public involvement for ethnic minority research: An urgent need for improvement. *Journal of the Royal Society of Medicine*, *114*(7), 347–350. https://doi.org/10.1177/0141076821994274.

Feustel, A. C., MacPherson, A., Fergusson, D. A., Kieburtz, K., & Kimmelman, J. (2020). Risks and benefits of unapproved disease-modifying treatments for neurodegenerative disease [article]. *Neurology*, *94*(1), E1–E14. https://doi.org/10.1212/wnl.0000000000008699.

Fletcher, S. (1992). Innovative treatment or ethical headache? Fetal transplantation in Parkinson's disease. *Professional Nurse*, *7*(9), 592–595.

Gillon, R. (1988). Ethics of fetal brain cell transplants. *British Medical Journal (Clinical Research Ed.), 296*(6631), 1212–1213. https://doi.org/10.1136/bmj.296.6631.1212.

Gorzynska, O., McGoohan, K., & Velayudhan, L. (2022). Patient and caregiver experiences of participating in Parkinson's disease clinical trials: A systematic review of qualitative studies. *Archives of Clinical Neuropsychology, 37*(3), 654–676. https://doi.org/10.1093/arclin/acab083.

Grauman, Å., & Drevin, J. (2022). Perceptions on using surplus embryos for the treatment of Parkinson's disease among the Swedish population: A qualitative study. *BMC Medical Ethics, 23*(1), 15. https://doi.org/10.1186/s12910-022-00759-y.

Halpern, J., Paolo, D., & Huang, A. (2019). Informed consent for early-phase clinical trials: Therapeutic misestimation, unrealistic optimism and appreciation. *Journal of Medical Ethics, 45*(6), 384–387. https://doi.org/10.1136/medethics-2018-105226.

Harding, E., Robinson, P., Wilson, J., Crutch, S. J., & Mummery, C. J. (2021). Injections of hope: Supporting participants in clinical trials. *BMJ, 375*, e066851. https://doi.org/10.1136/bmj-2021-066851.

Hellmers, N., Obeng-Aduasare, Y., de Melo-Martín, I., & Henchcliffe, C. (2016). Future needs for informed consent in stem cell clinical trials in neurodegenerative diseases. *Neural Regeneration Research, 11*(1), 83–85. https://doi.org/10.4103/1673-5374.169632.

Hirpara, D. H., Cleghorn, M. C., Sockalingam, S., & Quereshy, F. A. (2016). Understanding the complexities of shared decision-making in cancer: A qualitative study of the perspectives of patients undergoing colorectal surgery. *Canadian Journal of Surgery, 59*(3), 197–204. https://doi.org/10.1503/cjs.013415.

Hoffer, B. J., & Olson, L. (1991). Ethical issues in brain-cell transplantation. *Trends in Neurosciences, 14*(8), 384–388. https://doi.org/10.1016/0166-2236(91)90168-t.

Hug, K. (2008). Motivation to donate or not donate surplus embryos for stem-cell research: Literature review. *Fertility and Sterility, 89*(2), 263–277. https://doi.org/10.1016/j.fertnstert.2007.09.017.

Jensen, T. L., Gøtzsche, C. R., & Woldbye, D. P. D. (2021). Current and future prospects for gene therapy for rare genetic diseases affecting the brain and spinal cord. *Frontiers in Molecular Neuroscience, 14*, 695937. https://doi.org/10.3389/fnmol.2021.695937.

Kefalopoulou, Z., Politis, M., Piccini, P., Mencacci, N., Bhatia, K., Jahanshahi, M., et al. (2014). Long-term clinical outcome of fetal cell transplantation for Parkinson disease: Two case reports. *JAMA Neurology, 71*(1), 83–87. https://doi.org/10.1001/jamaneurol.2013.4749.

Kehagia, A. A., North, T. K., Grose, J., Jeffery, A. N., Cocking, L., Chapman, R., et al. (2022). Enhancing trial delivery in Parkinson's disease: Qualitative insights from PD STAT. *Journal of Parkinson's Disease*. https://doi.org/10.3233/JPD-212987.

Kemet, S. (2019). Insight medicine lacks-the continuing relevance of Henrietta lacks. *The New England Journal of Medicine, 381*(9), 800–801. https://doi.org/10.1056/NEJMp1905346.

Keränen, T., Pasternack, A., & Halkoaho, A. (2017). Hope for better treatment in participating clinical drug trials. *Duodecim, 133*(6), 587–591.

Kim, S. Y., De Vries, R., Holloway, R. G., Wilson, R., Parnami, S., Kim, H. M., et al. (2012). Sham surgery controls in Parkinson's disease clinical trials: Views of participants. *Movement Disorders, 27*(11), 1461–1465. https://doi.org/10.1002/mds.25155.

Kim, S. Y., Holloway, R. G., Frank, S., Beck, C. A., Zimmerman, C., Wilson, R., et al. (2006). Volunteering for early phase gene transfer research in Parkinson disease. *Neurology, 66*(7), 1010–1015. https://doi.org/10.1212/01.wnl.0000208925.45772.ea.

Kim, S. Y., Wilson, R. M., Kim, H. M., Holloway, R. G., De Vries, R. G., Frank, S. A., et al. (2012). Comparison of enrollees and decliners of Parkinson disease sham surgery trials. *Movement Disorders, 27*(4), 506–511. https://doi.org/10.1002/mds.24940.

Kunkel, D., Fitton, C., Roberts, L., Pickering, R. M., Roberts, H. C., Wiles, R., et al. (2017). A randomized controlled feasibility trial exploring partnered ballroom dancing for people with Parkinson's disease. *Clinical Rehabilitation, 31*(10), 1340–1350. https://doi.org/10.1177/0269215517694930.

Kwon, D. (2021). Failure of genetic therapies for Huntington's devastates community. *Nature, 593*(7858), 180. https://doi.org/10.1038/d41586-021-01177-7.

Lamore, K., Montalescot, L., & Untas, A. (2017). Treatment decision-making in chronic diseases: What are the family members' roles, needs and attitudes? A systematic review. *Patient Educ Couns, 100*(12), 2172–2181. https://doi.org/10.1016/j.pec.2017.08.003.

Largent, E. A., & Karlawish, J. (2020). Rescuing research participants after Alzheimer trials stop early: Sending out an SOS. *JAMA Neurology*, 77(4), 413–414. https://doi.org/10.1001/jamaneurol.2019.4974.

Largent, E. A., Karlawish, J., & Grill, J. D. (2018). Study partners: Essential collaborators in discovering treatments for Alzheimer's disease. *Alzheimer's Research & Therapy, 10*(1), 101. https://doi.org/10.1186/s13195-018-0425-4.

Lau, Y. H., Podlewska, A., Ocloo, J., Gupta, A., Gonde, C., Bloem, B. R., et al. (2022). Does ethnicity influence recruitment into clinical trials of Parkinson's disease? *Journal of Parkinson's Disease*. https://doi.org/10.3233/jpd-213113.

Long, K. N. G., Kim, E. S., Chen, Y., Wilson, M. F., Worthington, E. L., Jr., & VanderWeele, T. J. (2020). The role of hope in subsequent health and well-being for older adults: An outcome-wide longitudinal approach. *Global Epidemiology, 2*, 100018. https://doi.org/10.1016/j.gloepi.2020.100018.

Lowy, F. H. (1989). Fetal tissue transplantation: Time for a Canadian policy. *CMAJ, 141*(12), 1227–1229. https://www.ncbi.nlm.nih.gov/pmc/articles/PMC1451552/pdf/cmaj00205-0021.pdf.

Lyons, S., Salgaonkar, S., & Flaherty, G. T. (2022). International stem cell tourism: A critical literature review and evidence-based recommendations. *International Health, 14*(2), 132–141. https://doi.org/10.1093/inthealth/ihab050.

Massett, H. A., Mitchell, A. K., Alley, L., Simoneau, E., Burke, P., Han, S. H., et al. (2021). Facilitators, challenges, and messaging strategies for Hispanic/Latino populations participating in Alzheimer's disease and related dementias clinical research: A literature review. *Journal of Alzheimer's Disease, 82*(1), 107–127. https://doi.org/10.3233/jad-201463.

Mathur, S., DeWitte, S., Robledo, I., Isaacs, T., & Stamford, J. (2015a). People with Parkinson's design clinical trials charter to improve communication about, recruitment to, and retention in Parkinson's clinical trials. *Movement Disorders, 30*(10), E4–E5. <Go to ISI>://WOS:000360759500011.

Mathur, S., DeWitte, S., Robledo, I., Isaacs, T., & Stamford, J. (2015b). Rising to the challenges of clinical trial improvement in Parkinson's disease. *Journal of Parkinson's Disease, 5*(2), 263–268. https://doi.org/10.3233/jpd-150541.

Mathur, S., & Stamford, J. (2021). Bringing advanced therapies for Parkinson's disease to the clinic: The patient's perspective. *Journal of Parkinson's Disease, 11*(s2), S141–s145. https://doi.org/10.3233/jpd-212650.

Meinders, M. J., Donnelly, A. C., Sheehan, M., & Bloem, B. R. (2022). Including people with Parkinson's disease in clinical study design and execution: A call to action. *Journal of Parkinson's Disease, 12*(4), 1359–1363. https://doi.org/10.3233/jpd-223190.

Merola, A., Kobayashi, N., Romagnolo, A., Wright, B. A., Artusi, C. A., Imbalzano, G., et al. (2021). Gene therapy in movement disorders: A systematic review of ongoing and completed clinical trials. *Frontiers in Neurology, 12*, 648532. https://doi.org/10.3389/fneur.2021.648532.

Moore, S. F., Guzman, N. V., Mason, S. L., Williams-Gray, C. H., & Barker, R. A. (2014). Which patients with Parkinson's disease participate in clinical trials? One Centre's experiences with a new cell based therapy trial (TRANSEURO) [article]. *Journal of Parkinsons Disease, 4*(4), 671–676. https://doi.org/10.3233/jpd-140432.

Nishimura, A., Carey, J., Erwin, P. J., Tilburt, J. C., Murad, M. H., & McCormick, J. B. (2013). Improving understanding in the research informed consent process: A systematic review of 54 interventions tested in randomized control trials. *BMC Medical Ethics*, *14*, 28. https://doi.org/10.1186/1472-6939-14-28.

Peddie, V. L., Porter, M., Counsell, C., Caie, L., Pearson, D., & Bhattacharya, S. (2009). 'Not taken in by media hype': How potential donors, recipients and members of the general public perceive stem cell research. *Human Reproduction*, *24*(5), 1106–1113. https://doi.org/10.1093/humrep/den496.

Pfeffer, N. (2008). What British women say matters to them about donating an aborted fetus to stem cell research: A focus group study. *Social Science & Medicine*, *66*(12), 2544–2554. https://doi.org/10.1016/j.socscimed.2008.01.050.

Pietrzykowski, T., & Smilowska, K. (2021). The reality of informed consent: Empirical studies on patient comprehension-systematic review. *Trials*, *22*(1), 57. https://doi.org/10.1186/s13063-020-04969-w.

Planner, C., Bower, P., Donnelly, A., Gillies, K., Turner, K., & Young, B. (2019). Trials need participants but not their feedback? A scoping review of published papers on the measurement of participant experience of taking part in clinical trials. *Trials*, *20*(1), 381. https://doi.org/10.1186/s13063-019-3444-y.

Pollack, A. (2005). *Parkinson's patients suing Amgen over drug* (p. 1). The New York Times. 27/04/2005.

Prasad, E. M., & Hung, S. Y. (2021). Current therapies in clinical trials of Parkinson's disease: A 2021 update. *Pharmaceuticals (Basel)*, *14*(8). https://doi.org/10.3390/ph14080717.

Ravina, B., Swearingen, C., Elm, J., Kamp, C., Kieburtz, K., & Kim, S. Y. (2010). Long term understanding of study information in research participants with Parkinson's disease. *Parkinsonism & Related Disorders*, *16*(1), 60–63. https://doi.org/10.1016/j.parkreldis.2009.05.007.

Reddy, P., Martinez-Martin, P., Brown, R. G., Chaudhuri, K. R., Lin, J. P., Selway, R., et al. (2014). Perceptions of symptoms and expectations of advanced therapy for Parkinson's disease: Preliminary report of a patient-reported outcome tool for advanced Parkinson's disease (PRO-APD). *Health and Quality of Life Outcomes*, *12*, 11. https://doi.org/10.1186/1477-7525-12-11.

Rehncrona, S. L., Lund, W., & Neurotransplantation, G. (2006). Comparison between embryonic dopamine cell transplantation and subthalamic DBS for treatment of PD. *Movement Disorders*, *21*, S688. <Go to ISI>://WOS:000241058002363.

Riner, A. N., Girma, S., Vudatha, V., Mukhopadhyay, N., Skoro, N., Gal, T. S., et al. (2022). Eligibility criteria perpetuate disparities in enrollment and participation of Black patients in pancreatic cancer clinical trials. *Journal of Clinical Oncology*, *40*(20), 2193–2202. https://doi.org/10.1200/jco.21.02492.

Rochester, L., & Carroll, C. (2022). Implications of research that excludes under-served populations. *Nature Reviews. Neurology*. https://doi.org/10.1038/s41582-022-00688-9.

Santos Garcia, D., de Deus Fonticoba, T., Suarez Castro, E., Borrue, C., Mata, M., Solano Vila, B., et al., & Coppadis Study, G. (2019). Non-motor symptoms burden, mood, and gait problems are the most significant factors contributing to a poor quality of life in non-demented Parkinson's disease patients: Results from the COPPADIS study cohort. *Parkinsonism & Related Disorders*, *66*, 151–157. https://doi.org/10.1016/j.parkreldis.2019.07.031.

Shineha, R., Inoue, Y., Ikka, T., Kishimoto, A., & Yashiro, Y. (2018). A comparative analysis of attitudes on communication toward stem cell research and regenerative medicine between the public and the scientific community. *Stem Cells Translational Medicine*, 7(2), 251–257. https://doi.org/10.1002/sctm.17-0184.

Snyder, R. A. (2022). Clinical trial eligibility criteria: A structural barrier to diversity in clinical trial enrollment. *Journal of Clinical Oncology*, *40*(20), 2183–2185. https://doi.org/10.1200/jco.22.00537.

Sood, A., Prasad, K., Chhatwani, L., Shinozaki, E., Cha, S. S., Loehrer, L. L., et al. (2009). Patients' attitudes and preferences about participation and recruitment strategies in clinical trials. *Mayo Clinic Proceedings*, *84*(3), 243–247. https://doi.org/10.1016/S0025-6196(11)61141-5.

Stoehr, K., Pazira, K., Bonnet, K., Schlundt, D., Charles, D., & Hacker, M. (2022). Deep brain stimulation in early-stage Parkinson's disease: Patient experience after 11 years. *Brain Sciences*, *12*(6). https://doi.org/10.3390/brainsci12060766.

Sturkenboom, I. H., Graff, M. J., Borm, G. F., Veenhuizen, Y., Bloem, B. R., Munneke, M., et al. (2013). The impact of occupational therapy in Parkinson's disease: A randomized controlled feasibility study. *Clinical Rehabilitation*, *27*(2), 99–112. https://doi.org/10.1177/0269215512448382.

Sun, J., & Roy, S. (2021). Gene-based therapies for neurodegenerative diseases. *Nature Neuroscience*, *24*(3), 297–311. https://doi.org/10.1038/s41593-020-00778-1.

Swift, T. L. (2012). Sham surgery trial controls: Perspectives of patients and their relatives. *Journal of Empirical Research on Human Research Ethics*, 7(3), 15–28. https://doi.org/10.1525/jer.2012.7.3.15.

Tabrizi, S. J., Estevez-Fraga, C., van Roon-Mom, W. M. C., Flower, M. D., Scahill, R. I., Wild, E. J., et al. (2022). Potential disease-modifying therapies for Huntington's disease: Lessons learned and future opportunities. *Lancet Neurology*, *21*(7), 645–658. https://doi.org/10.1016/s1474-4422(22)00121-1.

Tarrant, C., Jackson, C., Dixon-Woods, M., McNicol, S., Kenyon, S., & Armstrong, N. (2015). Consent revisited: The impact of return of results on participants' views and expectations about trial participation. *Health Expectations*, *18*(6), 2042–2053. https://doi.org/10.1111/hex.12371.

The Economist Intelligence Unit. (2018). *The innovation imperative: The future of drug development.* https://druginnovation.eiu.com/wp-content/uploads/2019/05/Parexel-innovations-in-drug-development-part-1_V14.pdf.

Unsworth, D. J., Mathias, J. L., Dorstyn, D. S., & Koblar, S. A. (2020a). Are patient educational resources effective at deterring stroke survivors from considering experimental stem cell treatments? A randomized controlled trial. *Patient Education and Counseling*, *103*(7), 1373–1381. https://doi.org/10.1016/j.pec.2020.02.012.

Unsworth, D. J., Mathias, J. L., Dorstyn, D. S., & Koblar, S. A. (2020b). Stroke survivor attitudes toward, and motivations for, considering experimental stem cell treatments. *Disability and Rehabilitation*, *42*(8), 1122–1130. https://doi.org/10.1080/09638288.2018.1517193.

Valadas, A., Coelho, M., Mestre, T., Guedes, L. C., Finisterra, M., Noronha, A., et al. (2011). What motivates Parkinson's disease patients to enter clinical trials? *Parkinsonism & Related Disorders*, *17*(9), 667–671. https://doi.org/10.1016/j.parkreldis.2011.05.023.

Vawter, D. E., & Gervais, K. G. (1995). Ethical and policy issues in human fetal tissue transplants. *Cell Transplantation*, *4*(5), 479–482. https://doi.org/10.1016/0963-6897(95)00034-u.

Weernink, M. G. M., van Til, J. A., Groothuis-Oudshoorn, C. G. M., & Ijzerman, M. J. (2017). Patient and public preferences for treatment attributes in Parkinson's disease [article]. *Patient-Patient Centered Outcomes Research*, *10*(6), 763–772. https://doi.org/10.1007/s40271-017-0247-7.

Whone, A., Luz, M., Boca, M., Woolley, M., Mooney, L., Dharia, S., et al. (2019). Randomized trial of intermittent intraputamenal glial cell line-derived neurotrophic factor in Parkinson's disease. *Brain*, *142*(3), 512–525. https://doi.org/10.1093/brain/awz023.

FSC
www.fsc.org
MIX
Papier | Fördert gute Waldnutzung
FSC® C083411

Kolibri 360 GmbH
Ferdinand-Jühlke-Straße 7
99095 Erfurt, Deutschland
produktsicherheit@kolibri360.de